VOLUME IX

CHEMICAL EXPERIMENTATION UNDER EXTREME CONDITIONS

Edited by Bryant W. Rossiter

VOLUME X

APPLICATIONS OF BIOCHEMICAL SYSTEMS IN ORGANIC CHEMISTRY, in Two Parts

Edited by J. Bryan Jones, Charles J. Sih, and D. Perlman

VOLUME XI

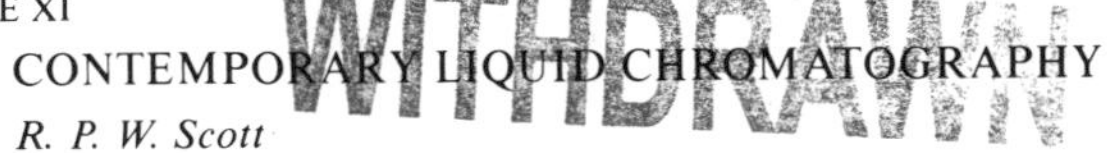

CONTEMPORARY LIQUID CHROMATOGRAPHY

R. P. W. Scott

VOLUME XII

SEPARATION AND PURIFICATION, Third Edition

Edited by Edmond S. Perry and Arnold Weissberger

VOLUME XIII

LABORATORY ENGINEERING AND MANIPULATIONS, Third Edition

Edited by Edmond S. Perry and Arnold Weissberger

VOLUME XIV

THIN-LAYER CHROMATOGRAPHY, Second Edition

Justus G. Kirchner

VOLUME XV

THEORY AND APPLICATIONS OF ELECTRON SPIN RESONANCE

Walter Gordy

VOLUME XVI

SEPARATIONS BY CENTRIFUGAL PHENOMENA

Hsien-Wen Hsu

VOLUME XVII

APPLICATIONS OF LASERS TO CHEMICAL PROBLEMS

Edited by Ted R. Evans

TECHNIQUES OF CHEMISTRY

ARNOLD WEISSBERGER, *Editor*

VOLUME XVII

APPLICATIONS OF LASERS TO CHEMICAL PROBLEMS

TECHNIQUES OF CHEMISTRY

VOLUME XVII

APPLICATIONS OF LASERS TO CHEMICAL PROBLEMS

Edited by

TED R. EVANS

Research Laboratories
Eastman Kodak Company
Rochester, New York

A WILEY-INTERSCIENCE PUBLICATION

JOHN WILEY & SONS

New York • Chichester • Brisbane • Toronto • Singapore

Library of Congress Cataloging in Publication Data
Main entry under title:

Applications of lasers to chemical problems.

(Technique of chemistry, ISSN 0082-2531; v. 17)
"A Wiley-Interscience publication."
Includes index.
1. Lasers in chemistry. I. Evans, Ted R. II. Series.

QD61.T4 vol. 17 [QD715] 540s [541.3′5] 82-1904
ISBN 0-471-04949-2 AACR2

Printed in the United States of America

10 9 8 7 6 5 4 3 2 1

CONTRIBUTORS

TED R. EVANS

Research Laboratories, Eastman Kodak Company, Rochester, New York

A. M. RONN

Department of Chemistry, Brooklyn College of the City University of New York, New York, New York

JACK WILSON

Laboratory for Laser Energetics, University of Rochester, Rochester, New York

JOHN C. WRIGHT

Department of Chemistry, University of Wisconsin, Madison, Wisconsin

INTRODUCTION TO THE SERIES

Techniques of Chemistry is the successor to the Technique of Organic Chemistry Series and its companion—Technique of Inorganic Chemistry. Because many of the methods are employed in all branches of chemical science, the division into techniques for organic and inorganic chemistry has become increasingly artificial. Accordingly, the new series reflects the wider application of techniques, and the component volumes for the most part provide complete treatments of the methods covered. Volumes in which limited areas of application are discussed can be easily recognized by their titles.

Like its predecessors, the series is devoted to a comprehensive presentation of the respective techniques. The authors give the theoretical background for an understanding of the various methods and operations and describe the techniques and tools, their modifications, their merits and limitations, and their handling. It is hoped that the series will contribute to a better understanding and a more rational and effective application of the respective techniques.

Authors and editors hope that readers will find the volumes in this series useful and will communicate to them any criticisms and suggestions for improvements.

ARNOLD WEISSBERGER

Research Laboratories
Eastman Kodak Company
Rochester, New York

PREFACE

In 1950 it was stated that "the most important tool bequeathed to the twentieth century was the photographic plate" [1]. Certainly one of the most important scientific tools discovered since that time has been the laser. Any device whose applications include welding, price tag scanning at the local supermarket, rangefinding, holography, and fingerprint detection, to mention just a few applications, must be a remarkable invention.

Chemists have adopted the laser for a wide variety of applications. Still the full scope of its utility cannot yet be measured. This book is intended to examine some of the applications of lasers to chemical problems.

Since the basic physics of lasers is covered in many general texts on physics and optics it is not included in this volume. The techniques concerned with picosecond lasers and measurements have been covered in a recent volume of this series [2].

The first chapter in this book is an introduction to basic laser types and techniques. The principles of each laser and the physical and chemical features necessary for understanding its mode of operation have been outlined. The second chapter covers the use of lasers in analytical chemistry. Since many analytical techniques involve spectroscopy it is not surprising that lasers have found wide and rapid application in this field. The sensitivity, selectivity, and limits of detection have been improved several orders of magnitude by the substitution of lasers for conventional light sources. The nonlinear Raman techniques and multiphoton spectroscopy are possible only because of the high intensities available from lasers. The final two chapters cover the areas of photochemical application of ultraviolet and visible lasers (Chapter III) and infrared lasers (Chapter IV). The use of ultraviolet and visible lasers, in place of conventional light sources, has allowed a much greater control of wavelength and extreme shortening of the pulse duration. Using the high intensities produced by lasers, both unusual excited states and high concentrations of excited states can be prepared. In the infrared, tens of photons may be absorbed by a reagent leading to a vibrationally hot molecule surrounded by a cold medium. This has spawned the field of infrared laser photochemistry, which could not exist without lasers. Although the hope that bond-selective chemistry could be induced has not yet been realized, isotope selective chemistry is now a reality.

In the past two decades the use of lasers in chemistry has greatly expanded. Analytical applications, isotope separations, and flash photolysis are the most important applications at this time. In the next two decades these uses are expected to be further refined and new, probably unexpected, applications will be found.

I would like to thank Dr. R. J. Lovett, University of North Dakota, for comments on parts of the manuscript.

1. H. Shapley, *Sci. Amer.*, **183,** 24 (1950).
2. S. C. Pyke and M. W. Windsor, in *Chemical Experimentation Under Extreme Conditions*, B. Rossiter, Ed., *Techniques of Chemistry*, Vol. IX, 1980, Chapter V.

Rochester, New York
April 1982

TED R. EVANS

CONTENTS

Chapter I

LASER SOURCES

Jack Wilson

The Standard Oil Company (Ohio)
3092 Broadway Ave
Cleveland, Ohio

1 INTRODUCTION

There is an almost bewildering array of lasers available today. Power levels available extend from milliwatts to megawatts. Pulse lengths range from picoseconds to continuous (usually referred to as c.w., which is an abbreviation for continuous wave). Wavelengths span the spectrum from ultraviolet to far infrared. At first glance, it appears that the chemist should start making a laser selection based on the wavelength category, since photons with sufficient energy to break a chemical bond are found only in the visible and ultraviolet region of the spectrum. However, lasers are now available with sufficient intensity to make multiphoton processes very probable, and it is possible to dissociate molecules with photons having wavelengths of 10.6 μm [1,2]. Indeed, for industrial application, use of long-wavelength photons may be preferable, since such photons can be produced efficiently, and hence economically.

Lasers are usually considered to be devices that produce photons at a fixed wavelength. This is true of most solid and gaseous state lasers and somewhat restricts their utility for chemical applications. Dye lasers, which operate mainly in the visible and near ultraviolet, and F-center lasers and tunable diodes, which operate in the infrared, can, however, be tuned continuously over a certain wavelength range. This continuous tuning permits matching of the laser wavelength to a resonance of the irradiated compound, thereby increasing the probability of an interaction.

The output of a laser falls within a narrow bandwidth, and so lasers have the property of producing very nearly monoenergetic photons. This is clearly their advantage over other light sources. Moreover, the laser output can be produced in a very short time. Mode-locking techniques, which are described below, make possible laser pulse lengths of the order of picoseconds. Pulse lengths of this duration permit kinetics measurements on a time scale impossible to achieve by other techniques.

Continuous-wave lasers are available with very high powers, for example, 20 kW is available commercially from a carbon dioxide laser. However, c.w. lasers have less obvious application to chemistry than do pulsed lasers, since in effect they are rather expensive heat sources.

The various major types of lasers available are given in Table 1. Each of these is described in more detail below. First, however, techniques of wavelength and temporal conversion, common to all lasers, are discussed. The difference between coherent output (in which the waves are all in phase) and amplified spontaneous emission is also explained.

Safety

The high power available from lasers can have undesirable consequences. Laser light in the visible or near infrared falling on the eye is focused to an ex-

Table 1 Major Types of Lasers Available

Laser Medium	*Pumping Technique*	*Wavelength(s)(nm)*	*Typical Pulse Lengths (μsec)*	*Pulsed Energy (J)*	*Maximum c.w. Power (W)*
Gas lasers					
Carbon dioxide	Low-pressure discharge, high-pressure electron beam stabilized and photoionized discharges	10,600 9,600	0.1–100	1–1000	15,000
Carbon monoxide	Electric discharge	5000–7000	10–1000	0.01	20
Argon ion	Low-Pressure discharge	351–528	6	4×10^{-5}	20
Krypton ion	Low-pressure discharge	450–799	0.02	10^{-6}	6
Helium-neon	Low-pressure discharge	633	—	—	0.02
Nitrogen	Fast pulsed discharge	337	0.01	0.001–0.01	—
HF	Photoionized discharge	2600–3500	1	1	30
XeF, KrF, ArF	Photoionized discharge	352,248,193	0.02	0.1–0.3	—
Solid-state lasers					
Nd	Optical flashlamp	1,060	10^{-5}–10^{3}	10^{-3}–10^{3}	1000
Ruby	Optical flashlamp	694	0.02–10^{3}	0.02–400	—
F-Center	Ar, Kr, or dye laser	2200–3300	—	—	0.05
Diodes					
GaAs/GaAlAs	Electric	800–904	0.10	10^{-6}	0.02
Lead based	Electric	2800–30,000	0.02	4×10^{-6}	0.0005
Liquid lasers					
Dye	Optical	400–960	0.003–1	10^{-3}–1	1

tremely small spot on the retina, causing damage that can be permanent and is definitely unpleasant [3]. Infrared light is absorbed in the cornea, possibly causing damage. It is vital therefore that workers with high-power lasers wear protective goggles that block the laser wavelength. This is true even if the main beam seems to be out of harm's way—it is incredibly easy for stray reflections to appear all over the laboratory. Permissible exposure levels have been established by the American National Standards Institute [4].

Another hazard in working with lasers is the high voltage frequently used in the power source. Obviously, all high-voltage components should be enclosed.

2 LASER PROPERTIES

The coherence and intensity of laser beams make possible nonlinear effects in certain media. These effects can be used to double, triple, or quadruple the laser frequency or to generate other frequencies in parametric amplifiers. The laser frequency can be down-converted by Raman processes. These techniques may be useful in matching a laser to an experiment. Except for the Raman process, these techniques require a coherent beam, so it is important to know whether the laser produces a sufficiently coherent beam or not.

Coherent Output

Coherent laser output is produced by a cavity mode. The cavity is the name given to the pair of mirrors placed at opposite ends of the active (i.e., amplifying) medium. A mode is a configuration of the electromagnetic wave inside the cavity that reproduces itself after a round trip within the cavity. It has been shown that approximately 100 round trips are required for a mode to be established [5]. Consequently, if the pulse length is too short to permit this number of trips, it may be difficult to establish a mode. Figure 1*a* shows a typical laser arrangement, namely, a pair of mirrors at either end of an active medium. In an established mode, the radiation in the mode stimulates the excited particles in the active medium to emit in phase with the mode, and hence the radiation becomes coherent. The output beam is approximately diffraction limited based on the width of the laser beam.

A simple, approximate, expression for the width D of the laser beam can be obtained by considering the case of a plane and a curved mirror, in which the radius of the curved mirror is equal to the mirror separation L. The rays reflect back on themselves, that is, form a mode, if they strike each mirror normally. This occurs when the diffraction angle, α_D, given by

$$\alpha_D = \lambda/D$$

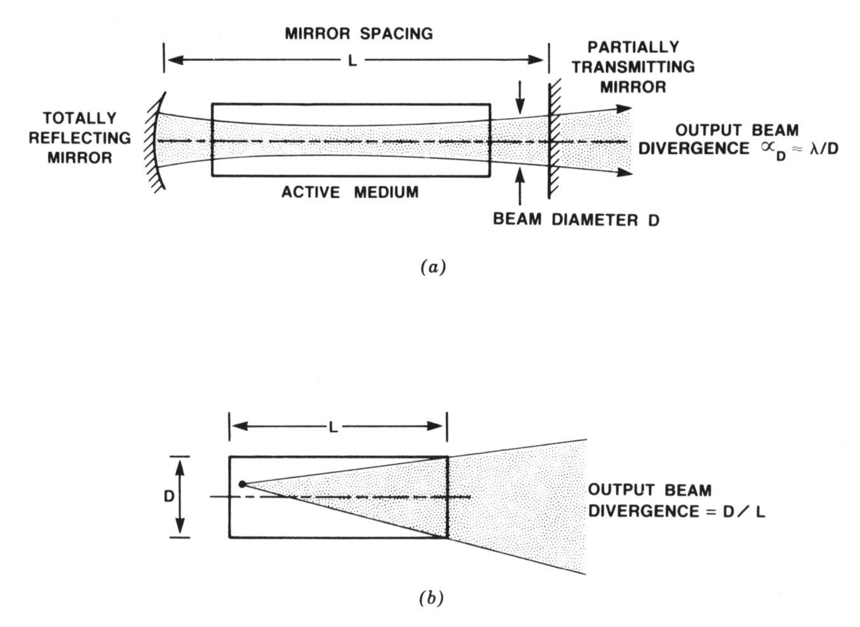

Fig. 1 (*a*) Coherent beam produced by a typical laser cavity, consisting of two mirrors at each end of the active medium. One mirror is fully reflective, and the other, the output mirror, is a partial reflector. (*b*) Amplified spontaneous emission from a high-gain laser.

in which λ is the laser wavelength and D is the laser beam diameter, equals the geometrical angle, α, given by

$$\alpha = D/L$$

Equating the two angles gives

$$D = \sqrt{\lambda L}$$

The laser output is then a beam of width D, diverging at an angle α_D or greater. For a more complete description of laser modes, see Ref. 6. If the width of the active medium is significantly greater than the mode size D just calculated, multimode operation can take place, in which several modes lase simultaneously. To extract energy in one mode from a large-aperture active medium, an unstable cavity, as described by Siegman [7], is sometimes used.

Some lasers have a gain so high that spontaneous emission from one end of the active medium can be sufficiently amplified in passing through the medium that it extracts most of the available laser energy at the other end. The remaining gain is insufficient to establish a mode. The output in this case consists of temporally incoherent light, known as amplified spontaneous emission, within a solid angle defined by the enclosure of the active medium (Fig. 1*b*). However, note that this output is still quite directional and can be useful for some purposes.

Harmonic Generation

In some crystals, the index of refraction varies in the presence of a strong electric field, that is, the index of refraction n takes the form

$$n = n_0 + \pi \chi E$$

in which χ is the second-order nonlinear susceptibility of the crystal, E is the electric field, and n_0 is the constant part of the index of refraction. The variable term becomes appreciable at intensities approaching *1* MW/cm^2, a value achievable with lasers. The effect of the nonlinear index is to generate a wave at twice the laser frequency; this wave is called the second harmonic. Thus if a laser beam of appropriate intensity is sent into a crystal, the emergent beam has components both at the original frequency ν and at 2ν. By solving Maxwell's equations [8], one can show that in a crystal of length L, the efficiency, η, of producing second-harmonic radiation is

$$\eta = \frac{I(2\nu)}{I(\nu)} = \frac{128\pi^5\chi^2L^2I(\nu)}{n^3c\lambda^2}\left[\frac{\sin(\Delta kL/2)}{\Delta kL/2}\right]^2 \sin^2\theta$$

in which c is the speed of light, $I(\nu)$ and $I(2\nu)$ are intensities at the respective frequencies, λ is the wavelength at the original frequency, θ is the angle between the direction of the light and the crystal axis, and $\Delta k = 4\pi[n_2 - n_0]/\lambda$ is the wave vector mismatch between the input beam and the second harmonic beam (n_2 is the index of refraction of the crystal at the frequency of the second harmonic). The above equation holds if the efficiency η is low, less than about 10%. It shows that efficient generation is achieved only if Δk is small, that is, when the indices of refraction at the two frequencies are equal. This equality can be achieved in birefringent crystals for which there are different indices of refraction depending on whether the polarization of the beam is perpendicular to the crystal axis (ordinary ray) or has a component along it (extraordinary ray). The index of refraction of the extraordinary ray depends on the angle θ, whereas that of the ordinary ray does not. If the input laser beam is polarized as an ordinary ray, the second harmonic is generated as an extraordinary ray. The angle θ can be altered until the indices at the two frequencies are equal. This is called type I angle tuning of the crystal. Type II angle tuning is achieved at a different angle when the input beam has two components, one an ordinary ray and one an extraordinary ray, and the second harmonic is an extraordinary ray. Type II conversion is usually more efficient than type I. A different technique for equalizing the indices at the two frequencies is temperature tuning—heating the crystal to the point at which the ordinary index of the input beam equals the extraordinary index of the second harmonic. In temperature tuning the crystal axis is normal to the input beam and the polarization; in fact, it is type I tuning in which the angle θ has been taken to the limit of 90°. Figure 2 shows the angle θ at different input frequencies for three different common doubling crystals, KDP, ADP, and CDA, and also the temperatures for temperature tuning.

What is not indicated above is the sensitivity to the angle θ. The range of angle, $\Delta\theta$, over which second-harmonic generation occurs is referred to as the acceptance angle. It is a function of the crystal length, such that the product $(\Delta\theta)L$ is a constant. Values for various crystals are given in Table 2. For a typical crystal length of 1 cm, the acceptance angle of KDP is only 500 μra-

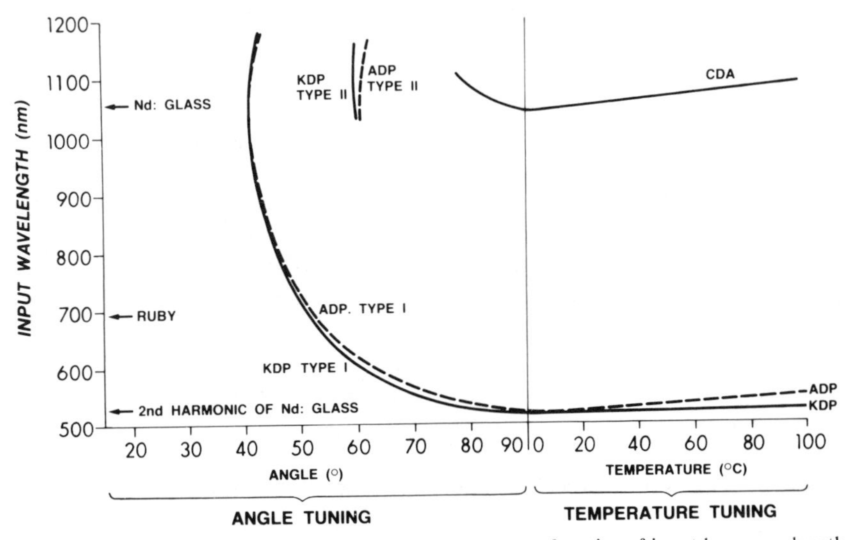

Fig. 2 Tuning angles and phase-matching temperatures as a function of input laser wavelength for frequency doubling in KDP, ADP and CDA crystals.

Table 2 Acceptance Angles for KDP, ADP and CDA Crystals at 1.06μm Input Wavelength

Crystal	$(\Delta\theta)$ *L (mradian-cm)*
KDP type I	0.55
KDP type II	0.8
ADP type I	0.5
CDA type I	5.0

dians. The angle sensitivity manifests itself in two ways. First, the crystal axis must be accurately tuned to the laser direction, and second, the input laser beam must be parallel to within less than the acceptance angle if significant conversion is to occur. These requirements make amplified spontaneous emission with its divergent output unsuitable for doubling.

Doubling is obviously useful for shifting the frequency of a laser to a region that may be more appropriate for the experiment at hand. For example, dye lasers function most efficiently in the visible region of the spectrum. However, by doubling the output of a dye laser, a tunable ultraviolet laser can be achieved.

A variation of doubling is summing. If the doubled output from a crystal, together with the remaining undoubled input are passed into an appropriately oriented crystal, the two waves can add to give the third harmonic of the original beam.

Parametric Oscillators

When the nonlinear susceptibility χ was introduced above, it was implicitly assumed to be a scalar quantity. Actually, it is a tensor quantity, which means that the input laser beam can generate radiation at frequencies other than 2ν. If the conversion crystal is placed inside a cavity resonant to a particular frequency, a wave at this frequency builds up. Actually two frequencies, ν_S and ν_I, the signal wave and the idler wave, result, and are related by

$$\nu = \nu_S + \nu_I$$

The crystal has to be tuned so that the index-matching relationship

$$k - k_S - k_I = 0$$

in which k is the wave vector at the input laser frequency, $k = 2\pi n/\lambda$, and k_S, k_I are similar quantities for the signal and idler frequencies.

By varying the angle of the crystal, the frequency of the signal and idler waves can be changed. Consequently, this is another technique for producing different, and in this case, tunable wavelengths from the original laser beam. However, since ν_S and ν_I are obviously smaller than ν, the tunable wavelengths are longer than the original wavelength. This technique is thus mainly used to produce tunable infrared radiation. For a review article, see Ref. 9.

Raman Down-Conversion

If a laser beam is passed through a medium with its own characteristic Raman frequency ν_c (as, e.g., a vibrational frequency), some of the laser light is converted to a new frequency, called the Stokes frequency ν_S, given by

$$\nu_S = \nu - \nu_c$$

At high laser intensities, this can become a stimulated process, and very high conversion efficiencies into the Stokes frequency can result. The Raman medium must have a Raman active excitation mode. Some media that have exhibited stimulated Raman scattering are given in Table 3, together with the frequency shifts (expressed in cm^{-1}) they generate. Loree et al. [10] have used this technique with rare gas halide pump lasers to generate a wide variety of ultraviolet and visible wavelengths. Djeu and Burnham [11], using electronic-transition Raman scattering in barium vapor, have demonstrated 80% photon conversion efficiency from the pump wavelength of 351 nm to an output wavelength of 585 nm.

Short-Pulse Generation

The cavity modes mentioned above can have different frequencies. These frequencies ν_n are such that an integral number of nodes n fits into the cavity, that is,

$$\nu_n = \frac{cn}{2L}$$

where L is the cavity length, c is the speed of light, and n is an integer. Ob-

Table 3 Materials that Have Exhibited Stimulated Raman Scattering

Material	*Frequency Shift* (cm^{-1})
Liquids	
Tetrachloroethylene	447
Carbon tetrachloride	459
Trichloroethane	640
Carbon disulfide	656
Nitrobenzene	1344
Acetone	2921
Water	3651
Solids	
α-Sulfur	216 470
Calcite	1084
Diamond	1332
Gases	
Nitrous oxide	774
Oxygen	1550
Nitrogen	2330
Deuterium	2991
Hydrogen	4155

viously the frequency spacing between modes is $c/2L$. Only those cavity modes whose frequencies lie within the bandwidth $\Delta\nu$ of the lasing transition have gain, so the number of cavity modes that can actually lase (see Fig. 3) is

$$m = \frac{\Delta\nu}{c/2L}$$

Under certain conditions, some, or all, of these modes can be in phase with each other. This phenomenon is known as mode-locking. Since the modes are in phase, the electric fields E_n can be added to give the total electric field E,

$$E = \sum_m E_n = \sum_m A_n \sin \frac{2\pi\nu}{c} n (x - ct)$$

and the intensity I,

$$I = E^2 = [\sum_m A_n \sin \frac{2\pi\nu}{c} n (x - ct)]^2$$

Here A_n is the stationary amplitude of the electric field E_n, x is distance, and t is time.

If one imagines the electric fields stretched out in space, that is, ignoring the mirrors, or, the equivalent, stretched out in time, and if the modes are all in phase at some point and time, by the definition of a mode, they are in phase again a distance $2L$ and a time $2L/c$ later. In between they cancel each other.

Consequently, the result of this addition and subtraction of electric fields is an intense burst of intensity, traveling back and forth in the cavity at the

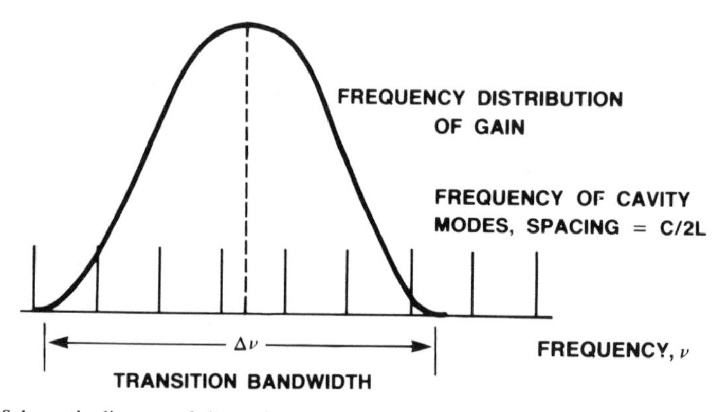

Fig. 3 Schematic diagram of the cavity modes and the laser transition bandwidth showing that several cavity modes may fall within the transition.

speed of light. The output of the mode-locked laser is a series of pulses separated in time by $2L/c$. The width of the pulse is determined by how many modes are contributing and the distance over which they can remain coherent, that is, the distance for which they become out of phase by one wavelength. This distance is simply

$$\frac{2L}{n_i - n_j}$$

for two modes of frequencies $\nu_i = n_i c/2L$ and $\nu_j = n_j c/2L$. The minimum distance is $2L/m$ if all the m modes within the bandwidth are contributing. The minimum pulse width t pulse then becomes

$$t_{\text{pulse}} = \frac{2L}{mc} = \frac{1}{\Delta\nu}$$

Thus if a large linewidth $\Delta\nu$ can be achieved, extremely short pulses can result.

As a numerical example, neodymium-glass lasers have a bandwidth greater than 10^{12} Hz and pulses of subpicosecond duration can be produced. In fact, it is sometimes necessary to put an etalon within the cavity of a neodymium-glass laser to reduce the bandwidth to generate a longer pulse.

Getting the modes to be in phase requires an optical switch within the cavity, which is open to let the intense pulse pass and closed in between. Two kinds are in common use, active and passive. The passive mode-locked switch is simply an absorbing medium that saturates at a certain intensity, passing high-intensity pulses and absorbing low-intensity pulses. For visible and near infrared lasers a dye is commonly used for this purpose. For carbon dioxide lasers, a cell filled with SF_6 is commonly used. Active mode-locked switches are crystals vibrating at ultrasonic frequencies, in resonance with the cavity length.

3 GAS LASERS

Gases can be convected very rapidly. Consequently, if the active medium is gaseous, it can be removed after each laser firing and replaced with fresh gas, ready for a new laser pulse. A high repetition rate can be achieved in this manner; an effective rate of 13 k Hz has been reported [12]. This ability to handle large quantitites of gases has led to high average powers from gas lasers—almost 100 kW has been achieved with a gas-dynamic carbon dioxide laser. Gases can be excited in large volumes by electrical discharge, an extremely efficient form of excitation, with efficiencies as high as 40% for carbon monoxide lasers. Thus gas lasers excel in efficient, high average power performance.

Electrically Powered Gas Lasers

The earliest electrically excited gas lasers employed a longitudinal gas discharge, at relatively low gas pressures (a few torr). Such discharges are wall stabilized. Flow is sometimes used in these low-pressure devices, but the objective is to remove impurities generated by the discharge that would terminate laser action if allowed to accumulate rather than enhance the power. A carbon dioxide laser working under these conditions might employ a 20-kV supply, with a distance between anodes of 1 m. The ratio of the electric field E to the pressure p determines the electron temperature, which itself affects the excitation and ionization rates in the discharge. Too low an electron temperature results in too little excitation of the desired active modes, and too high an electron temperature results in too much energy being invested in ionization. There is, therefore, an optimum value of E/p for best laser performance; for carbon dioxide lasers, this value is about 5 kV/cm-atm.

If one attempts to increase the output from a laser by increasing the pressure, the electric field must be increased correspondingly. This results in extremely high voltages if an axial field is used. Consequently, transverse electric fields, that is, fields perpendicular to the laser axis are used at higher pressures. Further problems arise at pressures around 1 atm. First, the optimum voltage for efficient operation may be insufficient to sustain a discharge. Second, a discharge at a voltage sufficient for breakdown is unstable and tends to break up into arcs of high current density. These problems are solved by providing an alternate source of electrons in the discharge. The discharge can then operate at the optimum voltage. Also, since the optimum voltage is too low to create a high ionization rate, the discharge is no longer unstable. The alternate source of electrons is an electron beam, ultraviolet light, or a secondary discharge fired in advance of the main discharge. The electron beam technique uses a beam of high-energy electrons, typically around 300 keV, passing through the discharge region to create secondary electrons by impact ionization. The technique can be pulsed or c.w. In the ultraviolet light technique, the discharge region is bathed in ultraviolet light from an additional discharge, frequently a surface discharge or flashboard. Electrons are created within the discharge volume by photoionization. Since ultraviolet light with the required intensity is only produced in pulsed discharges, this technique is limited to pulsed operation. It is used in several commercially available lasers.

A beam of high-energy electrons not only ionizes the gas through which it passes, but also excites it. Consequently, it is possible to pump some media with an electron beam alone. The advantage of this method is that levels with quite high excitation energy can be pumped, leading to short wavelength lasers. For example, molecular hydrogen has demonstrated laser action at

116 and 160 nm with direct electron beam pumping [13]. Although no commercially available lasers use this approach, suitable electron beam units can be purchased.

Pulsed high-pressure gaseous lasers employ short duration, high-current discharges. Such discharges generate copious quantities of electromagnetic noise. Good screening of the diagnostics, by locating them, or the laser, or both, inside screen rooms is virtually essential. Screens do not necessarily have to be elaborate; a wooden frame supporting an inside and outside layer of aluminum window screen can sometimes suffice.

Carbon Dioxide Lasers

Although the carbon dioxide laser has been pumped by optical techniques [14], by a gas-dynamic technique [15], and by transfer from chemical reactions [16], the pumping technique of major commercial importance is electrical [17]. Continuous lasers use low-pressure discharges and high-pressure electron beam sustained discharges; pulsed lasers use electron beam and ultraviolet preionization.

The importance of electrically powered carbon dioxide lasers stems from their efficiency, which can be as high as 20%. The gaseous medium employed in an electrical carbon dioxide laser is a mixture of carbon dioxide, nitrogen, and helium, roughly in the ratio 1:2:4. Nitrogen has a large cross section for vibrational excitation by electrons [18] by way of the process

$$e + N_2 \rightarrow N_2^- \rightarrow N_2(v) + e$$

In consequence, a large fraction of the energy of the discharge appears in nitrogen vibration. The $v = 1$ level of nitrogen is resonant with the ν_3 asymmetric stretch level (001) of carbon dioxide (see Fig. 4) and populates it by collisional vibration energy transfer. The symmetric stretch (100) and bending mode (020) levels are not readily populated by electrons or transfer and in fact are depopulated fairly rapidly by helium. Thus an inversion is generated between the (001) level and the (100) and (020) levels. Laser action can result between the upper level and either of the two lower levels, giving rise to output at 10.6 or 9.4 μm. The vibrational transitions actually comprise a large number of transitions between different rotational states of the upper and lower levels, and so many lines are possible.

In a simple cavity, laser action takes place on the line with the most gain, which is the P(18) transition. However, when a grating is used as one mirror of the cavity, a wavelength selective cavity can be produced, and laser action on a large number of lines results as the cavity is tuned to each particular line. Figure 5 shows the spectrum of laser transitions produced in this way.

As the pressure of the laser medium is increased, the individual rotational

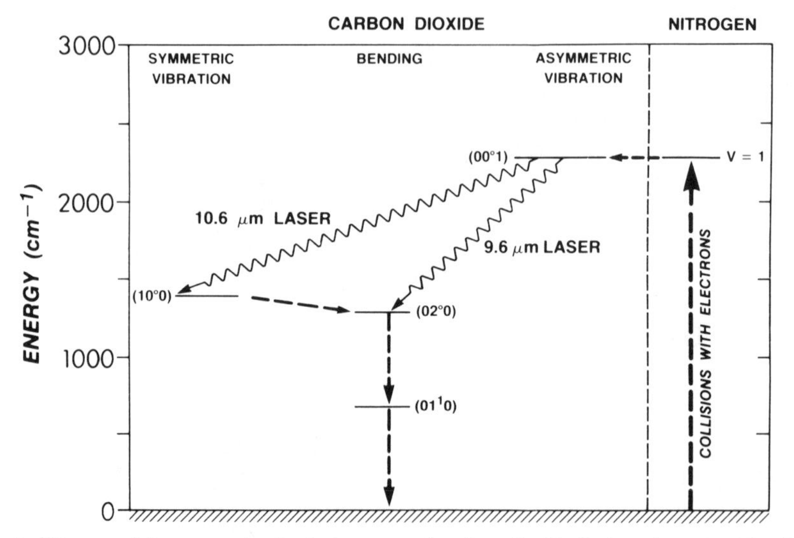

Fig. 4 Diagram of the energy levels of nitrogen and carbon dioxide that are important for the carbon dioxide laser.

lines become broader until at around 5 to 10 atm their width is equal to the spacing between transitions. At this point continuous tuning over the vibrational transition becomes possible (19).

By the technique of mode-locking described above, carbon dioxide lasers can generate short pulses. At 1 atm of pressure, the bandwidth of the carbon dioxide lines is around 10^{10} Hz, and subnanosecond pulses are possible. The mode-locking medium is typically sulfurhexafluoride [21] or carbon dioxide [22] used as a saturable absorber, or an active germanium crystal [23]. At pressures higher than 1 atm, pulses of 75-psec duration have been achieved [24]. An alternate technique for generating short pulses, called optical free induction decay, has resulted in 30-psec pulses [25]. Another approach is to use a reflective switch to isolate a short pulse. The reflectivity is generated in a pulsed mode by irradiating germanium with a short-pulse neodymium laser. Pulses as short as 5 psec have been produced this way [26].

A degree of spontaneous mode-locking can occur in the pulse of a carbon dioxide laser, and the output often contains rapid fluctuations in intensity. This can be masked if the detector-oscilloscope combination used to observe the pulses does not have sufficient temporal resolution. Devices that are commonly used for monitoring carbon dioxide lasers are mercury-cadmium-telluride infrared detectors, gold-doped germanium (which responds to the laser radiation even though it is apparently at a wavelength outside its region

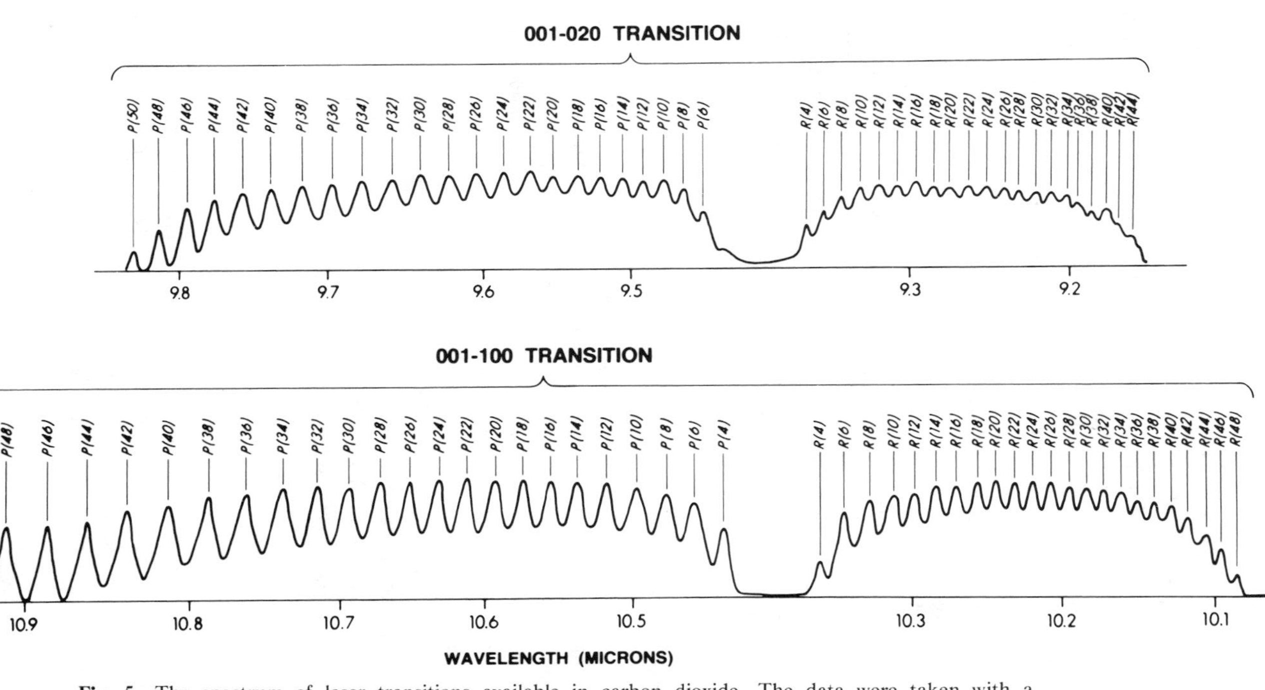

Fig. 5 The spectrum of laser transitions available in carbon dioxide. The data were taken with a Lumonics Model TEA-103 laser. Reproduced by courtesy of J. A. Nilson, Lumonics Research Limited.

of response), and photon-drag, pyroelectric, and hot-hole detectors. Pyroelectric detectors are offered as both power and energy detectors.

The gain of carbon dioxide lasers is sufficiently high that several transverse modes can exhibit laser action simultaneously. Frequently an aperture is inserted into the active medium and adjusted until only the lowest order mode is operating. If several modes are running, the laser output can have severe spatial nonuniformities.

The wavelength of the carbon dioxide laser is not transmitted by glass. Any transmitting optics must be of some other material, germanium, zinc selenide, and KCl or NaCl being commonly used. Germanium has a high index of refraction and consequently a large Fresnel reflection, which can be reduced by an antireflective coating. Salt is the cheapest material, but is, of course, hygroscopic. This presents no problem if a miniature heat lamp is directed at the salt so as to heat it to a temperature a little above that of the surroundings. Copper, gold, and silver all have reflectivities above 98% at 10 μm, and reflective optics of these materials are frequently used for carbon dioxide lasers.

Carbon Monoxide Lasers

Carbon monoxide lasers can be both pulsed and continuous. The usual mode of pumping is by some form of electrical discharge. Since a molecule of carbon monoxide is similar to a molecule of nitrogen, it can also be excited in vibration very efficiently by electrons. Unlike nitrogen, however, carbon monoxide can radiate on the vibrational transition, and it is on this transition that laser action is generated. Excitation by electrons puts energy into the vibrational mode and does not create an inversion between vibrational levels. Instead an inversion is created between two rotational levels of a P branch vibration-rotation transition by cooling the gas. The populations N_v, $N_{(v-1)}$ of the v and $(v - 1)$ vibrational levels are related by

$$\frac{N_v}{N_{(v-1)}} = \exp \frac{-E_v}{kT_v}$$

in which E_v is the vibrational spacing of carbon monoxide and T_v is the vibrational temperature, which will approach the electron temperature. Rotational transitions are dominated by collisions with the background gas and the population $N_{v,J}$ of the J rotational level of N_v is given by

$$N_{v,J} = N_v(2J + 1) \exp \frac{-BJ(J + 1)}{kT_R}$$

in which B is the rotational spacing and T_R is the rotational temperature of carbon monoxide. The rotational temperature is assumed to be close to the gas rotational temperature. It can be shown that even though $N_v < N_{(v-1)}$, an inversion can exist between $N_{v,J}$ and $N_{(v-1),(J+1)}$ provided that

$$\frac{T_v}{T_R} > \frac{E_v}{2(J+1)B}$$

Such an inversion is called a partial inversion. To achieve a partial inversion a low rotational temperature is necessary. Consequently, carbon monoxide lasers are typically cooled to liquid nitrogen temperatures. Carbon monoxide lasers have demonstrated the highest efficiency of all gas lasers, around 40% (27–29).

Laser action occurs from a variety of vibrational levels, and the laser output wavelength ranges between 5 and 6 μm.

Nitrogen Lasers

The nitrogen laser is a pulsed laser operating on the electronic *C–B* transition in nitrogen. It is most commonly excited in a very rapid transverse electrical discharge. In such a discharge, electrons preferentially excite the nitrogen $C(^3\Pi_u)$ state, since this is a direct Franck-Condon transition, rather than the $B(^3\Pi_g)$ state (Fig. 6). The $C(^3\Pi_0)$ state thus becomes inverted relative to

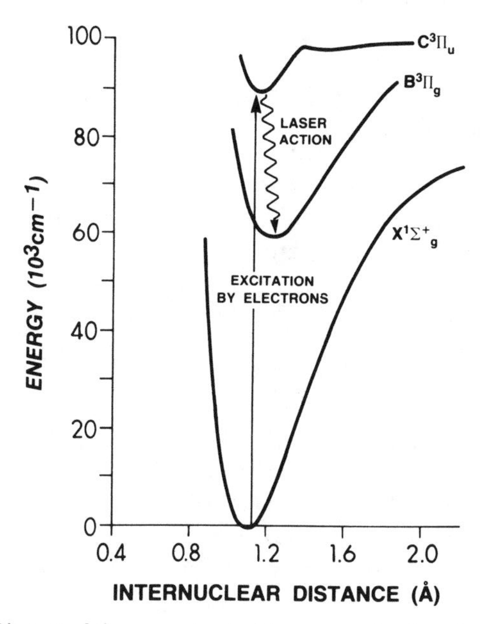

Fig. 6 Diagram of the energy levels of importance for the nitrogen laser.

the $B(^3\Pi_g)$ state, and laser action follows. The radiative lifetime of the v = 0 level of the $C(^3\Pi_c)$ state is 40 nsec [30]. This explains why a rapid discharge is required, since with a rapid removal rate by radiation, no significant $C(^3\Pi_u)$ population would be achieved unless the pumping rate were rapid. Laser action is on the (0–0) transition at 337.1 nm, and because of the high gain, is usually in the form of amplified spontaneous emission. Pulse lengths are typically around 10 nsec. Transitions to other vibrational levels of the $B(^3\Pi_g)$ state, namely, (0–1) at 357.6 nm, (0–2) at 380.4 nm, and (0–3) at 405.8 nm have shown lasing action if the energy is provided by transfer from excited argon. Long-pulse ($\sim$1 μsec.) laser action in nitrogen using electron beam pumping has been obtained by transfer of energy from argon [31].

Although the output from a single nitrogen laser is usually amplified spontaneous emission, with a high divergence angle, nitrogen laser emission with output at almost diffraction limited divergence can be achieved using an oscillator–amplifier combination [32]. The divergent oscillator output is focused by a lens onto a pinhole; the diameter of the pinhole corresponds to the desired degree of divergence. The light passing through the pinhole is collimated by a second lens and passed into the amplifier. The majority of the amplifier output is then in the collimated beam. This technique has also been used to obtain near-diffraction limited output from dye lasers [33]. An alternate technique for achieving near-diffraction limited ouput by means of an unstable cavity has also been demonstrated [34].

Detectors for nitrogen lasers are fast, ultraviolet-sensitive photodiodes. The output itself is invisible, but is manifested by blue fluorescence if a white card or other fluorescing material is placed in the beam.

A major use of nitrogen lasers is as a pump for dye lasers.

Ar^+, Kr^+ Ion Lasers

The rare gas ion lasers are excited in low-pressure discharges and can be either pulsed or continuous. The argon ion laser has various transitions between 334 and 529 nm, with major transitions at 488 and 514.5 nm. The krypton ion laser has transitions between 337 and 859 nm, with the major transition at 647 nm.

Units are available that produce all the lines from these various transitions.

Laser action occurs from various excited 4*p* levels of the argon and 5*p* levels of krypton ions. An energy level diagram for argon (Fig. 7) shows the multiplicity of levels involved [35]. The 4*p* levels themselves are roughly 20 eV above the ion ground state. Thus even in a discharge with high electron temperature, typically 2 to 4 eV, very few electrons have enough energy to excite these levels [36]. As a consequence the efficiency is not spectacular, 0.05% being typical. Nevertheless, argon ion lasers are produced with up to

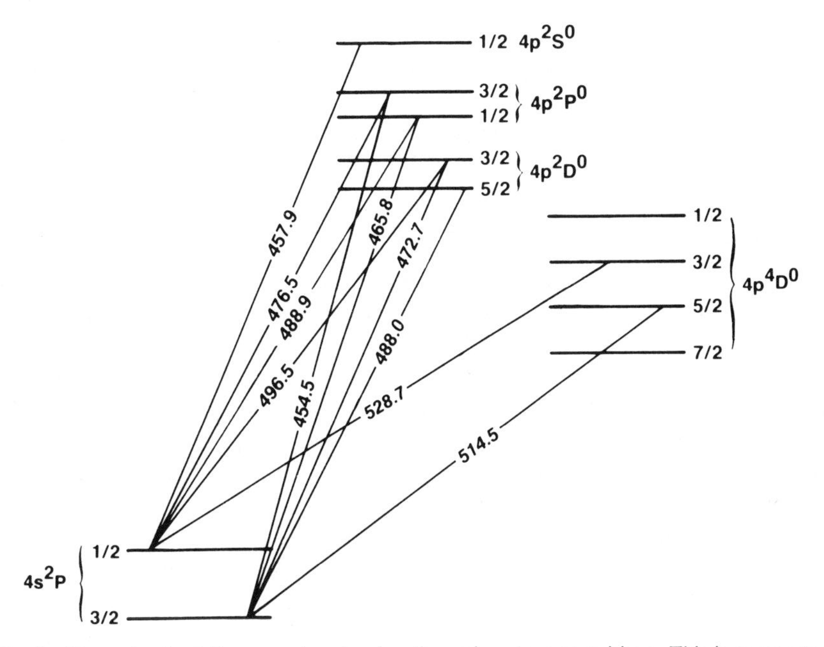

Fig. 7 Energy levels of the argon ion showing the various laser transitions. This is a corrected version of the diagram appearing in Ref. 35. The corrected diagram was kindly supplied by W. B. Bridges.

40 W of continuous output power, representing the highest continuous power available in the near ultraviolet and visible region of the spectrum. They are frequently used as the pump source for dye lasers to obtain high c.w. power tunable laser action. Ion lasers can also be made extremely stable for single-frequency operation and narrow linewidth. Linewidths of 10^{-4} cm^{-1} are achievable and find application in high-resolution spectroscopy. Mode-locking of ion lasers produces a train of short pulses whose duration can be as short as 5 psec. Using such a train of pulses to pump a dye laser results in subpicosecond tunable pulses.

Helium–Neon Lasers

Helium-neon lasers are low-power, low-pressure discharge lasers and operate at a wavelength of 632.8 nm. The maximum power available is 50 mW. Lower power units, with 1 or 2 mW output are remarkably inexpensive, costing around $100, and are used extensively as alignment devices.

The helium-neon laser was the first gas laser, and also the first c.w. laser [37]. The mixture is excited in a continuous discharge that produces the metastable 2*s* state of helium by electron impact. Collisional transfer to neon atoms results in population of the neon $3s$ levels, from which laser action results in population of the $2p$ levels (Fig. 8). The $2p$ states decay radiatively to the $1s$ levels of neon, which are metastable. However, if the diameter of the tube is large, the $2p$-$1s$ photons are reabsorbed (radiative trapping), limiting the effective decay rate of the $2p$ level, and hence the power output.

Electrochemical Gas Lasers

In practical terms, there is little difference between an electrically powered laser and an electrochemical laser. Indeed, at least one manufacturer provides a device that can be used for both types. As is described above, in an electrical laser the excited state that provides the laser energy is generated in an electrical discharge by electron impact. In an electrochemical laser the electron impact results in a specie, not necessarily excited, that subsequently reacts to produce a different specie in an excited state that then lases. In present forms the discharge-produced reactant requires several electron volts of energy, so that pulsed high-voltage transverse discharges are necessary to generate a high enough electron temperature to create the reactants.

Most electrochemical lasers use very reactive gases, that is, the halides, and particularly fluorine, as one component. The gas supply and exhaust system must be compatible with these gases. Full details of fluorine handling are given in Ref. 38, but briefly they are summarized below.

1. Metals should be copper, brass, aluminum, or steel.
2. Metal joints can be welded or brazed, but *not* soldered.
3. Nonmetals should be Teflon or KEL-F.

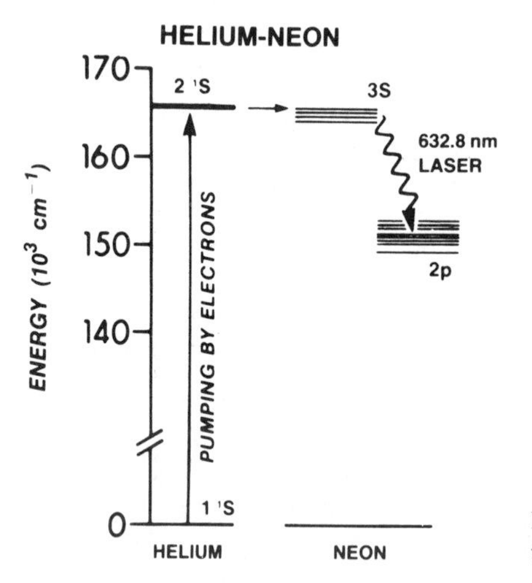

Fig. 8 The energy levels of helium and neon that contribute to the 632.8-nm transition.

4. O-rings should be avoided if possible, but if absolutely necessary, rings made of BUNA-N, lubricated as sparsely as possible with KEL-F grease, should be used.

5. Fluorine tanks should be in special, exhausted hoods, available from the gas supplier. Fluorine compatible regulators must be used.

6. Fluorine systems must be passivated before use. Passivation is achieved by letting the system stand overnight with a small amount of fluorine in it. This is repeated with increasing quantities of fluorine until the desired working concentration is reached and no change in pressure occurs on standing.

7. Exhausts should be to the outside atmosphere, at a safe place.

Fluorine is an extremely dangerous gas and should be treated with respect.

Rare Gas Halide Lasers

Rare gas halide lasers have been created in electron beam pumped discharges, ultraviolet preionized discharges, and electron beam stabilized discharges. For a review of rare gas halide lasers see Ref. 39. Commercially they are available in ultraviolet preionized discharges with pulsed energies up to about 200 mJ. In a research device, however, 350 J has been obtained from a KrF laser [40].

The gas mixtures used contain small amounts of rare gas ($\sim 10\%$ of Kr, 1% of Xe) and fluorine (0.2%) in a background of helium. The electrical characteristics of the gas are thus mainly determined by the helium, thereby permitting a high electron temperature. The high electron temperature means that there are electrons sufficiently energetic to dissociate fluorine and also to create excited rare gas atoms, and rare gas ions, that is,

$$e + F_2 \rightarrow F^- + F$$

$$e + R \begin{cases} \nearrow R^* + e \\ \searrow R^+ + 2e \end{cases}$$

where R is the rare gas. Two processes are then possible to produce excited rare gas fluorine molecules:

$$R^* + F_2 \rightarrow (RF)^* + F$$

$$R^+ + F^- + M \rightarrow (RF)^* + M$$

where M is any third body contributing to the recombination reaction. The state (RF)*, called an excimer, is a stable, electronically excited molecule, although it has a short radiative lifetime, typically of the order of 10 nsecs (41–43). Since the ground-state RF molecule is either unstable or so weakly bound that it is completely dissociated, there is automatically an inversion between the excited state of RF and the ground state, and laser action results. A potential energy diagram for KrF is given in Fig. 9, which shows

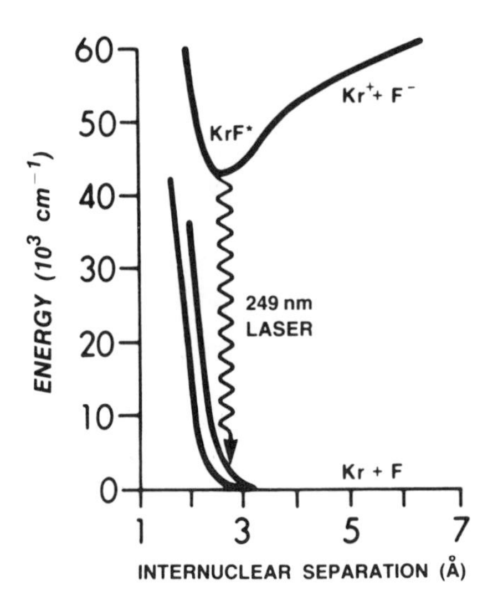

Fig. 9 Energy level diagram of krypton fluoride.

the above mechanisms. It appears that of the two mechanisms producing the rare gas fluoride, the reaction between the positive rare gas ion and the negative fluorine ion is the more important.

It might be expected that all the rare gases would combine with each of the halogens to form excimers. Many in fact do so, and Table 4 is a list of excimers that have generated laser action and the wavelength at which laser action occurs. As can be seen, this family of lasers covers an interesting range of the ultraviolet spectrum. Moreover, it is quite easy to generate other wavelengths through Raman shifting. Loree et al. [10] have demonstrated considerable coverage of the ultraviolet spectrum in this way. This class of lasers therefore seems particularly suitable for chemical applications.

One difficulty with current excimer lasers is that the output energy decays

Table 4 Raregas Halide Excimers That Have Produced Laser Action

Molecule	*Wavelength (nm)*	*Molecule*	*Wavelength (nm)*
Argon chloride	175	Xenon bromide	282
Argon fluoride	193	Xenon chloride	308
Krypton chloride	222	Xenon fluoride	352
Krypton fluoride	248		

with time if a static fill is used. Gas flow can be used to overcome this problem, but operation of the laser can then become very expensive!

The pulse length of present excimer lasers is around 25 nsecs. Fast photomultipliers with ultraviolet response are used to measure the power, and calorimeters can be used for measuring the energy. Longer pulse operation of excimer lasers has been reported [44, 45].

Hydrogen Halide Lasers

Hydrogen halide lasers are included as electrochemical lasers, since it is only in this form that they are available commercially. However, hydrogen halide lasers have been initiated optically, as well as chemically, by electron beams, and high-voltage and microwave discharges, and can be pulsed and c.w. Regardless of which initiating technique is used, the first step is the production of fluorine (or other halide) atoms. The important hydrogen–fluorine case is used here as an example. For more complete details on hydrogen fluoride and other chemical lasers, extensive reviews are available [46]. Once formed, the fluorine atoms react with hydrogen to form vibrationally excited hydrogen fluoride. The reaction is energetic enough to populate up to the third vibrational level in hydrogen fluoride:

$$F + H_2 \rightarrow HF(v = 1\text{-}3) + H$$

If the fluorine atoms are produced from fluorine molecules, a second reaction is possible, namely,

$$H + F_2 \rightarrow HF(v = 1\text{-}8) + F$$

with enough energy to populate up to the eighth vibrational level of hydrogen fluoride. Obviously these two reactions together constitute a chain reaction. If electrical initiation is used, the source of fluorine atoms is frequently sulfur hexafluoride (47), in which case there is no chain reaction. Use of sulfur hexafluoride presents somewhat of a problem in depositing sulfur throughout the apparatus unless the flow of the gas mixture is carefully configured. Since the product of the reaction is stable in the hydrogen halide lasers, there must be continual flow of gas to bring in fresh reactants. The hydrogen halide produced is a toxic gas and must be appropriately vented or neutralized.

The hydrogen halide lasers operate on vibrational-rotational transitions and are consequently in the infrared region of the spectrum. The wavelength ranges available with different halides, and also by substituting deuterium for hydrogen, are given in Table 5. The hydrogen chloride laser, using photolytic dissociation of the chlorine, was the first chemical laser invented [48].

The pulse length available depends on the technique used. As is stated above, hydrogen fluoride lasers can operate in a c.w. mode with continual flow of reactants. If a pulsed laser is made by reacting a static fill of gas, then

Table 5 Hydrogen Halide Molecules That Have Produced Laser Action

Molecule	*Wavelength (μm)*	*Molecule*	*Wavelength (μm)*
Hydrogen fluoride	2.6–3.5	Deuterium fluoride	3.6–5.0
Hydrogen chloride	3.5–4.1	Deuterium chloride	5.0–5.5
Hydrogen bromide	4.0–4.7	Deuterium bromide	5.6–6.4

the pulse length depends on the concentrations of the gases used, since a chemical reaction is involved. The shortest pulse length reported is 20 nsecs from a high-pressure electron beam initiated laser [49]. Commercial units generate pulse lengths around 1 μsec, although c.w. units are also available.

4 SOLID—STATE LASERS

Solid-state lasers, and liquid lasers, can contain an extremely high density of the excited atom and so produce considerable energy from relatively small volumes. The bandwidth of the active transition is usually high, and very short pulses can be produced by mode-locking. High intensities can be produced, sufficient in fact to damage the laser medium itself. The optical properties of solid-state lasers are excellent. The combination of properties of solid-state lasers makes them ideal for very high power, short-pulse operations. A disadvantage of solid-state lasers is that cooling takes place by conduction, which can be very slow in the glass media commonly used to contain the active atom, resulting in low average power operation. Efficiencies of solid-state lasers, except for the small diode lasers, are not very high.

Neodymium Lasers

The active medium in a neodymium laser consists of a host material containing neodymium ions as an additive; typical concentrations of the neodymium are 1 to 3% by weight. The host material is glass or a crystal of ytrium-aluminum-garnet (YAG), although a crystal of neodymium pentaphosphate has also been used [50]. The only pumping technique possible in these media is optical. Flashlamp pumping is used for pulsed operation, and high-intensity arc lamps are used for c.w. operation. The lamp spectrum is broad band, whereas the major absorption bands in the neodymium ion are less than 70 nm wide and are situated at roughly 520, 580, 740, 810, and 880 nm. Only the lamp output in these regions is useful for pumping the laser transition, and since this constitutes a small fraction of the total lamp output, this pumping technique is not very efficient. Neodymium laser efficiencies of 0.01% are common. After the neodymium ion has absorbed energy in the absorption bands, nonradiative transitions to the $^4F_{3/2}$ level occur, followed by laser action between the $^4F_{3/2}$ and $^4I_{11/2}$ level (see Fig. 10) at a wavelength of 1.06 μm. The radiative lifetime of the $^4F_{3/2}$ level is quite long, up to 900 μsec, depending on the neodymium concentration [51], and energy can be stored in the upper laser level for a relatively long time. Note, however, that at high neodymium concentrations, concentration quenching becomes important, reducing the lifetime of the level. The relatively long lifetime of the upper level has as a consequence that the flashlamps do not have to be of short duration; 100 to 200 μsec is typical. It is this storage property of the neodymium transition that makes it an excellent transition for high-power lasers—the pump energy is stored in the upper laser level until it is extracted by the high-power pulse being amplified. Neodymium-glass is used as the active medium in the extremely high power lasers constructed for thermo-

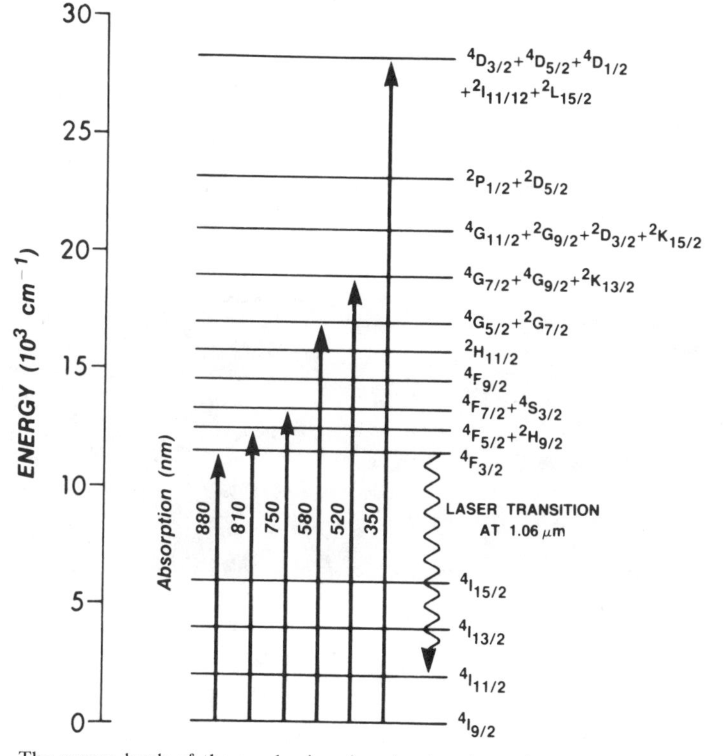

Fig. 10 The energy levels of the neodymium ion showing the various absorptions. The levels produced by absorption decay to the $^4F_{3/2}$ level, from which laser action occurs.

nuclear fusion research, namely, SHIVA, at Lawrence Livermore Laboratory, which has 20 beams and a maximum power of 30 TW, and OMEGA at the University of Rochester, with 24 beams and a maximum power of 12 TW.

The usual configuration of an optically pumped laser is a cylindrical rod surrounded by a cylinder of flashlamps. The axis of the rod is also the laser axis. The maximum rod diameter that can be manufactured practically is about 9 cm. Large diameters are desirable to reduce the intensity of the laser output for a given power or energy.

At the intensities created by neodymium-glass lasers, the nonlinear index of refraction of glass can become important. The index of refraction can be written as

$$n = n_0 + n_2 I$$

where n_0 is the ordinary, constant index of refraction, and the second term is the product of a coefficient n_2 and the intensity. Since n_2 has a typical value around 1×10^{-16} to 5×10^{-16} cm^2/W, this term becomes important for intensities above 10^{10} W/cm^2 and can result in defocusing of the beam and alteration of the pulse shape.

Neodymium lasers are amenable to mode-locking to produce very short pulses. A convenient arrangement to use is a combination of active and passive mode-locking, with a quartz crystal for the active mode locker, and a dye cell (containing Eastman 9740 or 9860 dye for short pulses and Kodak dye No. 14015 for pulses above 200 psec) for the absorbing passive mode locker (52). Figure 11 shows a mode-locked train produced in a neo-

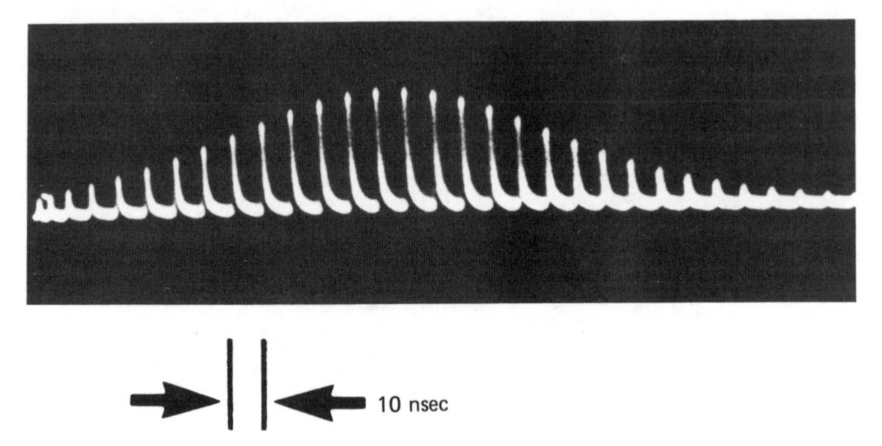

Fig. 11 A mode-locked train generated by a neodymium-glass laser. The detector used here did not have the resolution to demonstrate the pulse duration; however, the pulses are actually about 100 psec long.

dymium-glass laser. The bandwidth of the transition depends on the host material. In silicate glass it is 5×10^{12} to 6.5×10^{12} Hz, allowing pulses as short as 0.2 psecs to be produced; in YAG the bandwidth is 1.2×10^{11} Hz, limiting the shortest pulse to 8 psecs. The narrower bandwidth in YAG results in a higher gain, which is useful in small devices. There is also a difference of wavelength between the neodymium-glass lasers at 1.06 μm and the neodymium-YAG lasers at 1.064 μm. New, different types of glass, that is, phosphate and fluorophospate, are under investigation for improvements in performance of neodymium lasers.

Commercial units are available with up to 10^3 J of pulsed energy and with c.w. power up to 10^3 W. Doubling, tripling, and quadrupling accessories are readily available.

Ruby Lasers

Ruby was the first laser medium to be discovered (53). The laser action takes place in Cr^{3+} ions present as an impurity in the host crystal of Al_2O_3. Typical Cr^{3+} concentrations are around 0.05%. The Cr^{3+} absorbs flashlamp radiation into the wide bands resulting from transitions from the $^4A_2$ ground state and the $^4F_1$ and $^4F_2$ levels. Rapid relaxation to the ^2E level follows. This ^2E level has a lifetime of about 5 msec [54] and consequently can develop a considerable population. Laser action occurs between the ^2E level and the ground state (Fig. 12). Note that more than half the ions present must be pumped out of the ground state before an inversion can be achieved. Conse-

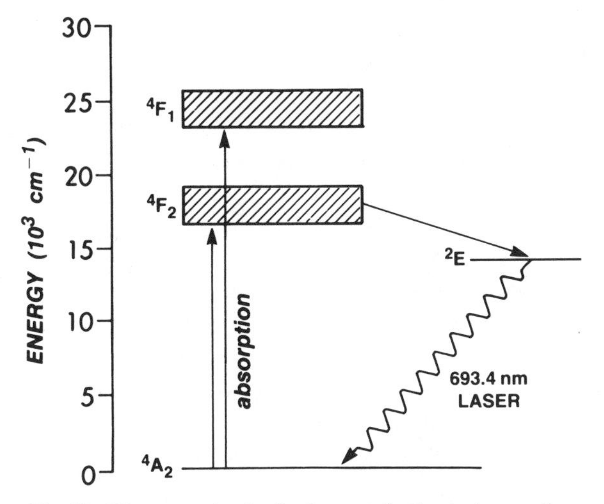

Fig. 12 The energy levels of ruby contributing to laser action.

quently, the ruby laser exhibits a fairly high threshold energy (i.e., energy input to achieve marginal laser action) and low efficiency.

F-Center Lasers

By bombarding alkali halide crystals with electrons, F_2^+ color centers can be created. On pumping these crystals with a visible laser, which can be krypton ion, argon ion, or a dye laser, tunable laser output is possible in the 2.2- to 3.1-μm wavelength region. The exact wavelength region available depends on the crystal used. In the commercial model, crystals are held at liquid nitrogen temperature to minimize the rate of disappearance of color centers.

Output powers of 240 mW at an efficiency of 9% have been obtained from potassium chloride, but practical values are about an order of magnitude lower. Linewidths of 1 MHz are possible, making the laser ideal for spectroscopic applications in the 2.2- to 3.1-μm region.

Semiconductor Lasers

Electrically powered semiconductor diodes are available that produce either c.w. or pulsed output. In general these are small devices with low energy and power output, and a high divergence angle. They are very efficient, efficiencies up to 80% having been achieved. Diodes of ternary compositions containing lead possess exceptional tuning ability and, depending on the composition, can be tuned between about 3 and 30 μm (55). The linewidth of such lasers is less than 10MHz, so that they form excellent sources for high-resolution spectroscopy in the infrared spectral region. A commercial spectrometer based on a tunable diode is available.

Diode lasers at 880 nm have been used to pump neodymium-glass lasers, since the Nd^{3+} ion has an absorption at 880 nm (56). The result is an efficient Nd^{3+} laser. Unfortunately, the energy available from diodes is small, around 10^{-4} J, and the cost of the diodes is sufficiently high that making a large array to generate a large pump energy is prohibitively expensive at the moment.

5 DYE LASERS

Dye lasers are used mainly in the liquid phase, although vapor phase dye lasers have been demonstrated. Pumping of the liquid phase dye lasers is optical, with a wide range of optical pumps being used, including linear and cylindrical flashlamps; nitrogen lasers; ruby lasers, direct and frequency doubled; neodymium lasers, direct, frequency doubled, and quadrupled for pulsed dye lasers; and argon and krypton ion lasers for c.w. dye lasers. Vapor phase dye lasers have been pumped by nitrogen lasers [57], and experiments with electrical discharge pumping are in progress [58]. An excellent survey of dye lasers is found in Schäfer (59).

Pulse energies up to 150 J have been produced (60), although an energy of 1 J is more common. Power levels for c.w. laser output are typically less than 1 W, although 20 W has been achieved (61). Dye lasers convert the laser pump energy to output energy very efficiently, with conversion efficiencies up to 50% being possible.

The importance of dye lasers, particularly in chemical applications, is the wide range of wavelengths available with dyes, covering the ultraviolet to the infrared, with each dye exhibiting continuous tuning over its working range, which can be 100 nm wide. Because of the colors generated, dye lasers are also incredibly beautiful.

The tunability of dye lasers stems from the broadness of the electronic energy levels in the dye molecules. Because the molecules are large, there are many vibrational and rotational levels associated with each electronic energy level whose widths overlap sufficiently to create one broad energy level. A set of energy levels for a typical dye is shown in Fig. 13. Absorption from the lowest level of the singlet state takes place to any of the levels of the excited singlet state. From there molecules relax rapidly, in less than a picosecond, to the lowest level of the excited state. Laser action occurs from that level to one of the levels of the ground electronic state. Since the vibration-rotational states of the ground electronic level cover a broad range of energy, the laser is tunable over this range. The mechanism described indicates that the absorption by dyes occurs at a wavelength shorter than the emission wavelength.

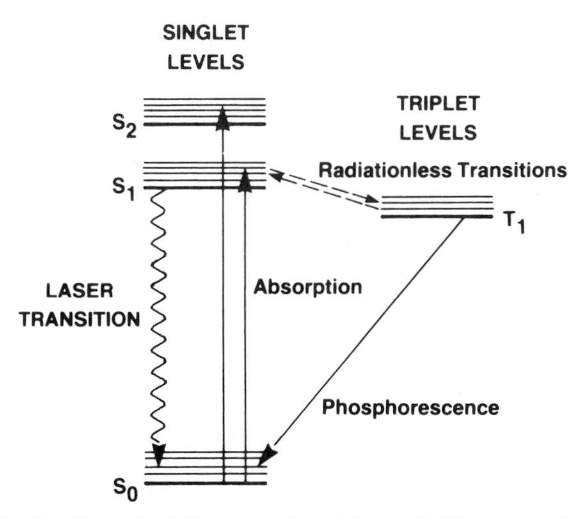

Fig. 13 Energy level diagram typical of dyes exhibiting laser action.

This is the case, as is seen in the absorption-emission spectrum of Rhodamine 6G shown in Fig. 14. Typical radiative lifetimes of the excited singlet are 1 to 10 nsecs, so dye lasers are not storage media. A loss mechanism from the excited singlet state is radiationless transition to the triplet state, which is usually metastable.

An individual dye does not cover the whole visible spectrum. Different dyes are available that cover this spectrum; Fig. 15 shows the spectral regions in which different families of dyes operate, together with the pumping sources that can be used.

In general, flashlamp-pumped dye lasers produce more energy, with longer pulse lengths (up to several hundreds of microseconds), than laser-pumped dye lasers. However, the long pulse length permits buildup of the triplet state, which severely limits the gain due to removal of molecules from the singlet system and absorption in the triplet system. Consequently, only a limited number of dyes exhibit laser action by flashlamp pumping. This situation is improved somewhat by using coaxial flashlamps, for which the inductance can be very low and the pulse length short, that is 1 to 2 μ-sec. Since the transfer time to the triplet state is of the order of 100 nsec, pumping by lasers with pulse lengths of the order of 10 to 20 nsecs results in insignificant transfer to the triplet state, and hence high gain and power. From the above, it might be thought that c.w. operation of dyes would be impossible. This is not true if some way can be found to remove the triplet state. Such removal is achieved by adding chemicals that quench the triplet state to the singlet state or by flowing the dye through the active region.

Dye lasers have an extremely large bandwidth, of the order of 10^{13} to 10^{14} Hz. Thus in principle they could produce pulses as short as 10^{-14} sec. While

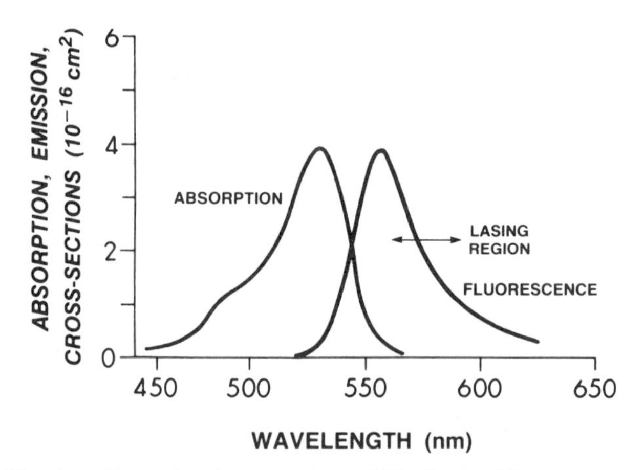

Fig. 14 The absorption and emission spectrum of Rhodamine 6G, a common laser dye.

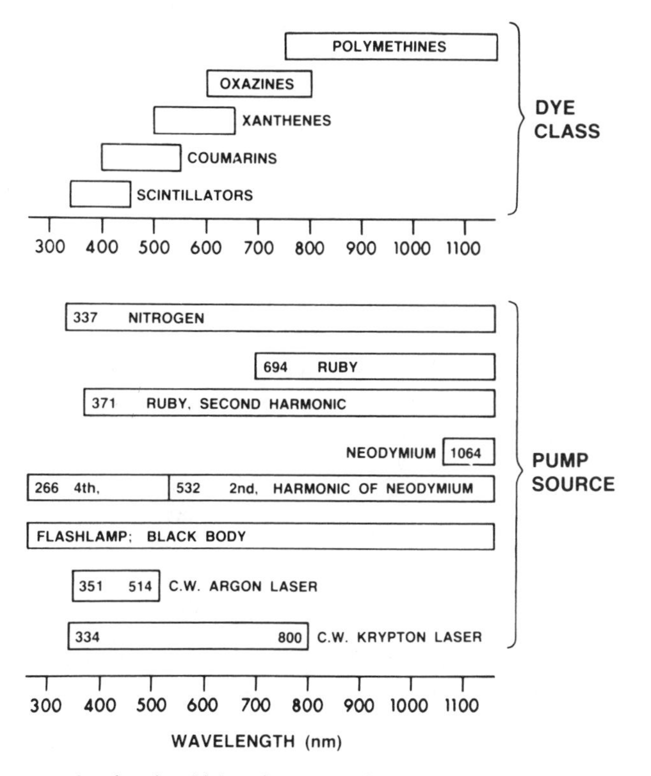

Fig. 15 The spectral regions in which various types of dyes operate, and appropriate pumps for these regions.

pulses this short are not readily achievable, dye lasers have been used for generation of subpicosecond pulses. In one arrangement this was realized by pumping with a mode-locked argon ion laser, producing a continuous train of 15 to 20 psec pulses (62, 63). It should be noted that measurement of pulse lengths this short is not trivial. Photodetector-oscilloscope combinations are usually limited to a resolution of about 100 psec. Instead one must use an ultrafast streak camera or two-photon fluorescence (64, 65).

6 COST AND AVAILABILITY

The objective of this chapter is to describe general types of lasers available, rather than individual models, and prices. Such information on individual models would rapidly become obsolete. Nevertheless, the scientist wishing to use a laser needs to know if there is a model available that will deliver the energy and power he requires, at a price that will not exceed his budget. The most convenient source of this information is the *Laser Focus Buyer's Guide* (66). This annual publication lists all the models of lasers commercially available, together with the laser energy, power, pulse length, repetition rate, beam divergence, and price. In addition, other items of laser-related equipment, such as frequency doubling crystals, detectors, optical components, and electro-optical devices are listed.

References

1. R. V. Ambartzumian and V. S. Lethokov, *Acc. Chem. Res.* **10,** 61 (1977); and in *Chemical and Biochemical Applications of Lasers*, Vol. 3, C. B. Moore Ed., Academic, New York, 1977, p. 167.
2. J. P. Aldridge, J. W. Birely, C. D. Cantrell, and D. C. Cartwright, in *Laser Photochemistry, Tunable Lasers, and Other Topics*, Physics of Quantum Electronics, Vol. 4, S. F. Jacobs et al, Eds., Addison-Wesley, Reading, MA, 1976.
3. C. D. Decker, *Laser Focus*, **13** (8), 6 (August 1977).
4. *Safe Use of Lasers*, ANSI Standard, Z-136.1, 1973, American National Standards Institute, 1430 Broadway, New York, NY 10018; see also *Laser Safety Guide*, 1975, Laser Institute of America, 4100 Executive Park Drive, Cincinnati, OH 45241.
5. A. G. Fox and T. Li, *Bell Syst. Tech. J.*, **40,** 453 (1961).
6. H. Kogelnik and T. Li, *Proc. IEEE*, **54,** 1312 (1966).
7. A. E. Siegman, *Proc. IEEE*, **53,** 277 (1965).
8. D. A. Kleinman, *Phys. Rev.*, **128,** 1761 (1962).
9. S. E. Harris, *Proc. IEEE*, **57,** 2096 (1969).
10. T. R. Loree, R. C. Sze, and D. L. Barker, *Appl. Phys. Lett.*, **31,** 37 (1977).
11. N. Djeu and R. Burnham, *Appl. Phys. Lett.*, **30,** 473 (1977).
12. J. Wilson, *Appl. Phys. Lett.*, **8,** 159 (1966).
13. R. T. Hodgson, *Phys. Rev. Lett.*, **25,** 494 (1970).
14. I. Wieder, *Phys. Lett.*, **13,** 759 (1967).
15. E. T. Gerry, *IEEE Spectrum*, **7,** 51 (1970).
16. T. A. Cool, R. R. Stephens, and T. J. Falk, *Int. J. Chem. Kinet.*, **1,** 495 (1969).
17. C. K. N. Patel, *Appl. Phys. Lett.*, **7,** 290 (1965).
18. G. J. Schulz, *Phys. Rev.*, **155A,** 988 (1964).
19. N. G. Basov, E. M. Belenov, V. A. Danilychev, O. M. Kerimov, A. S. Podsosonnyc, and A. F. Suchkov, *Sov. J. Quantum Electron.*, **4,** 1119 (1975).
20. A. J. Alcock, K. Leopold, and M. C. Richardson, *Appl. Phys. Lett.*, **23,** 562 (1973).
21. R. Fortin, F. Rheault, J. Gilbert, M. Blanchard, and J. L. Lachambre *Can. J. Phys.*, **51,** 414 (1973).
22. A. F. Gibson, M. F. Kimmitt, and C. A. Rosito, *Appl. Phys. Lett.* **18,** 546 (1971).

23. O. R. Wood, R. L. Abrams, and T. J. Bridges, *Appl. Phys. Lett.*, **17,** 376 (1970).
24. A. C. Walker and A. J. Alcock, *Opt. Commun.*, **12,** 430 (1974), also A. J. Alcock and A. C. Walker, *Appl. Phys. Lett.*, **25,** 299 (1974).
25. H. S. Kwok, E. Yablanovitch, *Appl. Phys. Lett.*, **30,** 158 (1977).
26. S. A. Jamison and A. V. Nurmikko, *Appl. Phys. Lett.*, **33,** 598 (1978).
27. G. L. McAllister, V. G. Draggoo, and R. G. Eguchi, *Appl. Opt.*, **14,** 1290 (1975).
28. M. M. Mann, *AIAA J.*, **14,** 549 (1976).
29. M. J. W. Bowness and R. E. Center, *J. Appl Phys.*, **48,** 2705 (1977).
30. L. W. Dotchin, E. L. Chupp, and D. J. Pegg, *J. Chem. Phys.*, **59,** 3960 (1973).
31. L. Y. Nelson, G. J. Mullaney, and S. R. Byron, *Appl. Phys. Lett.*, **22,** 79 (1973).
32. A. Schneiderman and I. Itzkan, *J. Appl. Phys.*, **41,** 2253 (1970). This reference describes the application of this technique to the neon laser, which, because of its narrower bandwidth output, is more suitable for interferometry. However, the same authors also used the technique with a nitrogen laser.
33. I. Itzkan and F. W. Cunningham, *J. Quant. Electron.*, **QE-8,** 101 (1972).
34. G. C. Thomas, G. Chakrapani, and C. M. L. Kerr, *Appl. Phys. Lett.*, **30,** 633 (1977).
35. W. B. Bridges, *Appl. Phys. Lett.*, **4,** 128 (1964).
36. W. B. Bridges, A. N. Chester, A. S. Halstead, and J. V. Parker, *Proc. IEEE*, **59,** 724 (1971).
37. A. Javan, W. B. Bennett, Jr., and D. R. Herriott, *Phys. Rev. Lett.*, **6,** 106 (1961).
38. H. W. Schmidt, *Handling and Use of Fluorine and Fluorine-Oxygen Mixtures in Rocket Systems*, National Aeronautics and Space Administration, NASA SP-3037, 1967.
39. C. A. Brau, "Rare Gas Halogen Excimers," *Excimer Lasers*, C. K. Rhodes, Ed., Springer-Verlag, New York, 1979, Chap. 4.
40. R. Hunter, 7th Winter Colloquium on High Power Visible Lasers, Park City, Utah, February 16–18, 1977.
41. P. J. Hay and T. H. Dunning, *J. Chem. Phys.*, **69,** 2209 (1978); and *J. Chem. Phys.*, **69,** 134 (1978).
42. J. G. Eden and S. K. Searles, *Appl. Phys. Lett.*, **30,** 287 (1977).
43. R. Burnham and S. K. Searles, *J. Chem. Phys.*, **67,** 5967 (1977).
44. L. F. Champagne, J. G. Eden, N. W. Harris, N. Djeu, and S. K. Searles, *Appl. Phys. Lett.*, **30,** 160 (1977).
45. B. Forestier and B. Fontaine, *High-Power Lasers and Applications*, K. L. Kompa and H. Walther, Eds., Springer-Verlag, New York, 1978, p. 53.
46. The literature on chemical lasers is immense. For bibliography see (a) R. W. F. Gross and J. F. Bott, *Handbook of Chemical Lasers*, Wiley, New York, 1976; (b) A. N. Chester, *Proc. IEEE*, **61,** 414 (1973); (c) K. L. Kompa, *Chemical Lasers, Topics in Current Chemistry, Vol. 37*, Springer-Verlag, New York, 1973.
47. H. Pummer and K. L. Kompa, *Appl. Phys. Lett.*, **20,** 356 (1972).
48. J. V. V. Kasper and G. C. Pimentel, *Phys. Rev. Lett.*, **14,** 352 (1965).

49. R. A. Gerber and E. L. Patterson, *J. Appl. Phys.*, **47,** 3524 (1976).
50. J. Wilson, D. C. Brown, and W. K. Zwicker, *Appl. Phys. Lett.*, **33,** 614 (1978).
51. E. Snitzer, *Appl. Opt.*, **5,** 1487 (1966).
52. W. Seka and J. Bunkenburg, *J. Appl. Phys.*, **49,** 4 (1978).
53. T. H. Maiman, *Nature*, **187,** 493 (1960).
54. T. H. Maiman, *Phys. Rev. Lett.*, **4,** 564 (1960).
55. K. J. Linden, K. W. Nill, and J. F. Butler, *IEEE J. Quant. Electron.*, **QE-13,** 720 (1977).
56. S. R. Chinn, H. Hong, and J. W. Pierce, *Laser Focus*, **12** (5), 64 (1976).
57. P. W. Smith, *Opt. Acta*, **23,** 901 (1976).
58. G. Marowsky, R. Cordray, F. Tittel, B. Wilson, and J. W. Keto, *J. Chem. Phys.*, **67,** 4845 (1977).
59. F. P. Schäfer, Ed., *Dye Lasers*, Springer-Verlag, New York, 1977.
60. V. A. Alekseev, I. V. Antonov, V. E. Korobov, S. A. Mikhnov, V. S. Produkin, and V. B. Skortsov, *Sov. J. Quant. Electron.*, **1,** 643 (1972).
61. J. Jethwa, F. P. Schäfer, and J. Jasny, *IEEE J. Quant. Electron.*, **QE-14,** 119 (1978).
62. J. M. Harris, R. W. Chrisman, and F. E. Lytle, *Appl. Phys. Lett.*, **26,** 16 (1975).
63. R. K. Jain and J. P. Heritage, *Appl. Phys. Lett.*, **32,** 41 (1978).
64. A. J. Maria, W. H. Glenn, M. J. Brienza, and M. E. Mack, *Proc. IEEE*, **57,** 2 (1969).
65. M. A. Duguay, *IEEE J. Quant. Electron.*, **QE-6,** 725 (1970).
66. *Laser Focus Buyer's Guide*, published annually by Advanced Technology Publications, Inc., 1001 Watertown St, Newton, MA 02165.

Chapter **II**

APPLICATIONS OF LASERS IN ANALYTICAL CHEMISTRY

John C. Wright

Department of Chemistry
University of Wisconsin-Madison
Madison, Wisconsin

1 INTRODUCTION

The rapid development of the laser and the associated technologies has caused a revolution in the field of spectroscopy. New techniques have been developed that were not possible without the laser and old techniques have been improved and expanded in their capabilities. Spectroscopy in all of its forms is one of the most frequently used analytical methods for chemical analysis and one would expect the laser to have as large an impact in this area. Except for the field of Raman spectroscopy, which the laser made practical, the laser has not yet been used in any substantial way for analytical determinations. Partially, the dormancy of the laser can be attributed to the existence of well-established techniques that are capable of performing the necessary measurements at a lower cost or with greater simplicity. There is a great inertia in replacing these techniques. More importantly, however, methods are still being developed that utilize many of the unique capabilities of the laser in a broad analytical way. It is the responsibility of the analytical chemist or scientist functioning as an analytical chemist to develop a marriage between the laser and the chemical system that is responsive to the two individual personalities. In this chapter, we explore a number of promising possibilities.

Absorption spectroscopy is perhaps the most widely used analytical tool. Infrared spectroscopy is extensively utilized for gas analysis and organic structural identification, while UV-visible spectroscopy is extensively used for quantitative analysis. The laser has proven especially valuable in the infrared, where the high-intensity and narrow bandwidths have solved the long-standing problem of poor detector sensitivity and low source intensity. Applications in UV-visible spectroscopy have been more limited because the broader absorption lines characteristic of this region for condensed phase samples do not require spectrally narrow sources. Nevertheless, the laser has made possible a number of new absorption techniques that are capable of measuring extremely small absorbances.

Fluorescence spectroscopy is an analytical method that is applied in many areas of science. It can be divided into atomic fluorescence and molecular fluorescence. The fluorescence intensity can be increased by increasing the excitation intensity up to the point of saturation. In atomic fluorescence, the laser is a convenient source for exciting narrow lines of atomic transitions with high intensities. At the higher intensities that are available, factors other than the excitation source intensity limit the use of atomic fluorescence. The present status of this work is discussed in some detail. Likewise in molecular fluorescence spectroscopy, the technique is not limited by the excitation intensity, but by the broad-band nature of the molecular transitions themselves. The methods currently under study for overcoming these problems are discussed later.

The very high peak power of the laser has resulted in a completely new detection method in which the ionization that results from a single photon transition to an excited state nearer the ionization continuum or from a multiple photon transition into the ionization continuum is measured directly. This method has provided single atom detection capabilities and it promises to have a wide applicability for analytical measurements both by itself and in combination with other methods such as mass spectroscopy.

The laser has opened an entirely new field of nonlinear spectroscopy. Since it is a coherent light source, the laser can be focused to spot sizes limited only by diffraction. The electric field from the light in the focal spot can easily approach or exceed the electric fields within or between molecules, even with readily available lasers. When the light intensities approach the molecular bonding energies, there can no longer be a linear relationship between the electric field of the electromagnetic wave and the polarization induced in the material. In this regime, numerous nonlinear effects become important. In particular, it becomes possible to drive molecular vibrations with light and thereby stimulate Raman scattering. This ability has given birth to a number of coherent Raman spectroscopies that are described here in depth. We do not discuss conventional Raman spectroscopy in this chapter because it is widely understood and appreciated.

There are numerous other areas of potential interest in analytical chemistry that cannot be included because of space limitations. These topics include high-resolution techniques such as saturation spectroscopy [1-2], two-photon spectroscopy [3-4], polarization spectroscopy [5-7], and various double resonance methods [8] and scattering techniques, such as Doppler velocimetry [9-12] and Rayleigh scattering. It is suggested that the serious reader follow these areas as well. Other applications have no doubt escaped this author's attention or interest and it is sobering to think that many important applications can probably not even be forseen at the present time. This Chapter therefore concentrates on some areas that appear particularly promising at present.

2 ABSORPTION SPECTROSCOPY

An absorption measurement requires either that one measure the change in the intensity of an excitation source as it passes through an absorbing material or that one measure effects of the absorption (e.g., thermal heating). Both of these techniques have been advanced substantially by development of laser-related methods. If one is measuring a change in transmitted light, it is important to have a low-noise, reproducible measurement of the light intensity. Laser sources provide high-photon fluxes that can both improve the shot noise contribution to the measurement error and overcome the problems of working at low light levels with insensitive detectors. The directionality of a laser can permit different approaches, such as intracavity absorption techniques and long path lengths for remote monitoring. If one is measuring effects of the absorption, the high flux levels of the laser can increase the number of photons absorbed and therefore the effects associated with absorption.

Direct Absorption Methods

The laser offers little advantage over blackbody sources for conventional analytical UV-visible absorption spectroscopy. In later sections, we see how special techniques can give the laser some unique capabilities in this spectral region. The laser does offer some impressive advantages for infrared absorption spectroscopy, however. Infrared transitions are much sharper than those encountered in UV-visible spectroscopy. Gas phase measurements at low pressures have linewidths determined by Doppler broadening and are typically 10^{-3} cm^{-1} at room temperature [13]. It is therefore more important to have good resolution and narrow bandwidths for an infrared measurement, particularly for a gaseous sample. The intensity of light from a blackbody source is always decreased as the passband of the monochromator is decreased since spectral brightness must be constant. The low brightness of blackbody sources in the infrared and the poor detectivities of infrared detectors quickly impose limits on the resolution that can be achieved by conventional infrared instrumentation. A laser source provides a new approach to this problem. The bandwidth of a laser is not improved by throwing away energy at unwanted wavelengths as must be done with conventional approaches, but instead by changing the cavity to concentrate the power in the bandwidth desired. High intensities and high resolution are therefore compatible for coherent sources. The sources used for infrared laser spectroscopy are the semiconductor diode lasers [14], the optical parametric oscillator [15], the color center laser [16, 17], the spin-flip Raman laser [18], and the many different kinds of fixed frequency gas lasers [19]. All these lasers can be operated with a narrow bandwidth and a high enough intensity to make detector noise a minor contribution to the measurement error.

Many applications for infrared absorption spectroscopy depend on laser infrared sources, including atmospheric pollution monitoring [13], combustion profiling, explosives detection [20] (by detecting volatile organics from the explosives), and isotope analysis [21] (particularly as it might be used for dating). This short list of applications represents a much larger number that are under study but it serves to illustrate the analytical requirements. Suppose, one would like to make a gas phase measurement at low concentration levels at atmospheric pressure. A gaseous species at atmospheric pressure undergoes collisional broadening of the vibrational transitions [22]. Figure 1 shows the infrared spectrum of NH_3 at low and high pressures [23]. The broadening that is evident causes a loss in selectivity for specific gases and a loss in sensitivity since the absorption coefficient at peak absorbance becomes lower. Typically, the linewidth of room temperature gases is about 10^{-3} to 10^{-2} cm^{-1} at low pressures and broadens to about 0.05 to 0.5 cm^{-1} at atmospheric pressure [22]. A semiconductor diode laser has been commonly used for performing infrared measurements in both the high- and low-pressure regimes. Diode lasers typically have bandwidths of 10^{-4} cm^{-1}, more than sufficient for infrared applications [14].

A semiconductor diode laser can be used for making a laboratory infrared absorbance measurement down to about 10^{-4} absorbance units (lower values are possible) [13]. Most gases have absorption coefficients of 10 to 100 cm^{-1} atm^{-1}. Over a 1-m path length, one could detect concentrations of about 10^{-7} to 10^{-8} atm or 10 to 100 ppb. The coherence and directionality of a

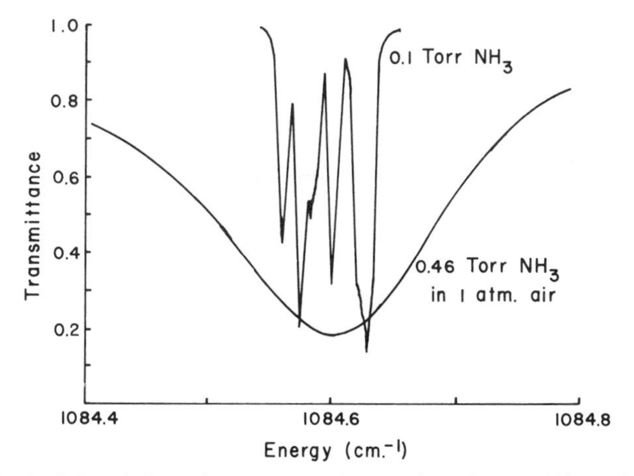

Fig. 1 Infrared absorption spectrum of ammonia at 0.1 and 750 torr [23].

diode laser allows one to make absorbance measurements over longer path lengths of several kilometers. Atmospheric fluctuations over these path lengths limit the smallest measurable absorbance to about 10^{-3} [22]. Over a 1-km path, one could then detect concentrations of about 10^{-9} to 10^{-10} atm or 1 to 0.1 ppb. Some typical values for absorption measurements are given in Table 1 for different pollutant gases. The realization of these detection limits depends on the spectral overlaps between the lines of the analyte and background gas.

There are difficulties in using the diode lasers over wide spectral regions [14]. It is possible to tailor the composition of a diode laser to work anywhere in the region from 330 to 3700 cm^{-1} (2.7 to 30 μm). Unfortunately, any one diode laser can only be tuned over about 50 cm^{-1}. To obtain wavelengths out of the 50-cm^{-1} range, a different diode laser must be constructed. (Some diode lasers can be tuned over 300 cm^{-1} [24].) Even within the 50-cm^{-1} range of a given diode laser, one can tune continuously only 1 to 2 cm^{-1} by varying the diode current before the laser hops to a different mode [14]. This problem is not a severe difficulty because one is usually interested in one specific transition that is free of interference from other absorptions. One can also use double beam or multiple wavelength techniques to overcome spectral interference and remain working with one line [13, 22].

A key method for improving the infrared measurement is to use derivative spectroscopy [13, 22]. The wavelength of the laser is modulated at a frequency and the AC component of the transmitted intensity is measured. If the modulation frequency is higher than the noise frequencies, noise contributions are eliminated and the ability to measure small changes in intensity can improve markedly. Atmospheric fluctuations, for example, occur on a time scale of 10^{-2} sec or longer, so that if one modulates the laser at 10^4 Hz, the fluctuations are effectively frozen and do not contribute to the signal [22].

Table 1 Predicted Minimum Detectable Concentration [23] for Different Gases of Atmospheric Pressure over a 1-km Path Length in the Atmosphere

Gas	$\bar{\nu}$ *(cm^{-1})*	*k (cm^{-1}atm^{-1})*	*Predicted Minimum Concentration (ppb)*
Freon-11 (CCl_3F)	847	110	0.27
Freon-12 (CCl_2F_2)	921	275	0.11
Vinyl chloride	940	11	2.8
Ethylene	950	42	0.71
Ozone	1052	22	1.3
Ammonia	1085	93	0.33

Three basic methods are used to measure atmospheric pollution by infrared laser spectroscopy [13]: (1) Sample the atmosphere and analyze the sample in the laboratory. (2) Measure the concentrations directly in the air over a long path length by locating the detector at the end of the path or placing a mirror there to return the beam to a detector. (3) Rely on scattering from atmospheric particulates or distant objects to return the beam to a detector. The first method has the advantage that intracavity absorption or optoacoustic detection methods can be used for enhanced sensitivity and the sample can be measured at lower pressures for enhanced selectivity even though the path length of the measurement is shorter. The first method has all the disadvantages of an indirect measurement made at a different place and time from those of the situation of interest. Both the second and third methods allow real time monitoring of atmospheric conditions. The second method requires placing either a mirror or detector at the end of the measurement path and therefore restricts the situations where it might be used. To eliminate the remote station, a previously developed method called light detection and ranging (LIDAR, in analogy to RADAR) was adopted. LIDAR used particulate and aerosol concentrations to generate Mie backscattering [25]. In the adaptation, one sends out two different wavelengths, one wavelength where an analyte gas absorbs and one where it transmits. Then by examining the backscattered signal, one can measure the differential absorption between the two wavelengths, which is determined by analyte gas concentration. This differential absorption LIDAR is called DIAL [26, 27]. If the laser source is pulsed, it is also possible to obtain accurate ranging information about the spatial variations of analyte concentration by simply looking at the backscattering with different time delays. A time resolution of 300 nsec provides a spatial resolution of 90 m. A more extensive description of these methods is available elsewhere [13, 27, 28].

Optoacoustic Spectroscopy

A technique with expanding applications has joined the set of tools used by the analytical spectroscopist [29]. Spectroscopy has long been concerned with the light emitted by objects after they have been excited. More commonly, however, the excited states do not relax by light emission but release their energy as thermal energy. Even if a sample emits light, often the light is absorbed by the walls of its surroundings and is converted to heat also [13]. By modulating the excitation source, one also modulates the thermal relaxation and sound waves can be generated. Sensitive microphones can detect this sound and thereby measure the absorption directly. One therefore eliminates the problem of trying to measure a small absorption by observing a small change in transmitted light intensity. Now if the intensity of the excitation source is increased, the absorption and the acoustic signal also increase. One is then limited in measuring small absorptions only by the sensitivity of the microphone, the intensity of the source, and any background absorption that could interfere. The laser is a good excitation source for this application because it can provide high intensities and excellent resolution. The spectral response of the instrument can even be calibrated by replacing the sample with carbon black or samples of known spectral characteristics.

Optoacoustic spectroscopy has been applied to gas, liquid, and solid samples. Gas samples are measured directly by placing the sample gas within a chamber that is coupled to a microphone [30, 31]. A buffer gas is often used to make the acoustical properties of the chamber insensitive to variations in samples. Condensed phase samples can be measured by placing them within a similar chamber where acoustical contact between the heated area on the sample and the microphone is established by the buffer gas [29]. The losses involved in this arrangement lower the inherent sensitivity. More direct contact can be achieved by other methods. The microphone can be mounted directly on a solid if its properties are suitable. Robin eliminated the microphone altogether and evaporated a lead or indium film on the surface of a ruby sample and cooled it to the superconducting transition of the film [32]. At the superconducting transition, the thin film resistance becomes incredibly sensitive to small variations in temperature, such as that produced by light absorption. Liquid samples can also be measured by direct contact with the microphone [33–38]. The microphone can be placed directly in the liquid sample or the sample container can be made from a piezoelectric ceramic that acts as a microphone. This approach can cause contamination of both the measuring system and the sample. An alternate arrangement is to mount a microphone through suitable acoustical contacts to a standard absorption cuvette. Voightman et al., for example, glued a microphone onto a quartz disk, which was attached in turn to a cuvette through a droplet of glycerol [39].

In cases where the temperature changes caused by light absorption are transferred to the microphone by an intervening phase, such as optoacoustic spectroscopy of opaque solids with a buffer gas in the detection chamber, the bulk absorption of the sample may not be measured and the surface contributes the major fraction of the signal [40–43]. The depth of sample that contributes a signal depends upon the modulation frequency and the thermal properties of the sample [40]. The time delay between the generation of heat in the bulk and its transmission to the surface is described by a thermal conductivity. If the modulation frequency is significantly higher than the rate associated with thermal conduction from a given depth to the surface, the thermal modulation amplitude is small [40, 41]. The sampling depth is therefore a function of the modulation frequency and the sample thermal conductivity. The spectra from solid samples are also complicated by the wavelength dependence of the reflection coefficient, which contributes to the observed optoacoustic spectrum by preventing the light from reaching the sample.

The microphones used for optoacoustic spectroscopy are quite sensitive, typically providing 1 V/torr of signal [44]. With FET preamplifiers directly attached to the microphone and modern amplifiers and lock-in or boxcar detectors, signals as low as 0.2-μV can be detected. These signals correspond to pressure changes of only 2×10^{-7} torr or to absorptions of as little as 10^{-9} J from a pulsed laser [44]. If the pulse itself contains 1 mJ, an absorbance of 10^{-6} can be measured [33, 39, 44]. The minimum detectable absorption can be lowered by using more sensitive microphones or more powerful lasers than those used for the illustration [33, 44]. As the laser power is increased, one eventually reaches the point where the background absorption obscures the signal of interest [13]. For gas and liquid phase samples, the background is usually caused by absorption in the cell windows. The windows cause two contributions—direct absorption in the window and scatter from the windows, followed by absorption at the walls and detectors. Both contributions can be diminished by using high-quality windows, acoustically baffling the windows, or using windowless cells (at least for gas phase samples). If a pulsed laser is used, the direct absorption from the window produces a signal that arrives at a different time than that from the gas or liquid. Temporal resolution can then discriminate against window absorption [44]. Absorbance measurements down to 10^{-9} have been reported for gas phase samples [13]. For a gas phase infrared measurement, this would correspond to 1 to 10 ppt. Liquid sample measurements have been done for β-carotene analysis [38], Cd trace analysis [37] using complexation with dithizonate, and a variety of other important solutions [39]. Detection limits of about 10 to 20 pg/ml are typically obtained. Absorptions of 10^{-7} cm^{-1} are measurable [33]. For solid samples, the background is caused by light scattering from the sample surface followed by absorption at the cell walls. The background is therefore much higher for solids, and low-concentration measurements become difficult [40].

The upper limit for absorption measurement by optoacoustic spectroscopy occurs when the amount of absorption is an appreciable fraction (10%) of the excitation intensity and the response begins to saturate. One is then limited in the highest absorbance measurable to about 10^{-1} [13]. Thus for gas phase samples, this technique possesses an impressive linear dynamic range of 8 orders of magnitude.

The signal levels can be increased by using resonantly enhanced cells [45]. If the modulation frequency matches natural acoustical resonances of the cell, the sound wave amplitude can be enhanced. One can also change the buffer gas to optimize the coupling between the sample and the microphone

[46, 47]. The buffer gas also influences the acoustically resonant frequencies of the cell. Figure 2 shows the enhancement in the optoacoustic signal from methane that occurs as a function of modulation frequency for different buffer gases [47]. It is clear that the buffer gas has a profound influence on the resonant frequencies and amplitudes. Thomas et al. [47] point out that the acoustically resonant frequencies can be used as a fingerprint of the buffer gas for qualitative information about the buffer gas.

The linewidths obtainable by optoacoustic spectroscopy can be narrower than the absorption linewidth if Doppler-free methods are used. Marinero and Stuke split a dye laser beam into two counter-propagating beams that were separately chopped at frequencies ω_1 and ω_2 [48]. Both were focused into a sample where they interacted with gaseous I_2 molecules moving at different velocities. When the dye laser was tuned to a position in the Doppler broadened line-profile of the I_2 transition, one beam was resonant with a

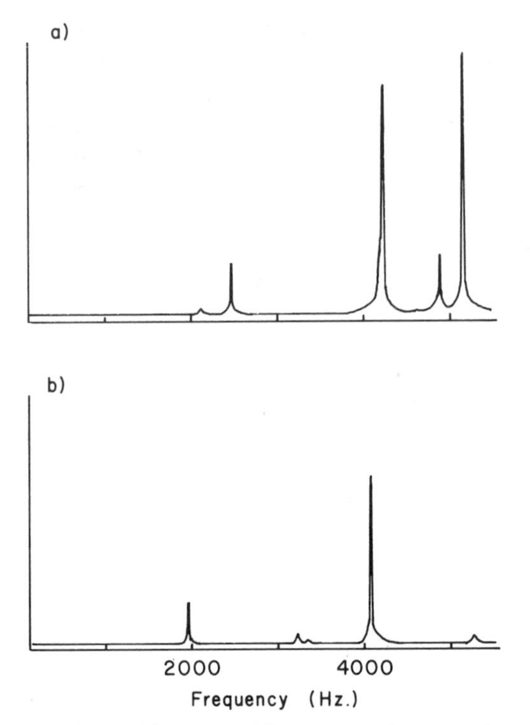

Fig. 2 (*a*) An optoacoustic signal is generated in a pure methane sample at 760 torr and the signal amplitude is monitored as the modulation frequency is changed [47]. Acoustical resonances appear corresponding to the natural resonances of the cell. (*b*) The same experiment is repeated, but the methane has been diluted by 760 torr of N_2. The different properties of N_2 cause a shift of all the resonant frequencies.

velocity subset of the molecules while the other beam (propagating in the opposite direction) was resonant with a different subset that had the same velocity magnitude but the opposite direction. At line center, however, both beams were resonant with the same subset, namely, the molecules that were standing still relative to the beams. If the laser beams were intense enough to cause nonlinearity in the molecular excitation, one would obtain an optoacoustic signal at the sum and difference frequencies ($\omega_1 + \omega_2$) and ($\omega_1 - \omega_2$) corresponding to the interaction of the two beams with the nonlinear molecule as the intermediary. Other positions within the line-profile would not produce such a signal because the two beams would not have a common subset of molecules to cause the interaction between the beams. Thus a Doppler-free spectrum could be achieved.

Thermal Lensing Spectroscopy

The heating effects associated with the absorption of light create not only pressure waves that are detected in photoacoustic spectroscopy, but also differences in the index of refraction that change the propagation of light through the medium [49, 50]. When the light beam is Gaussian in shape, as is the TEM_{00} mode of a laser, the index of refraction profile has a similar shape and acts as a lens. For most materials, heating causes a decrease in the index of refraction (dn/dT is negative) and the resulting lens is a negative one that causes a propagating TEM_{00} beam to diverge. This thermal lens was first observed by Gordon and co-workers [51] and has since been used in many applications to measure very low absorbances [52–59].

When a laser with a Gaussian beam profile is first allowed to pass through a sample, heating occurs across the beam and the thermal lens is established. As the heating continues, the thermal lens grows stronger according to the relationship

$$f(t) = f(\infty)\left(1 + \frac{t_c}{2t}\right) \tag{1}$$

where $f(t)$ is the time dependent focal length of the thermal lens, t is elapsed time since the establishment of the lens, and t_c is the time constant for the buildup of the lens [57]. Equilibrium is reached when the heat deposited in any volume element by the laser equals the heat leaving that volume element by thermal conduction. The time constant can be written as

$$t_c = \frac{\rho C_p w^2}{4k} \tag{2}$$

where ρ is the sample density, C_p is the specific heat, w is the size of the laser beam in the medium, and k is the thermal conductivity of the sample. At equilibrium, the focal length $f(\infty)$ becomes

$$f(\infty) = \frac{\pi k w^2}{2.303 PA \, dn/dT} \tag{3}$$

where P is the power of the laser (watts), A is the absorbance of the sample, and dn/dT is the variation of the index of refraction with temperature [49]. This approximation is good when the absorbance is small and the sample is nonfluorescent so all absorbed energy goes into heat. Strong thermal lenses are created when the laser power per unit area is high and/or when the absorbance is high so a large amount of heat is deposited. The coefficient dn/dT should also be large so the heat is effectively translated into an index of refraction change. At the same time, a small thermal conductivity causes higher temperatures in the absorption region and a stronger thermal lens. Nonpolar solvents are particularly attractive because they have low thermal conductivities and large values for dn/dT [58].

The effects of the thermal lens are usually quantitated by measuring the intensity of the laser beam at the very center of the beam profile ($I_{\text{beam center}} = I_{bc}$) [58]. To maximize the thermal lens effect and minimize the time constant, the laser is focused into the sample. The ratio of the beam as it propagates to and from the focal spot can be described by the equation

$$w^2 = w_o^2 \left[1 + \left(\frac{Z}{Z_c} \right)^2 \right] \tag{4}$$

where w_o is the minimum radius at the focal spot or the beam waist, Z is the distance from the lens, and Z_c is the confocal parameter that describes the length of the focal volume [60]. Both w_o and Z_c can be written in terms of the wavelength of light and the $f/\#$ associated with the cone of light from the lens to the focus:

$$w_o = \frac{\lambda(f/\#)}{\pi}$$

$$Z_c = \frac{2\lambda(f/\#)^2}{\pi} \tag{5}$$

$$Z_c = 2(f/\#)w_o = \frac{2\pi w_o^2}{\lambda}$$

If the beam waist of the laser is in the center of a sample cuvette, no thermal lens is observed by measuring I_{bc}. The situation is analogous to placing a lens at the focus of a second lens. If the sample is moved closer to the laser and away from the waist, the thermal lens introduces additional divergence in the laser and causes the waist to move back from the laser. In the far field, the laser beam appears smaller and I_{bc} becomes larger. However, if the sample is moved away from the laser and away from the beam waist, the thermal lens introduces additional divergence in an already diverging beam. The beam grows larger in the far field and I_{bc} becomes smaller. If the sample is moved too far from the beam waist, the thermal lensing is weakened by the lower power densities of the laser outside the focal region. It has been shown that the optimum position is one confocal parameter from the beam waist [58]. Dovichi and Harris measured this optimum as shown in Fig. 3*a* where the change in $I_{bc}(\infty)$ when the equilibrium thermal lens is established from its value $I_{bc}(0)$ before the thermal lens forms is shown as a function of the distance from the lens [59]. The thermal lens effect vanishes when the sample is at the beam waist and maximizes one confocal parameter on either side. An explicit expression can be written for the light intensity when the sample is one confocal parameter away:

$$\frac{I_{bc}(\infty) - I_{bc}(0)}{I_{bc}(0)} = \frac{-2.303P(dn/dT)A}{\lambda k} \tag{6}$$

This expression assumes the sample length is narrow compared with $2Z_c$. It is very similar to a Beer's law expression, where the right-hand side of the equation is simply $2.303A$. The additional factors for the thermal lens case have been combined into an enhancement factor E [49, 58]:

$$\frac{I_{bc}(\infty) - I_{bc}(0)}{I_{bc}(0)} = 2.303EA \tag{7}$$

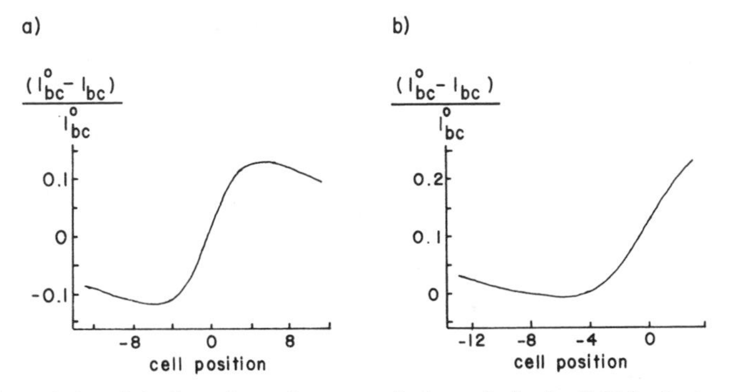

Fig. 3 The variation of the intensity at the center of a beam in the far field I_{bc} for two different cases. (*a*) The beam is focused at different positions relative to the center of a cell. (*b*) Two cells are positioned on either side of the focused beam—one at one confocal parameter from the focus and the other movable on the other side of the focus. The confocal parameter is 5.5 for both cases. I^0_{bc} is the intensity at the beam center before a thermal lens can be established. (After reference 59).

where

$$E = \frac{P(dn/dT)}{\lambda k}$$

The enhancement factor depends on the laser power and the thermal properties of the sample. It is important to realize that E can be larger or smaller than unity, that is, one can obtain less signal than with a direct Beer's law measurement. To understand this, it is necessary to realize that $I_{bc}(\infty)$ and $I_{bc}(0)$ represent intensities obtained after the beam passes through the sample at two different times—before and after the thermal lens is established. The amount of absorption taking place is the same at all times.

The analytical applications of thermal lensing have been investigated most extensively by Dovichi and Harris [57-59]. They focus a single laser with a stable TEM_{00} beam profile and good amplitude stability to create the thermal lens. The lens is measured by placing a pinhole in the middle of the far-field beam profile. Others have used two lasers, an intense laser to establish the thermal lens and a stable laser to probe and measure it [50]. Several different measurement methods were used by Dovichi and Harris. In one the laser is chopped and the intensity I_{bc} is measured before and after the thermal lens is established [48]. Equation 6 can then be used to establish the sample absorbance after suitable standardization. Alternatively, a series of measurements can be made as the thermal lens builds up in accordance with (1) and (3) [57]. By fitting the entire signal with the time constant and EA, a greater precision can be achieved because of the increase in information. Suitable calibration procedures can fix the time constant needed in the fit so that only EA is a parameter for an unknown sample. The sample absorbance can then be obtained by standardization with known absorbances. This approach is similar to that used in other kinetic methods [61].

Dovichi and Harris used the multiple measurement method with a 160-mW argon laser (514.5 nm) and found that a CCl_4 blank had an absorbance of 5×10^{-6} for a 1-cm path length [57]. The actual change in the laser intensity $[I_{bc}(0) - I_{bc}(\infty)]/I_{bc}(0)$ was 4×10^{-3}. Carbon tetrachloride, however, has a low thermal conductivity and a high dn/dT, so with the relatively high laser power used, a large enhancement factor (E) of 868 was achieved. CCl_4 is also a favorable solvent because it has a very low absorbance in the visible. Most other common solvents have higher values because of absorption by high overtone vibrational transitions [34, 50]. Dovichi and Harris could measure smaller changes in the blank absorbance because of analyte absorption so an absorbance of 7×10^{-8} was found at the detection limit [57]. With only a two-point measurement of $I_{bc}(0)$ and $I_{bc}(\infty)$, an absorbance of 2.2×10^{-6} would represent the detection limit. Thus the multiple measurement technique allows a significant improvement in the precision and detection limit for thermal lensing methods.

Dovichi and Harris have developed a rather ingenious method for correcting the thermal lensing signal for absorption in the blank [59]. A sample cuvette placed before the beam waist caused the far-field beam size to contract, while one placed after the beam waist caused it to expand. Thus two identical cuvettes placed symmetrically on either side of the beam waist cancel the effects of each other, assuming the same amount of light passes through each. Any small differences in the cuvettes or the amount of light passing through each can be compensated by moving one cell relative to the other so they do exactly cancel. Figure 3*b* shows how the thermal lensing signal varies as one cell is translated relative to the beam waist. The signal goes to zero at the position where the two cells cancel each other's effects. Having thus nulled the system, one can fill one cell with a sample and the thermal lensing signal depends on the difference between the two cells. The procedure is very simple yet provides an excellent method for background subtraction. $Fe^{2+}-1,10$ phenanthroline complexes have been detected at absorbances of 4×10^{-7}, corresponding to Fe^{2+} concentrations of 5×10^{-11} *M* [59].

Intracavity Absorption

If an absorbing sample is placed within the cavity of a laser, one finds the output power of the laser can be changed dramatically even for weak absorption [62–67]. This effect is the basis of the intracavity absorption technique. An example application is shown in Fig. 4. A dilute solution of a rare earth ion was placed inside the cavity of a ruby laser pumped dye laser that was operated in a broad-band mode with dielectric mirrors [68]. The output of the dye laser was passed through a monochromator and focused onto a photodiode array to display the spectral profile of each laser output pulse. The rare earth solution has reasonably narrow absorption features that are superimposed on the normal laser output. The absorptions are quite pronounced in this figure although the actual absorbances measured directly were small. For example, the solution absorption shown in Fig. 4*b* corresponds to 0.096 *M* $Eu(NO_3)_3$, which has an absorbance of 0.006 at the peak. An enhancement of about 35 is obtained by using the intracavity method. Considerably larger enhancements are possible for other laser configurations.

A number of physical phenomena contribute to the enhancement of intracavity absorption [69–71]. If a sample is inside the cavity, there are multiple passes through the sample because of the resonator structure of the cavity. The number of passes, and therefore the enhancement factor, increases as the reflectivity of the output mirrors is increased. The enhancement also

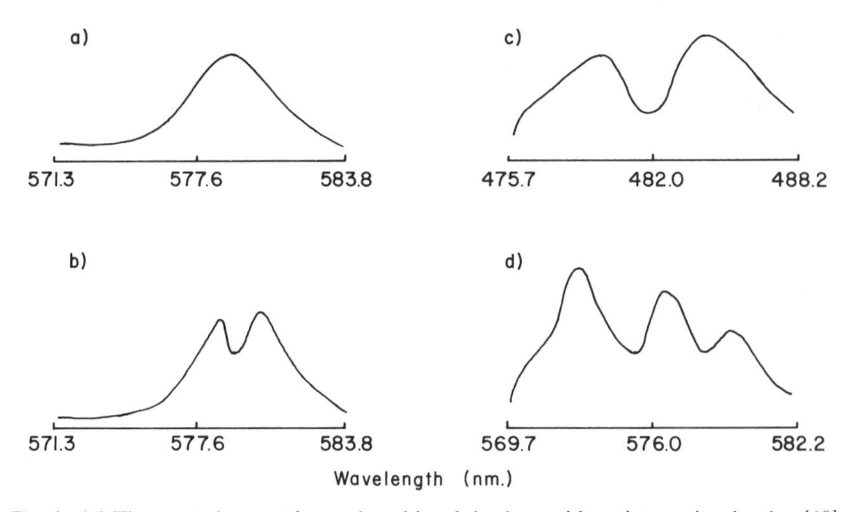

Fig. 4 (*a*) The spectral output from a broad-band dye laser with no intracavity absorber [68]. (*b*) The spectral output from the laser when a 0.096 *M* $Eu(NO_3)_3$ solution was placed inside the cavity. (*c*) 0.0037 *M* $Pr(NO_3)_3$. (*d*) 0.008 *M* $NdCl_3$.

depends on how close the laser is operated to threshold. The condition for lasing states that the gain of the laser must exceed the losses in the cavity. If the laser operates near threshold where the gain is only slightly larger than the losses, the output power becomes a very sensitive function of the losses. If the losses are increased by the insertion of a weakly absorbing sample, the lasing can be actually terminated. Although the enhancement factor becomes very favorable when one is operating near threshold, the requirements on laser stability also become particularly severe because one must distinguish between intensity changes caused by sample absorption and those caused by fluctuations in the laser gain or cavity losses. Even scattering from dust particles behind the coatings of the laser end mirrors or from the edges of mirror mounts can produce perceptible effects near threshold [59].

If the absorption lines of the sample are sharper than the gain bandwidth of the laser, the absorption enhancement also depends on the competition among cavity modes [71]. Each mode in a cavity has a characteristic gain and loss that depends on absorption at the wavelength of that mode, as well as several additional factors. The lowest loss mode in the cavity is the mode that begins to lase. In doing so, it begins to deplete the population inversion that caused it to lase. Several effects can then occur. If the other modes share the same population inversion (i.e., the gain of these other modes is determined by the same set of molecules in the lasing medium), the oscillating mode prevents the other modes from reaching threshold. In the other extreme, if the other modes obtain their gain from different parts of the lasing medium and if there is no excitation transfer within the lasing medium, the oscillating mode does not affect the ability of the other modes to reach threshold. If there is excitation transfer, energy can flow to the mode that has the lowest losses and is oscillating. One therefore has a mode competition that depends on the amount of communication among modes. The introduction of additional losses into a cavity at the wavelengths of an absorbing sample can cause a mode that had competed favorably to become unable to do so. Its energy can be transferred to other modes that have less loss. This can cause dramatic changes in the intensity within modes. The magnitude of the effect depends on how close mode losses are to each other and how much communication occurs among modes. If the sample absorption occurs over bands that are broader than the gain profile, then the active modes all experience similar losses and no mode competition occurs.

The pulse width of a laser can strongly affect the enhancement factor expected for intracavity work [73]. If the pulse is short, the number of passes is limited by the time the laser is active. Additionally, the pulsed lasers are high-gain lasers that operate most reliably far above threshold. There is also not always sufficient time for the excitation energy in a mode to flow to the other modes with lower losses. The c.w. lasers therefore are more sensitive to the quenching effects and provide larger enhancements [73, 74]. They also have greater amplitude stability, so the effects of intracavity absorption can be more readily measured.

Thermal lensing is an important contribution to cavity losses [75]. This loss mechanism is time dependent as opposed to the other factors that cause loss. The time dependence can be utilized to discriminate between thermal lensing and direct sample absorption. If the dye laser excitation source is chopped or pulsed, the output can be watched as a function of time to obtain the intensity before and after the thermal lens is established.

Gas phase spectra are particularly well-suited for intracavity absorption because the lines are much sharper than the gain profile of a c.w. dye laser, running in a broad-band configuration. Hill and co-workers [72] constructed an enclosed dye laser that could be evacuated to remove any contaminant gases and then refilled with a gas sample of interest. The dye laser had highly reflective cavity mirrors. The output beam was scanned with a high-resolution monochromator to identify where absorption occurred. The workers were successful in measuring the extremely weak absorption lines of O_2 that occur around 630 nm. The minimum detectable absorbance was estimated to be about 10^{-8} for a 1-cm path length. Hill and co-workers estimated that an enhancement of 3×10^4 was achieved. The observed line-shapes were not Lorentzian and this was attributed to nonlinearity between the intracavity absorption and the sample absorption. Such nonlinearities are a problem when mode competition becomes an important mechanism for enhancement. Present theories are inadequate for quantitatively predicting the functionality [72].

The identification and quantitation of intracavity absorption is simplified for sharp-lined spectra such as those of gases in which the transitions appear as lines on a background where absorption is not occurring. The task becomes more difficult if the sample spectrum is broad and there are no spots in the laser output profile where absorption is not occurring. One must then substitute a blank for the sample to measure how much intracavity absorption occurs. The comparison is difficult to quantitate because the laser characteristics must not change for the two measurements. Shirk et al. have had success in achieving good measurements for this analytically important situation [75]. It was first necessary to stabilize the c.w. dye laser so the spatial beam profile and power were constant. A Pockels cell was placed in the cavity in order to have a loss that could be electronically ajusted. The argon ion laser used to excite the dye laser was chopped and the dye laser output was measured immediately after it was turned on to avoid the effects of thermal lenses. When the sample solution was introduced in an intracavity position, the dye laser output decreased because of the intracavity absorption. Instead of quantitating the intracavity absorption by measuring the output power with the sample and a blank in the cavity, Shirk and co-workers instead adjusted the Pockels cell voltage so that the output power with a blank inside the cavity matched the output power with the sample inside the cavity. This procedure achieves its quantitation by a substitution of controllable losses with a Pockels cell for the losses with a sample. Since the dye laser is operated at the same conditions for the blank and sample measurements, complicating effects associated with changes in the dye laser characteristics with power are eliminated. Excellent long-term reproducibility can be achieved with this procedure. The greatest sensitivity was achieved when the dye laser operated very near threshold, but there was also the greatest noise and instability near threshold. The dynamic range for an absorption also becomes limited near threshold because small increases in absorption can extinguish the laser. The dye laser was usually run 10 to 30% above threshold powers to get acceptable stability and dynamic range. Under these conditions, an absorbance of 5×10^{-5} could be measured at the detection limit [75, 76]. This instrument was used to detect iron at levels of 0.45 ng/ml, a value limited by the blank. Levels of 90 pg/ml could have been detected against the blank [76].

Hohimer and Hargis have used intracavity techniques for isotope specific measurements of $^{127}I_2$ and $^{129}I_2$ [77]. If a gas phase I_2 sample is placed within a dye laser cavity operated in a broadband mode, the dye laser output is attenuated at the narrow absorption lines of the individual I_2 isotopes. To monitor the laser output only at the wavelengths of a particular I_2 isotope, a cell containing this isotope was inserted into the output beam and the resulting fluorescence was monitored. That fluorescence was only excited at the wavelengths where the specific isotope in the sample cell absorbed, so the fluorescence could be used as a measure of how much light was absorbed in the cavity by the specific isotope. A high concentration of the I_2 isotope in the sample resulted in a low fluorescence signal from the monitoring cell. This technique was used to selectively measure both $^{127}I_2$ and $^{129}I_2$ isotopes at concentrations of 5×10^{11} molecules/cm^3.

3 FLUORESCENCE SPECTROSCOPY

Lasers are a well-suited light source for the traditional absorption and emission spectroscopic measurements. The bandwidth achievable with a laser is superior to the linewidth of any spectroscopic transition of current analytical interest or the band pass of even the highest resolution monochromator. The power within that bandwidth remains very high so that detection limits can be improved in many situations. For an absorption measurement, the improvement arises if the detector and electronics are insensitive to light and contribute appreciable dark noise or if shot noise from the finite number of photons per second is important. For an emission measurement, the improvement arises for a variety of reasons. It is true that the emission intensity can be increased to the limit imposed by saturation, but the major factors for the improvement with laser excitation are concerned with a variety of more practical considerations. In this section, we concentrate on the influence of the laser in the many different kinds of emission or fluorescence measurements important for analytical spectroscopy.

The high powers and narrow bandwidths of modern lasers should permit development of analytical techniques that have very low detection limits and high selectivity for particular analytes. The selectivity of an analytical technique becomes particularly important to practicing analytical chemists as they work at lower and lower concentrations where the number of other elements or molecules present begins to skyrocket. It is often found that an analytical technique is limited by its selectivity or susceptibility to interference and not by its ability to measure low concentrations. The susceptibility to interference is not unrelated to the question of detection limits. If there are components in a mixture that cause interference, the mixture can be diluted sufficiently to eliminate the interference if the method permits detection at the resulting lower concentration. It is the responsibility of the analytical chemist to develop new chemical and physical methods that can exploit the unique characteristics of the laser for analytical measurments, particularly its high intensity and narrow bandwidth. In the following discussions, we examine the performance of different laser analytical techniques in terms of their promise in providing selective analyses and low detection limits. We do not emphasize other important parameters such as analysis speed, small sample determination, time resolution, dynamic range, equipment cost, complexity, and reliability because they can change markedly as these fledgling techniques are developed.

The fluorescence intensity that can be generated from an arbitrary sample can be described very simply if we are willing to make a number of assumptions. Since the prime concern in this chapter is to present all the important characteristics of laser analytical methods in a clear framework and not to provide a general description of all possible systems, we adopt a simple model for a laser excited fluorescence experiment. The analyte in our sample has only two energy levels that can be appreciably populated: level 1, which has a population of N_1 cm^{-3} and is the ground state, and level 2, which has a population of N_2 cm^{-3}, a lifetime of τ_2 sec, and a radiative quantum efficiency of η and is the only excited state that can be populated. We allow the possibility that level 2 can relax to levels other than the ground state, but we assume these other levels decay so quickly to the ground state that their populations may be neglected. This last assumption allows us to consider nonresonant fluorescence and simplifies the mathematics considerably. Its validity becomes important only when we consider saturation phenomena and even there it only affects calculations for the amount of saturation. Because such calculations involve many other assumptions for systems of analytical interest, this approximation does not hinder the description. We also assume that the concentrations are too low to absorb incoming photons or fluorescence photons (pre- and postfilter effects). The fraction of the total fluorescence from level 2 that occurs at the particular transition we are observing is designated by the branching ratio β. The spontaneous fluorescence produced by the sample at the transition observed is

$$S_{\text{fluor.}} = \beta\eta\tau_2^{-1}N_2V \tag{8}$$

while the signal detected is

$$S_{\text{out}} = \beta\eta\tau_2^{-1}N_2V\Omega_{\text{coll.}}\eta_{\text{det.}} \tag{9}$$

where S_{out} = output signal in counts/sec
V = sample volume under observation
$\Omega_{\text{coll.}}$ = efficiency of the optical system including monochromator and filters
$\eta_{\text{det.}}$ = efficiency of detecting and counting fluorescence photons.

Assuming the excitation intensity is not attenuated as it propagates through the sample, the rate of change of level 2 is

$$\frac{dN_2}{dt} = \frac{\sigma I}{h\nu}(N_1 - N_2) - \tau_2^{-1} N_2 \tag{10}$$

where σ = optical cross section for the transition (cm^2)
I = laser intensity (W/cm^2)
$h\nu$ = energy of a laser photon (joules).

The first term represents the laser excitation rate from the ground state (the difference $N_1 - N_2$ is required to account for stimulated emission), while the second term represents all the radiative and nonradiative relaxation processes that depopulate the excited level. If we assume the laser excites the sample for a time much longer than τ_2, the population of level 2 reaches a steady-state value where $dN_2/dt = 0$. If we also require that the total population $N = N_1 + N_2$, (10) can be solved for N_2.

$$N_2 = \frac{\sigma I N}{2\sigma I + h\nu\tau_2^{-1}} \tag{11}$$

Substituting (11) into (9), we obtain

$$S_{\text{out}} = \frac{\beta\eta\tau_2^{-1}\sigma I N V \Omega_{\text{coll.}}\eta_{\text{det.}}}{2\sigma I + h\nu\tau_2^{-1}} \tag{12}$$

There are two important limiting cases for this expression. At low excitation intensities, the expression becomes

$$S_{\text{out}}^{low} = \frac{\beta\eta\sigma P N l \Omega_{\text{coll.}}\eta_{\text{det.}}}{h\nu} \tag{13}$$

where P = laser flux (W) and l = path length (cm). This last expression is the familiar one applicable to conventional fluorescence spectroscopy. There is a linear relationship between output signal and concentration in this region (assuming the pre- and postfilter effects are negligible). The output depends directly on the fluorescence quantum efficiency, the fraction of the total fluorescence we are examining, and the instrumental efficiencies. The output also depends linearly on the average power but does not depend on the area of the sample illuminated. One cannot continue to increase the output signal by increasing the excitation intensity, however, because one reaches the second limiting case.

The second limiting case occurs at very high excitation intensities, where the rate of stimulated emission exceeds the normal excited state relaxation rate ($2\sigma I \gg h\nu\tau_2^{-1}$). Then

$$S_{\text{out}}^{\text{high}} = \frac{\beta\tau_{\text{rad.}}^{-1} N V \Omega_{\text{coll.}}\eta_{\text{det.}}}{2} \tag{14}$$

where $\tau_{\text{rad.}}^{-1}$ = radiative or natural lifetime of level 2($= \eta\tau_2^{-1}$). This equation expresses the saturation condition where the output signal has reached its highest possible value. One can see from (11) that the excited and ground states are equally populated at saturation (if the two states had different degeneracies, the saturation populations would be proportional to the state degeneracy). Since the absorption depends on the difference in the ground- and excited-state populations, the absorbance goes to zero at saturation. This condition is shown dramatically in an experiment performed by Hosch and Piepmeier [78]. Their measurement of the absorption of the Na D lines in a flame is shown in Fig. 5 as a function of the laser power density. As the power density approaches saturation, there is a decrease in the absorption until it vanishes. Note also that the absorption linewidth appears to increase at higher powers because the peak of the absorption line is more easily saturated than the wings of the line. Since the excited state population has reached its limiting value, the output intensity depends on the rate of radiative relaxation at the transition we are monitoring and the total number of analyte species in the excited volume, NV. Note that although the output signal is independent of excitation intensity at this point, the attainment of

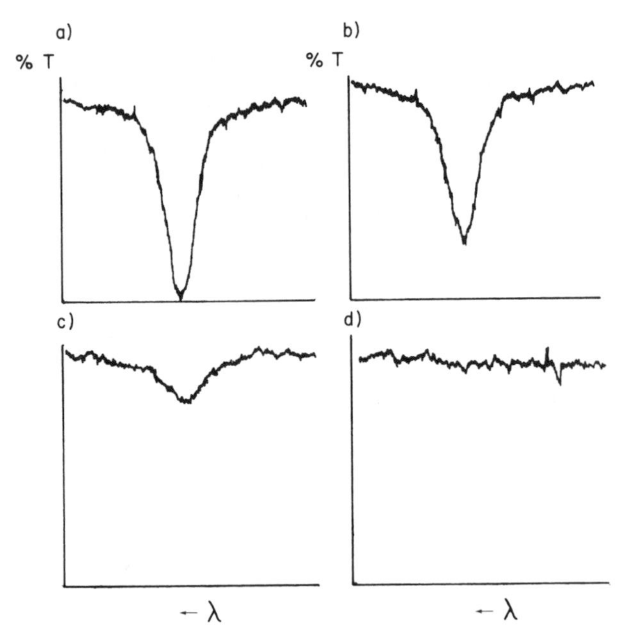

Fig. 5 The absorption measured by a dye laser for the Na 589.0-nm transition in a flame is shown at four different power levels. The decrease in the absorption occurs because of saturation of the Na transition by the laser. (*a*) 0.2 kW/cm^2. (*b*) 3 kW/cm^2. (*c*) 42 kW/cm^2. (*d*) 500 kW/cm^2.

the saturation conditions depends on the power density (W/cm^2). To achieve saturation, the laser excitation can be focused into the sample in order to force the stimulated rate to exceed the normal relaxation rates. However, if a laser with a given energy output is focused more than is necessary to achieve saturation, the output signal diminishes because the excited volume is smaller. Despite the decrease in signal, there are several analytical advantages to working in saturation [78–84]. The output signal becomes independent of excitation intensity (as long as it is high enough to saturate) and corrections for laser fluctuations are not necessary. The biggest advantage is that the output signal does not depend on the quantum efficiency. Interferences could be present in the sample that would normally quench the fluorescence, but under saturation conditions no quenching occurs assuming the power densities are high enough to overcome the faster relaxation rate caused by quenching. Experimentally, there is a trade-off between choosing a degree of laser focusing that will ensure insensitivity to quenching and maintaining a large excitation volume for maximum signal. In addition, one must remember that saturation occurs within a given volume and if the sample is excited below saturation outside of that volume, it becomes important to ensure that the optical system examines only the saturated excitation volume.

There are several other benefits to be derived from operation in the saturation regime [79, 83]. Since the sample ceases to absorb light at saturation, the laser excitation is not attenuated as it travels through the sample, thus reducing the importance of the prefilter effect at high analyte concentrations. Additionally, if the area of the saturated region extends through a large part of the sample toward the detection optics, the absorption of resonance fluorescence (the postfilter effect) is also diminished. Thus the working curve maintains linearity to higher analyte concentrations and improves the dynamic range of the method.

The quantitative prediction of fluorescence in the saturation regime is complex because of many factors. We already know that the lineshape of a transition changes as saturation is approached because the wings of a line absorb less than the peak and therefore do not reach saturation at the same rate. Additionally, the increased relaxation rate caused by stimulated emission shortens the lifetime of the excited level and causes additional lifetime broadening, which in turn changes the optical cross section [85]. At higher analyte concentrations near saturation, where the laser becomes attenuated in traversing the sample, the degree of saturation decreases as the beam passes further into the sample. Thus the attenuation becomes a complex relationship among saturation, spontaneous emission, and path length in the material [86]. The prediction of saturation threshold is also complicated by the need to know the relaxation rates operative within the inhomogeneous envelope of a transition [79, 80, 84]. If a very narrow-band laser is used, the analyte species within the inhomogeneous linewidth that are in resonance (i.e., are within the homogeneous linewidth of an individual analyte species) with the laser can be saturated. If there is no excitation transfer between analyte species within the inhomogeneous envelope, a "hole is burned" in the absorption profile and in the ground-state population. If there is rapid transfer between analyte species, the "hole becomes filled" and the entire population can be saturated. Thus the rate of transfer between species is an important parameter in predicting saturation. In the gas phase, collision rates usually determine the rate of transfer, while in condensed phases both collisions and/or energy transfer can contribute.

These benefits of working at saturation are offset when one is looking at resonance fluorescence because the scattered light does not saturate [79]. Thus if one has a laser of a given power that is being focused to reach saturation, the scattered light does not change, but the output signal actually goes down as saturation conditions are reached because smaller excitation volumes are being sampled. Similarly, if one is increasing the laser power in order to reach saturation, keeping the focal parameters constant, the scattered light increases in proportion to laser power, but the output signal approaches a constant value at saturation. When resonance fluorescence is being monitored, the noise level of the experiment is determined by the scattering, and therefore the S/N ratio is degraded by working in the saturation regime.

One can calculate some typical values for reaching the saturation regime. For atomic transitions where favorable absorption cross sections are $\sim 10^{-12}$ cm^2 and lifetimes are 10^{-9} sec, the condition for saturation ($2\sigma I \sim h\nu\tau^{-1}$) is met when photon fluxes of 5×10^{20} photons/(sec)(cm^2) (or for 400 nm light, 250 W/cm^2) are reached. Such fluxes can be readily achieved. For molecular transitions with absorption cross sections of $\sim 10^{-16}$ cm^2 and lifetimes of 10^{-12} sec, the condition for saturation is met when photon fluxes of 5×10^{27} photons/(sec)(cm^2) (or for 400 nm light, 2.5×10^9 W/cm^2) are reached. This value of power could result in sample destruction. For the LaPorte forbidden transitions of lanthanide and actinide ions in condensed phases where absorption cross sections of 10^{-19} cm^2 and lifetimes of 10^{-12} sec are typical, the condition for saturation is not met until photon fluxes of 5×10^{30} photons/(sec)(cm^2) (or for 400-nm light, 2.5×10^{12} W/cm^2) are reached and sample destruction is assured. If the level excited has a longer lifetime (and many rare earth ions have levels with lifetimes of milliseconds), the saturation condition can be more easily met.

It is of interest to estimate how low a concentration can be measured by laser excited fluorescence techniques. To provide insight into the important aspects of such estimates, we consider the experiment performed by Fairbanks et al. [87] in 1975 and still widely quoted. These researchers measured absolute Na concentrations of 100 atoms/cm^3 using resonance fluorescence in a Na vapor cell. A tunable, narrow-band dye laser was attenuated with neutral density filters to avoid saturation. A power of 3 μW (9×10^{12} photons/sec) was used to excite a Na vapor over a 2-cm path length in a cell that was well baffled to eliminate scattered light. An $f/1.8$ optical system (collection efficiency of about 2%) imaged the fluorescence onto a photomultiplier that we assume had a detection quantum efficiency of 10%. The Na D line has an optical cross section of 8.3×10^{-12} cm^2 and we assume unit quantum efficiency. If these values are substituted into (13), one finds

$$S_{\text{out}}^{\text{low}} = 0.3N \tag{15}$$

The minimum detectable concentration depends on how low an output signal can be measured in the presence of the scattered light. The scattered light at the photomultiplier tube in this experiment was reduced by 3×10^{-10} from the incident light intensity (2.7×10^3 photons/sec). One should actually be able to see below the scattered light intensity if this scattered light is constant. Fairbank and co workers were able to get a factor of 30 lower than the scattered light by wavelength modulating the light at 1.5 Hz and by monitoring the output with a phase-sensitive detector. They were able to reach a lower limit of 90 photons/sec or 9 photoelectrons/sec. Substituting into (15), one sees it should be possible to observe 30 atoms/cm^3, a value close to the 100 atoms/cm^3 observed in the experiment. This value of 100 atoms/cm^3 corresponds to 1.7×10^{-19} moles/liter.

An experiment by Balykin et al. improved on this value [88]. By extensive baffling and use of two detectors operating in a coincidence configuration, it was possible to obtain a 10^{14} rejection of laser light, which then permitted single Na atom detection.

The same calculations could be performed for molecules. The primary difference is that the molecular transitions have absorption cross sections more than 5 orders of magnitude smaller than the Na transitions because of the many possible vibronic and rotational pathways for radiative transitions. One might therefore expect that the detection limits would also be poorer by a comparable factor.

Many factors could be improved in our example to obtain better detection limits. In fact, Schawlow has stated that if he knew the 100 atom/cm^3 would be quoted so extensively, he would have improved that value considerably [89]. The most obvious improvement is to measure nonresonance fluorescence where scattered light can be rejected by a single or double monochromator. The noise level would then be determined by background fluorescence or inelastic light scattering and/or by noise in the detection electronics. Higher power lasers can be used with expanded beams to obtain the optimum fluorescence levels while avoiding saturation. Pulsed lasers might have advantages in many instances if proper gated electronics is available to detect light only during the time fluorescence is present. Background light and electronic noise can then be strongly discriminated against. It is interesting to compare conventional Raman spectroscopy, which has been carefully optimized over the years, with the potential for laser-induced fluorescence spectroscopy. Raman scattering cross sections are typically 10^{-29} cm^2 and the best Raman instruments are capable of measuring concentrations of 10^{-4} *M* or higher. A typical atomic absorption cross section is 10^{-12} cm^2, while a typical molecular absorption cross section is 10^{-16} cm^2. By extrapolating, one should be able to measure atomic concentrations of 10^{-21} *M* or 0.6 atoms/cm^3 and molecular concentrations of 10^{-17} *M*. Let us now look at what actually has been accomplished in systems of practical analytical interest and compare the results with these estimates.

Laser-Excited Atomic Fluorescence

The traditional method of sample preparation for generating sharp optical transitions characteristic of the analyte species of interest is conversion to the gaseous atomic state. The gas phase atoms have very sharp and intense absorption lines that are well-suited to a laser's narrow bandwidth and high powers. Although there have been many studies that show the excellent selectivity and detection limits achievable with lasers, many of these have used simple vapor cells that are clearly unsuitable for analysis of ordinary samples. Analytical chemists have extensive experience in conversion of samples to atomic vapors because of the importance of that step in many of the traditional methods of analytical spectroscopy. Of the many techniques available, flame and furnace techniques have been most extensively used for sample atomization prior to laser excitation with some work on inductively coupled plasmas. We are not concerned in this chapter with techniques that use a laser itself to blast a sample into the gas phase. Excellent reviews are available elsewhere [90, 91].

Atomic Fluorescence in Flames

Laser excited atomic fluorescence was first demonstrated in flames by Denton and Malmstadt [91] and Fraser and Winefordner [92] in 1971 using a conventional flame burner and nebulizer. This early work was very encouraging. A linear working curve was obtained over 4 orders of magnitude and detection limits of 0.01 μg/ml were obtained for Ca and In despite a number of instrumental factors that needed improvement, including the wavelength coverage, the laser bandwidth (0.1 to 1 nm) and powers, the burners and nebulizer, the presence of scattered laser light, and interference from radiofrequency noise. With simple improvements in the electronics and burners, the detection limit was improved by about 5 to 0.002 μg/ml for In and in addition two-photon excitation was demonstrated in flames for Cd and Zn at 10-μg/ml concentrations [93]. Since then a number of studies have been performed that have expanded the number of atoms that can be determined by this method from 10 to 25 [94–100], the detection limits have been lowered by about 50 to 5000, and linearity is achieved over about 5 orders of magnitude in concentration [96]. A selection of typical detection limits is given in Table 2 along with values that would be obtained if an inductively coupled plasma source were used for the analysis.

The typical system that has been used in these experiments is shown in Fig. 6 [96]. A pulsed laser with high peak power is used for excitation with a beam expanding telescope, which increases the area of the laser beam through the flame to avoid saturation effects that would decrease the signal. The beam expander also includes a spatial filter that removes broad-band fluorescence from the dye emission. The laser bandwidth used in the best experiments of this type was still only about 0.03 nm, much broader than the typical Doppler linewidths in a flame (about 0.002 nm). The dye laser can be frequency doubled to permit UV excitation when necessary. Baffles, beam stops, and diaphragms are placed in the system at the laser input, as a laser beam stop, behind the flame in the field of the monochromator, and at the input to the monochromator in order to eliminate as much scattered light as possible when resonance fluorescence is being observed. Standard capillary burners are usually used with either air-acetylene or N_2O–acetylene flames. Ultrasonic nebulizers improve the efficiency markedly [99]. The monochromator is normally a low-resolution instrument with a focal length of about 0.35 m and an $f/6.8$ aperture since the resolution is obtained from the narrow bandwidth of the laser excitation. A photomultiplier tube that is designed for fast time response is used for fluorescence detection. The photomultiplier signals are measured by a standard boxcar integrator with a 50-Ω input impedance. The detection system is capable of 15-nsec aperture times. The very small duty cycle strongly discriminates against photomultiplier dark current, flame luminosity, and other background light that is not coincident with the laser firing. The discrimination against background luminosity is sufficiently good that wide slits can be used on the monochromator without difficulty [96].

When resonant fluorescence transitions are used, the detection limits are determined primarily by the Rayleigh and Mie scattering from particulates and gases in the flame [101]. It is well established that the use of nonresonant fluorescence transitions can eliminate that source of noise. The very low duty cycle of a gated detection system should discriminate so strongly against other sources of noise that single photon events from the laser excited fluorescence should be detectable. If one reaches that limit, one reaches the situation where the important analytical parameters that control the detection limits in flames are the efficiency of introducing the analyte species into

Table 2

Element	*Excitation* (nm)	*Fluorescence* (nm)	*Excitation for Optogalvanic Detection* (nm)	*Detection Limit for Laser Excited Fluorescence* (ng/ml)	*Detection Limit for Inductively Coupled Plasma* (ng/ml)	*Detection Limit for Optogalvanic Flame Spectroscopy* (ng/ml)
Ag	328.1	328.1	328.1	4[a]	4[b]	1[c]
Al	394.4	396.1	—	0.6[a]	0.2[d]	—
Ba	553.7	553.7	307.2	8[a]	0.01[d]	0.2[c]
Bi	306.8	306.8	306.8	3[a]	50	2[c]
Ca	422.7	422.7	300.7	0.01[e]	0.0001[d]	0.1[c]
Ce	371.6	394.2	—	500[g]	0.4[d]	—
Cd	228.8	228.9	—	8[a]	0.07[f]	—
Co	357.5	347.4	—	1000[a]	0.1[f]	—
Cr	359.3	359.3	301.8	1[a]	0.08[f]	2[c]
Cu	324.7	324.7	324.8	1[a]	0.04[f]	100[c]
Dy	364.5	353.6	—	300[g]	4[b]	—
Er	400.8	400.8	—	500[g]	1[b]	—
Eu	459.4	462.7	—	20[g]	1[b]	—
Fe	296.7	373.5	302.1	30[a]	0.09[d]	2[c]
Ga	403.3	417.2	287.4	0.9[a]	0.6[d]	0.07[c]
Gd	376.8	336.2	—	800[g]	7[b]	—
Hf	368.2	377.8	—	10[5b]	10[b]	—
Ho	405.4	410.4	—	150[g]	10[b]	—
In	410.4	451.1	303.9	0.2[a]	30[b]	0.006[c]
K	—	—	294.3	—	—	1.0[c]
Li	670.8	670.8	670.8	0.5[a]	—	0.001[c]
Lu	465.8	513.5	—	3000[g]	8[b]	—
Mg	285.2	285.2	285.2	0.002[e]	0.003[d]	0.1[c]

Mn	279.5	279.5	279.5	0.4[a]	0.01[b]	0.3[c]
Mo	390.3	390.3	—	12[a]	0.2[d]	—
Na	589.0	589.0	285.3	<0.1[a]	0.02[d]	0.05[c]
Nb	405.9	408.0	—	1500[h]	0.3[d]	—
Nd	463.4	489.7	—	2000[g]	10[b]	—
Ni	361.0	352.4	300.2	2[a]	0.2[d]	7[c]
Os	442.0	426.1	—	1.5×10^5[h]	—	—
Pb	283.3	405.8	280.2	0.2[e]	1[b]	0.6[c]
Pr	427.2	430.6	—	1000[g]	30[b]	—
Rh	369.2	350.2	—	150[h]	3[b]	—
Ru	372.8	349.9	—	500[h]	—	—
Sc	391.2	402.4	—	10[h]	3[b]	—
Sm	366.1	373.9	—	150[g]	20[b]	—
Sn	—	—	286.3	—	—	2[c]
Sr	460.7	460.7	—	0.1[e]	0.003[d]	—
Tb	370.3	350.9	—	500[g]	200[b]	—
Ti	365.4	365.4	—	2[a]	0.03[d]	—
Tl	377.6	377.6	291.8	4[a]	200[b]	0.09[c]
Tm	371.8	409.4	—	100[a]	7[b]	—
V	370.4	411.2	—	30[a]	0.06[d]	—
Yb	398.8	346.4	—	10[g]	0.02[d]	—

[a] S. J. Weeks, H. Haraguchi, and J. D. Winefordner, *Anal. Chem.*, **50,** 360 (1978).
[b] V. A. Fassel and R. N. Kniseley, *Anal. Chem.*, **46,** 1110A (1974).
[c] G. C. Turk, J. C. Travis, J. R. DeVoe, and T. C. O'Haver, *Anal. Chem.*, **51,** 1890 (1979).
[d] P. W. J. M. Boumans and F. J. de Boer, *Spectrochim. Acta*, **30B,** 309 (1975).
[e] J. J. Horvath, J. D. Bradshaw, J. N. Bower, H. S. Epstein, and J. D. Winefordner, *Anal. Chem.*, **53,** 6 (1981).
[f] K. W. Olson, W. J. Haas, Jr., and V. A. Fassel, *Anal. Chem.*, **49,** 632 (1977).
[g] N. Omenetto, N. N. Hatch, L. M. Fraser, and J. D. Winefordner, *Anal. Chem.*, **45,** 195 (1973).
[h] N. Omenetto, N. N. Hatch, L. M. Fraser, and J. D. Winefordner, *Spectrochim. Acta*, **28B,** 65 (1973).

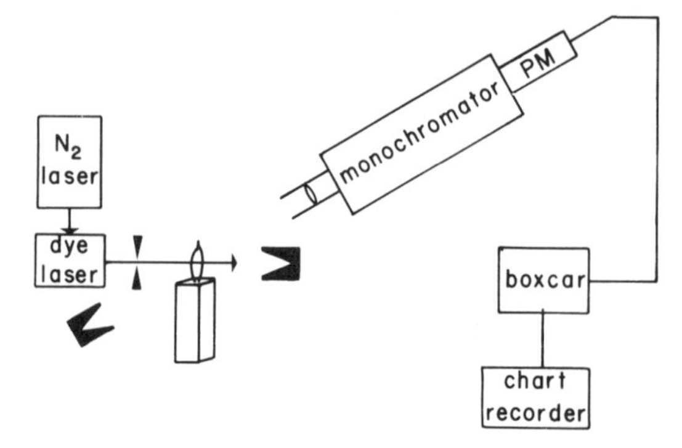

Fig. 6 The instrumental system used to record atomic flame fluorescence consists of a N_2 pumped dye laser whose beam is spatially filtered, expanded in area to avoid saturating the transition, and passed through the flame containing the analyte. Baffles are arranged to trap direct laser light and to prevent scattered light from reaching the monochromator. Fluorescence is detected with a conventional monochromator-photomultiplier and measured with a boxcar integrator.

the flame, the spatial structure of the analyte population in the flame, the atomization efficiency of the flame, and quenching.

The conversion of a sample to the gas phase and the spatial structure of a flame has been investigated in a very nice experiment performed by Smith and co-workers [102]. The fluorescence output is directly related to the absolute concentration in the saturation regime by (14), with experimentally determinable parameters comprising the proportionality constant. By calibrating the experimental system with a tungsten lamp standard of spectral irradiance, all the parameters could be determined. A focused c.w. dye laser was then used to saturate the sodium transition and the absolute fluorescence intensity was measured. The results were checked by a direct absorption measurement, which provided consistent results. The vertical profile of an air–acetylene flame was measured for an aspirated Na concentration of 1.2 ppm and the results are shown in Fig. 7. The ordinate in this figure has been changed to take the thermal equilibrium between the $^2P_{1/2}$ and $^2P_{3/2}$ states into account [102].

Once species are introduced into the flame, they must be converted to the atomic state in order to be observable in atomic fluorescence. The conversion efficiency β expressed as the ratio of the atomic form to all the flame species of the analyte was measured by a number of workers in different flames [103]. Typically values between 5 and 80% were obtained.

The quenching of fluorescence in a flame depends on the analyte species involved, the flame gases, and the flame temperature. Measurements of the quenching are difficult to obtain, but it is generally accepted that quenching

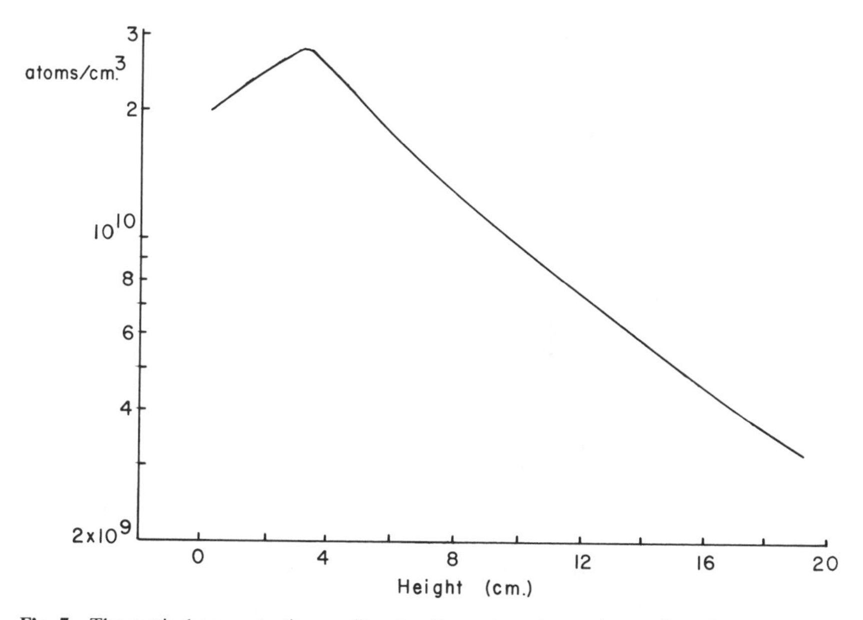

Fig. 7 The vertical concentration profile of sodium atoms in an air-acetylene flame is shown.

can reduce the fluorescence by 100 or more under typical conditions. The saturation intensity determined by Smith and co-workers can be used to obtain a value for the radiative quantum efficiency of Na emission in an air–acetylene flame using equation 8 of their paper [102]. A value of 1.3×10^{-3} is then obtained for the quenching.

Let us now combine this information to see what kinds of flame concentrations can be expected from a given sample. A 1.2-μg/ml solution produces total analyte concentrations of 6×10^{10} cm^{-3} at the optimum spectral position of the flame, of which only 10% on the average are in the atomic form and only 0.1% of these fluoresce under unsaturated conditions. If we accept 100 atoms/cm^3 as a detectable concentration without considering the quenching and atomization efficiencies, their incorporation would increase this value to 10^6 atoms/cm^3. If the experiment were done in a flame, this value would correspond to a solution concentration of about 20 pg/ml. The detection limits at the present time are essentially at this concentration [99]. As we saw earlier, the 100 atoms/cm^3 should be capable of being lowered to about 0.6 atoms/cm^3 or a detection limit of 100 fg/ml in the solution before aspiration. Additionally, the quenching can be eliminated by increasing the laser power to reach the saturation regime. Thus there remains good reason to believe the assertion that the present detection limits in laser excited atomic fluorescence in flames can be lowered into and below the pg/ml range [96].

If these improvements are possible, they will provide a technique with an outstanding dynamic range. Linear working curves are maintained up to concentrations of 10 μg/ml for most elements where absorption of the excitation and fluorescence light becomes important [96]. If one is working in the region of saturation, this upper limit can be increased because absorption is absent in this regime.

The accuracy and precision of the technique require correction for laser intensity fluctuations in both the short term and long term. Several papers have addressed this problem and it has been shown that with proper precautions, the relative standard deviation of the laser power measurement can be reduced to less than 0.2% [104]. Published values had not appeared for the standard deviation of the atomic fluorescence measurement when corrections were performed for the laser power fluctuations. The signal itself does contribute shot noise and fluctuates because of temporal changes in the flame. It is expected that the ultimate precision and accuracy will be found to be similar to those for flame emission or atomic absorption measurements since the controlling factors are similar.

The interferences of the method can be divided into those interferences characteristic of all flame spectroscopies and interference due to line overlaps. The latter interference can be largely eliminated because the linewidths of the absorption and fluorescence profiles are very sharp [96]. If the laser bandwidth and the monochromator bandpass are reasonably narrow, it would always be possible to select a line that does not overlap lines from other elements. Some interference might be expected if molecular species that have broad excitation and emission features form in the flame. The flame interferences include the familiar effects of molecular species formation in flames, particularly oxide formation, ionization interferences, and various physical interferences.

The high powers that are characteristic of pulsed lasers permit one to reach the saturation region without difficulty, although with c.w. lasers one can also reach the saturation region after focusing. Several experiments have explored the characteristics of working in saturation both theoretically and experimentally. The fluorescence intensity is shown in Fig. 8 as a function of the laser spectral power density [102]. At a laser power of 8×10^5 W/cm^2 nm, the fluorescence saturates and does not increase further with increases in laser intensity. Clearly if one is working at laser powers higher than this value, the fluorescence would be insensitive to laser power fluctuations. Figure 5 shows that the absorption coefficient vanishes as saturation is reached because the ground and excited states are reaching equivalent concentrations. As a result, the shape of the excitation line changes. The peak intensity of the excitation line is saturated while the wings of the line are unsaturated because of their lower absorption cross section. Thus the wings contribute in-

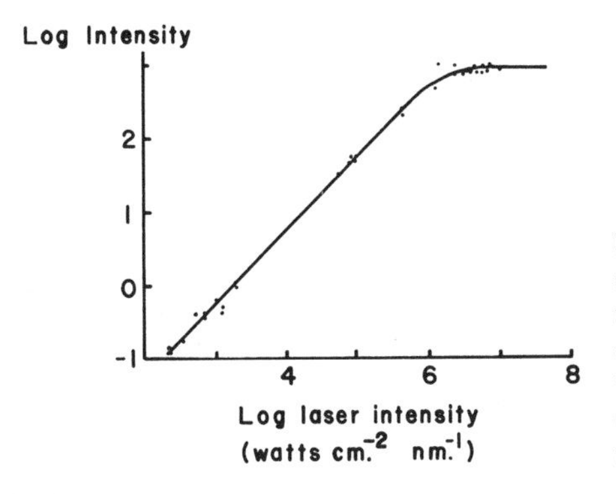

Fig. 8 The fluorescence intensity of Na emission in a flame is shown as a function of the excitation laser intensity. The fluorescence increases linearly with excitation intensity until it finally saturates and the excited- and ground-state atomic populations reach their statistical limits.

creasingly to the excitation profile and the line appears to broaden. Since the Lorentzian portion of a line profile dominates in the wings, the lifetime and collisional broadening contributions become important in describing the lineshape of an excitation transition that is saturated at its peak.

The vanishing absorption coefficient in saturation improves the dynamic range of the analytical working curve [81, 105]. The loss of absorption permits the excitation beam to propagate without attenuation, thus eliminating the prefilter effect. If the diameter of the excitation beam is sufficiently large to cause saturation of all the atoms viewed by the monochromator, the postfilter effect is also eliminated. These changes are shown in the working curve of Fig. 9 for the unsaturated regime and the saturated regime [81]. It is important to remember, however, that if saturation only occurs in the focused region of a beam, there can still be attenuation of excitation or fluorescence outside the focal region. Thus the spatial variation of saturation becomes an important parameter in defining the working curve.

Green et al. [106] have studied the advantages of using a c.w. dye laser as the excitation source in laser excited atomic fluorescence in order to circumvent some of the technical problems of a pulsed laser excitation, specifically the radiofrequency noise, the laser intensity fluctuations, the short aperture times required in the gated electronics, the laser bandwidths that are larger than the analyte linewidths, and the maintenance of a precise wavelength. The argon pumped dye laser they used had a power of about 100 W and a bandwidth of 0.003 nm (compared with a 0.002-nm Doppler width for a typical line). It was mechanically chopped at 2 kHz and the fluorescence was synchronously detected to eliminate flame background. A capillary burner was used with air-acetylene, hydrogen-air, or hydrogen-oxygen/argon. Fluorescence was detected with either filters in front of a photomultiplier or a 0.5-m monochromator. Both Na and Ba were studied since these elements could be excited with the laser. It was not possible to evaluate the detection

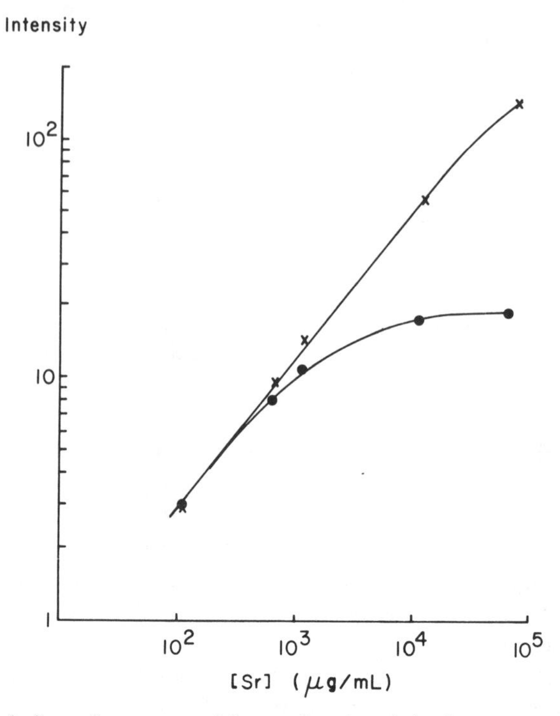

Fig. 9 The atomic flame fluorescence of Sr as a function of the Sr concentration introduced into the flame. The lower curve shows the working curve when the laser power is too low to saturate the Sr transition, while the upper curve shows the working curve when saturation is achieved. The extended linearity is caused by the elimination of the post- and prefilter effects when the transition is saturated and absorption is bleached.

limit for Na because it was exceeded by the laboratory contamination. It was found that Ba could be detected at 8 ng/ml in an air–acetylene flame (the same as that obtained in Table 2 for the same flame) and 2 ng/ml in a hydrogen–oxygen/argon flame. The dynamic range was also the same as that measured with pulsed lasers. The detection limit was determined by the fluctuations in the scattered light from the flame. The fluctuations have both a shot-noise component from the statistical nature of the light intensity of the scatter and a flicker component from the fluctuations of the scattering medium itself. The fluctuations in the scattered light were sufficiently small that they were able to detect signals about 22 times weaker than the scattered light intensity itself. It was shown that the flicker component was more important than the shot noise component and therefore could not be improved by going to higher excitation intensity to reduce the shot noise contribution. Since both c.w. and pulsed excitation have the flicker component of scattering as the limitation in measuring resonance fluorescence, it is not surprising that the same detection limits are obtained. By incorporating wavelength modulation into the dye laser as was done by Fairbanks et al. [87], the scattered light contribution can be rejected more effectively. Goff and Yeung [107] have found that detection limits could be lowered by 30, which is about the same factor that Fairbanks et al. could see below the scattered light level.

An important parameter in atomic fluorescence experiments is the correct setting of the laser wavelength. Since these linewidths are sharp, it is quite important to have excellent laser wavelength stability over a reasonably long term. That is a definite problem with many or most of the commercially available systems, either c.w. or pulsed. One attractive solution to this problem might be to use the optogalvanic effect that results from exciting the plasma of a hollow cathode lamp with the excitation laser to provide a feedback stabilization of the laser frequency [108]. An interesting alternative is the use of metal atom resonance line lasers [109]. These lasers consist of heated cells that contain a metal vapor and are excited by an excimer laser. They lase at the exact wavelength needed for atomic fluorescence.

Atomic Fluorescence in Furnaces

A second common method of converting a sample is the use of high-temperature furnaces or other nonflame cells to dry, vaporize, and atomize the sample. This technique has a number of well-known attractive features for laser excited atomic fluorescence that distinguishes it from flame techniques. One of the difficulties in using flames is the high dilution of analyte that occurs because of the high flow rates of flame constituents that both support the flame and transport the sample into the flame and the gaseous expansion that accompanies the combustion. Additionally, the flame environment is not always best suited for efficient emission because of collisional quenching processes and compound formation. Nonflame cells allow better control of the emission environment and permit very small sample volumes to be examined. Additional comparisons between flame and nonflame cells appear in many other references [110].

Several experiments have been performed with heated Na vapor cells that resemble nonflame cells [86–88, 111, 112]. There are important differences, however, between the heated vapor cell and the high-temperature furnaces and graphite cuvettes are required for analytical work. A graphite cuvette must be rapidly heated to the atomization temperature to establish a high atomic density in the vapor before the vapor diffuses out of the cell. In addition, the high temperature required for atomization produces blackbody radiation, which must be distinguished from fluorescence. Nevertheless, the Na vapor cell experiments have been valuable in studying the basic principles of the technique, determining the importance of different experimental variables, and defining the limits on the reproducibility, dynamic range, and detection limits of the analytical measurement.

As is described earlier, Fairbanks et al. [87] measured the resonance fluorescence from Na in a cell with a c.w. dye laser and obtained a linear correlation between fluorescence intensity and concentration over 9 orders of magnitude in concentration with a lowest detectable concentration of 100 atoms/cm^3. Gelbwachs et al. [112] modified this experiment slightly so that it was not necessary to look at the resonance fluorescence. Instead, they excited the D_1 transition of Na at 589.6 nm with a pulsed dye laser and observed the D_2 fluorescence at 589.0 nm with a ¾-m monochromator. By extrapolating their S/N ratio at 200 atoms/cm^3, they estimated that 10 atoms/cm^3 could be observed. Gelbwachs et al. also suggested this method be named SONRES for saturated optical nonresonant emission spectroscopy.

Brod and Yeung [86] used a pulsed laser to excite fluorescence from a Na vapor cell in their study of the influence of saturation on the important analytical variables. They found that the linearity of calibration curves was improved in the saturation regime, where the prefilter effect can be reduced or eliminated. In addition, they showed that the absorption profile of transitions and the fluorescence time dependence are strong functions of the degree of system saturation.

A number of experiments have been performed with pulsed laser excitation of fluorescence in graphite cuvettes under conditions that would be suitable for a typical analytical sample [113–116]. Pulsed lasers (10-nsec pulses) are particularly well-suited for fluorescence excitation from graphite cuvettes because short aperture times can be used with gated electronics to minimize detection of blackbody radiation from the cuvette. A typical experiment is described by Hohimer and Hargis, who determined thallium in aqueous solution [113]. Their instrumentation is sketched in Fig. 10. A N_2 laser pumped dye laser was frequency doubled to 276.8 nm to excite the $6^2P_{1/2} \rightarrow 6^2D_{3/2}$ transition of Tl. The laser energy was 0.15 μJ in a 5-nsec pulse and the bandwidth was 0.03 nm. The beam was directed into a vitreous carbon atomizer of conventional design. A light trap was added behind the sample region and baffles were added in front of the sample region to restrict the photomultiplier's field of view to the excitation region and to reduce scattered light and the contribution from blackbody radiation. The fluorescence at 352.9 and 351.9 nm was collected with an $f/2.8$ optical system and was imaged through interference filters onto a cooled photomultiplier. The photomultiplier signal was amplified and measured with a conventional boxcar integrator that was operated with an aperture time of 50 nsec. The laser wavelength and intensity could be constantly monitored by passing the beam emerging from the atomizer into a temperature stabilized Tl vapor cell. The ratio of the

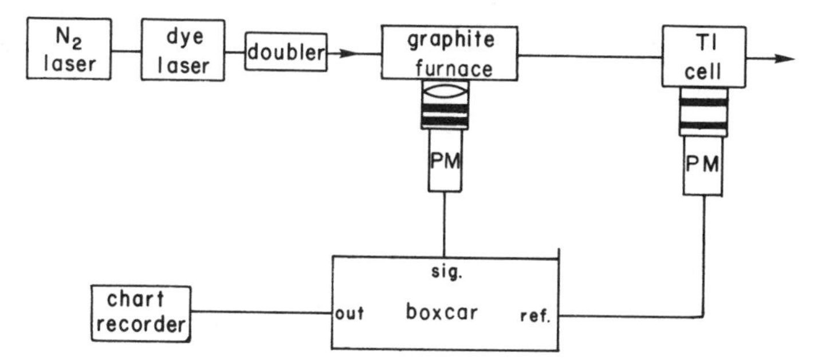

Fig. 10 A nitrogen-pumped dye laser can be used directly or with frequency doubled to excite Tl fluorescence from a sample that is atomized in a graphite furnace. The fluorescence is viewed through filters by a photomultiplier. The signal is corrected in a dual channel boxcar integrator using a reference signal that comes from fluorescence generated in a heated Tl oven. The ratioing procedure reduces intensity and wavelength fluctuations of the laser.

amplitude of each laser pulse measured by this vapor cell to the analytical signal was used to correct for laser fluctuations.

This experiment was able to measure Tl concentrations over 7 orders of magnitude down to a lowest concentration of 0.5 pg/ml or an absolute detection limit of 2.5×10^{-14} g. The standard deviation for the measurements was 9% at the lowest concentrations and improved to 4% at higher concentrations. The limiting noise was produced by dark current of the photomultiplier and noise associated with the boxcar integrator and other electronics. Since nonresonant fluorescence was monitored, scattered light did not contribute. Blackbody radiation from the graphite cuvette also did not contribute significantly both because of the short wavelength that was monitored and because of the short aperture time of the electronics. This situation suggests that the low detection limits can be lowered still further. Dark noise from a photomultiplier can be essentially eliminated by a short aperture time for the boxcar integrator. The detection limit should therefore scale directly with the power of the excitation laser. Commercial systems are presently available that produce 10 mJ of tunable UV radiation at 276.8 nm. If other factors do not arise to limit the detectability, one should be able to measure absolute amounts of 3.7×10^{-19} g of Tl.

Other experiments have shown comparable results for analysis of Pb, Fe, and Cs [107–109]. Bolshov et al. [115] frequency doubled a Nd:YAG pumped dye laser to excite the 283.31-nm transition of Pb and the 296.69-nm transition of Fe. The fluorescence was detected at 405.78 and 373.49 nm for the two elements, respectively. An oscilloscope was used to measure the intensities. With this system, these workers could measure concentrations of

2.5 pg/ml of Pb and 25 pg/ml of Fe and absolute amounts of 7.5×10^{-14} g of Pb and 7.5×10^{-13} g of Fe. The detection limits for these elements were determined by background fluorescence that was generated in the quartz windows of the graphite furnace. Similarly, Hohimer and Hargis [116] demonstrated limits of 1.2×10^{-12} g of Cs in a 20-pg/ml solution. In this case, the detection limit was determined by the intense blackbody radiation from the atomizer at the 852.1-nm transition that was monitored.

Molecular Fluorescence

If a laser is used to excite molecular fluorescence from compounds in a condensed phase, one is immediately faced with the problem it is necessary to excite into a broad absorption band instead of the sharp absorption lines encountered in atomic fluorescence. In this situation, it becomes much more difficult to distinguish the background signals from the fluorescence of the molecule of interest. Although the fluorescence intensity scales with the excitation intensity, the lower limit of detection does not improve if the background intensity also scales with the excitation intensity. Additionally, the high spectral powers characteristic of lasers are wasted if they simply excite a broad absorption band. Thus a tungsten lamp may have a brightness of only 0.01 W/(cm^2)(sterad)(nm) [117], but it can deliver more than 20 mW to a sample with a 50-nm absorption band. This excitation intensity excites a fluorescence intensity that is sufficiently high that the detection limit of a well-designed instrument is determined primarily by background emission and not fluctuations in the detector dark current [118–120]. Thus reduction in background levels requires careful attention to the factors that cause background noise when designing an experiment.

A number of factors contribute to background noise [118–120]. At very low background levels, the dark noise of the detectors and electronics becomes important. Usually this factor is not as important as extraneous fluorescence and Raman scattering. The solvent always contains impurities that fluoresce. Some commercial suppliers of solvents take particular care to furnish high-quality solvents with low fluorescence backgrounds. Nevertheless, the backgrounds can be seen easily. Additional purification can lower the background [118]. Exposure of the solvent to daylight or a strong light source for a long time can bleach some of the impurities so they no longer fluoresce [121]. All the optical components in the instrumental setup, including the optics, filters, and sample container, also fluoresce. The windows of the sample container are particularly troublesome because it is difficult to separate the sample and window fluorescence from each other. Raman scattering is a weak effect, but since the solvent concentration can be orders of magnitude higher than the sample concentration, Raman scattering from the solvent becomes an important limitation. Finally, the instrumentation must be capable of very high rejection of stray light so the much higher intensities of the excitation source are blocked from detection. In this regard, it is important to remember that lasers are not perfectly monochromatic. They all emit light at wavelengths distant from the lasing wavelength. The light that falls at the same wavelength as the fluorescence monitored must be blocked by appropriate filters.

The laser offers some specific advantages over other more conventional sources [122]. The high monochromaticity makes the laser easier to filter and it is easier to achieve high rejection ratios for stray light. In addition, the Raman scattering from the solvent falls in lines that are narrower than those that would be obtained with a broader band excitation source. Raman scattering can be spectrally distinguished from sample fluorescence by simply tuning to a fluorescence wavelength between the Raman lines of the solvent. The coherent nature of the laser allows it to be easily directed and to be focused to very small, diffraction-limited spot sizes. Small sample volumes can therefore be used. Finally, the high powers of the laser can provide high background levels that have a low shot noise component.

Several papers have examined the instrumental factors that control the detection limit [123, 124]. Ishibashi et al. have performed a rather complete study of the problem and have achieved a 5×10^{-14} M detection limit for fluorescein [123]. Once one reduces the extraneous fluorescence, the most important adjustment for obtaining low detection limits is choosing the optimum excitation and fluorescence wavelengths. This choice is dictated by the size of the Stokes shift between the absorption and fluorescence maxima. If there is a large shift, the excitation wavelength can be set to the absorption maximum and the fluorescence can be monitored at wavelengths outside the region of Raman scattering. If there is a small shift, however, the fluorescence signal falls in the same region as Raman scattering. One must then find an excitation wavelength where the absorption cross section is large, but at the same time places the Raman lines at wavelengths outside the region where the fluorescence is monitored. Ishibashi et al. prepared graphs of the signal-to-noise ratio versus fluorescence wavelength for a series of different excitation wavelengths [123]. The optimum conditions could then be easily identified.

Numerous studies have been performed on different compounds to determine the advantages of using laser excitation sources [121–148]. Representative detection limits are quoted in Table 3. They depend on the particular experimental configurations and the sample characteristics. The values are generally in the parts per trillion region. Several comparisons have been performed with conventional spectrofluorometers, but these are very dependent on the particular instrument [126, 143], its upkeep, and its resulting wavelength rejection capabilities. If one is careful to use a stable light source and

Table 3

Compound	*Detection Limit* (ng/ml)	*Reference*
9-Acridanone of flurazepam	0.5	138
Acridine	1.0	141
	0.03	143
Aflatoxins	0.0025	147
Anthracene	0.01	143
	0.0044	140
Benzene	19	140
Carprofen	2.5	138
Chlorophyll	0.001	143
Chrysene	0.03	143
Fluorescein	0.8	142
	0.20	143
	0.002	126
	2×10^{-5}	123
Fluoroanthene	0.01	143
	0.001	140
Phenanthrene	0.03	143
Pyrene	0.02	143
	0.0005	140
Pyrido[2,1-6]quinazoline	0.5	138
Quinine sulfate	1.0	138
	0.01	143
Rhodamine B	0.005	125
	0.0005	124
Rhodamine 6G	0.00062	147
Riboflavin	0.00047	144
	0.0006	123
Salicylic acid	1.0	138

adequate light rejection capability, the detection limits can approach those for a laser source [143]. A simple filter fluorometer, for example, has been described that uses a conventional tungsten lamp and has a 2×10^{-13} *M* detection limit for fluorescein (about 0.08 pg/ml) [119]. The choice of an excitation source for a fluorescence measurement should therefore be made with care and should include consideration of the optical systems required for processing the exciting light, the stability of the systems, the wavelength regions required, the chemical application of interest, and the cost, complexity, and reliability of the excitation source in the different possible systems. As the laser technology continues to develop more reliable and convenient sources with broader wavelength coverage and lower costs, the laser will begin to displace the conventional sources.

The direct molecular fluorescence experiment does not utilize many of the unique characteristics of modern lasers. For the laser to be a competitive analytical light source, either novel chemical methods must be developed that will make the molecular fluorescence measurement more compatible with the unique properties of the laser or novel laser methods must be developed that will provide new capabilities for fluorescence spectroscopy. Generally, one must seek methods that increase the selectivity of laser-excited molecular fluorescence, since this limits its range of applicability, its use for identifying compounds, and its lower detection limit. A number of techniques are now under study that deal with this problem.

Liquid Chromotography Detection with Laser Excited Fluorescence

One approach to providing the improved selectivity required for practical analysis problems is to combine the excellent separation capabilities of modern chromatographic techniques with the sensitivity of fluorescence methods [122, 149]. The illuminated sample area in the effluent from a high-pressure liquid chromatograph (HPLC) is typically 0.1×0.2 cm, much smaller than the illuminated area of a sample in a conventional fluorometer. Since the output of a laser can be conveniently focused into sample areas much smaller than the effluent volumes, there is no problem transferring all of the laser energy to the sample. Diebold and Zare developed a detection system where the flowing effluent from a LC column was suspended between the ends of the column tubing and a solid rod [149]. The suspended liquid formed a windowless volume into which the laser could be focused without having to pass through materials that might fluoresce. This system has been used with the 325-nm line of an 8-mW He–Cd laser and with the 488- and 514.5-nm outputs of a 0.5-W argon laser. With the He–Cd laser, a detection limit in the effluent of 1.8 pg/ml was obtained for the measurement of aflatoxin in agricultural samples. It was limited by background fluorescence. This limit corresponded to 7 fg of aflatoxin in the detection volume. Several additional determinations have been made using this system with good success [121, 136].

Yeung and Sepaniak have designed a different fluorescence detector for HPLC [122, 150–152]. A laser is focused within a capillary tube through which the effluent from the HPLC flows. A small optical fiber is introduced into the capillary tube so the end of the tube is close to the excitation volume of the laser. The fluorescence generated in the sample volume is collected by the fiber optic and is transferred to a monochromator for measurement. The shape of the effluent stream in the excitation region is confined in this design and does not depend on the solvent mixture used as the eluant. It is also less susceptible to bubbles from the degassing of solvents. This system has been used for a variety of samples, including aflatoxins, drugs, and coal-derived liquids. Detection limits have been in the 10-pg range.

Enzyme and Immunological Methods

A second approach to providing improved selectivity is combining the excellent detection capability of laser excited fluorescence with the excellent selectivity of many biological reactions, such as enzyme mediated and antigen-antibody reactions. Fluorescence immunoassay (FIA) procedures have been studied extensively in the clinical chemistry field with the expectation that FIA methods could replace many of the radioimmunoassay procedures, which have all the problems of radioactive materials associated with them [153].

The different FIA methods have been broadly classified into homogeneous and heterogeneous assay procedures [153]. For the heterogeneous methods, an antiserum to the specific antigen of interest is obtained and an excess of the antigen with a fluorescent tag attached is added [153–155]. The fluorescently labeled antigen attaches to the antibodies to form a large complex. Then any additional antigen that is introduced displaces some of the labeled antigen that is already attached to the antibody. The labeled antigen and the free antigen can both compete on roughly equal terms for the available antibodies. Thus when the antigen from an unknown sample is added to the mixture of antibodies and labeled antigen, the fluorescence from the antibody-labeled antigen complex decreases by an amount that depends on how much antigen was present in the unknown sample. The sensitivity of this assay procedure depends on the relative amounts of labeled antigen, free antigen, and antibody. For the lowest detection limit, the antibody and the labeled antigen are kept low, so low concentrations of free antigen can have a noticeable effect. These conditions are unsuitable if the antigen concentration is large because the labeled antigen–antibody concentration becomes very low and insensitive to the exact amount of free antigen. Consequently higher concentrations of antibody and labeled antigen must be used. The key to the success of these assays is distinguishing between the fluorescence of free labeled antigen and the labeled antigen–antibody complex.

Heterogeneous procedures rely on a physical separation of the two fluorescent species. Considerable success has been obtained in attaching antibodies to different solid supports such as polystyrene beads [153]. The competition between labeled and free antigen for the antibody sites on the support can then take place. The labeled antigen that does not attach to the support is washed off. If the fluorescence of the labeled antigen in the wash liquid is measured and compared with what it would have been if there were no free antigen present, the difference is directly proportional to the amount of free antigen. Alternatively, the fluorescence from the labeled antigen on the solid support can be measured. Any decreases in fluorescence from that when no free antigen was present are directly proportional to the amount of free antigen.

Lidofsky et al. carried out the separation of the labeled antigen from the labeled antigen–antibody complex using gel permeation HPLC with the laser-excited fluorescence detector [135]. They developed an assay procedure for insulin that has a detection limit of 0.4 ng/ml or 7×10^{-11} *M*. This detection limit is in the same range as the traditional radioimmunoassay procedure for insulin and it justifies the expectation that FIA methods can compete with present methods. Much work remains to be done on these procedures before they can become clinically useful. Seven samples contained so many fluorescent materials that the fluorescence from the labeled insulin–antibody complex was lost.

If the characteristics of the free labeled antigen and the labeled antigen–antibody complex fluorescences are sufficiently different and can be spectroscopically distinguished, it is not necessary to carry out the separation. Homogeneous RIA procedures are based on such differences [153, 156, 157]. Several physical phenomena have been utilized to distinguish the two species. As an example, we consider the "sandwich" assay, where the antigen can bond with two antibodies. Two sets of labeled antibodies are prepared—one set absorbs at high energy while the second set emits at low energy. Both sets are mixed with the solution containing the antigen to be measured. The antigen can bond with one antibody from each set (of course other combinations are possible). Now when the label that absorbs at high energy is excited, the excitation energy can be transferred to the second label, which has a lower energy level and can fluoresce. The energy transfer is efficient because the two labels are spatially close to each other, being bound in the same molecule. Such a transfer is not possible if the two labels are not bound to the same antigen. Direct excitation of the lower energy fluorescence is still possible, but this is less efficient, being off the absorption maximum.

A related approach labels both the antigen and the antibody with separate fluorescing molecules. One type of molecule is chosen to absorb at high energies while a second type is chosen to emit at low energies. Energy transfer can then occur between the absorbing molecule and the emitting molecule when the labeled antigen and antibody combine. The choice of the two types of molecules is restricted by the requirement that the emission profile of one overlaps the absorption profile of the other. Otherwise the energy transfer becomes inefficient.

Fluorescence depolarization techniques have been applied to homogeneous FIA procedures [153, 156]. If a molecule is excited by polarized light, its reemission is also polarized along the direction of the induced dipole moment. If the molecule changes its orientation between the excitation and emission events, the fluorescence polarization is lost. The polarization methods for a homogeneous assay rely on the differences in fluorescence depolarization between the labeled antigen and the labeled antigen–antibody complex. The complex is so much larger and heavier than the free labeled antigen that the depolarization effects are much smaller and the two can be distinguished on the basis of the depolarization.

The detection limits for the homogeneous and heterogeneous methods are determined by both the ability to separate the fluorescence of the complex from the fluorescence of the free labeled antigen and the ability to measure low concentrations of the labeled complex. For HPLC methods, laser excited fluorescence is measured at the exit from the column. Many of the other methods use standard fluorescence methods with cuvettes. Several groups have spread the mixtures on a transparent surface and illuminated the surface from below with a focused laser operating at the total reflection angle [158, 159]. The small focal volume ($\sim 25\ \mu m^3$) defined by the edges of the laser and the penetration depth from the total internal reflection is translated across the sample by moving the transparent surface. Whenever a labeled complex enters the focal volume, a sharp increase in the fluorescence is observed. Hirschfeld used this procedure to perform essentially single-molecule detection [159]. The intensity of the laser in the focal colume is about 2×10^5 W/cm^2 or 5×10^{23} photons/(sec)(cm^2), a value sufficiently high that bleaching occurs rapidly. Bleaching occurs because each time an organic molecule enters the excited state, there is a finite probability that it will undergo a photochemical reaction. If the absorption cross section is 10^{-17} cm^2, the molecule is reexcited about 5×10^6 times/sec. Even if the probability of reaction is only 0.1%, it is still clear that these light intensities will decompose the organic rapidly. Experimentally the mean lifetime for bleaching was observed to be about 2.5 msec [160], a time comparable to that required for a molecule to diffuse through the focal volume. Thus one observes pulses of fluorescence whenever a sample molecule is within the focal volume that corresponds to individual molecular entities.

In addition to the probability for excited state reaction, there is also a fixed probability for an excited organic molecule to fluoresce. The ratio of the two probabilities is a constant that is independent of excitation intensity or other quenching processes. If a molecule is illuminated with sufficient intensity and for a sufficient time to completely bleach it, the number of fluorescence photons emitted by that molecule is a constant determined by the ratio of fluorescence to reaction probabilities. The value is independent of any other quenching processes [160] that might occur, such as the concentration quenching caused by having many fluorescein molecules close to each other. It is the maximum number of photons a molecule can emit. A bonded set of dye molecules therefore emits a total number of photons during the life of those molecules that scales directly with the number of dye molecules and is independent of concentration quenching effects.

Enzyme-based methods also have a great deal of selectivity because of the selectivity of the enzyme. Imasaka and Zare [121] have developed a clever application of enzyme-mediated reactions that has resulted in methods for the measurement of a number of different materials and a very sensitive method for the determination of reduced nicotinamide adenine dinucleotide phosphate, NADPH, or NADP by enzyme amplification. Two enzyme-mediated reactions form the key to the procedures:

$$\text{glucose-6-phosphate} + \text{NADP} \xrightarrow{\text{glucose-6-phosphate dehydrogenase}} \text{NADPH} + \text{6-phosphoglucanolactone} \quad \text{(a)}$$

$$\text{L-glutamic acid} + \text{NADP} \xleftarrow{\text{glutamate dehydrogenase}} \text{NADPH} + \alpha\text{-ketoglutaric acid} \quad \text{(b)}$$

NADPH is naturally fluorescent under the conditions of the experiment, while NADP can be made fluorescent by adding concentrated base. Imasaka and Zare [121] excited fluorescence from samples contained in ordinary cuvettes with the 325-nm output of a He–Cd laser. Glucose-6-phosphate could be determined by running reaction (a) and measuring the fluorescence from the product NADPH. A detection limit of 2×10^{-9} *M* glucose-6-phosphate was obtained from a 1-ml solution. α-Ketoglutaric acid could be determined by running reaction (b) and measuring the fluorescence from the product NADP after addition of strong base. A detection limit of 4×10^{-9} *M* was obtained. In both cases, the detection limits were determined by background fluorescence from the distilled water and reagents.

Very low concentrations of NADP and NADPH can be measured by combining reactions (a) and (b) so an enzyme amplification scheme is obtained. Any NADP in the original solutions is reduced to NADPH by the first reaction. The NADPH produced in the first reaction can then be oxidized back to NADP by the second reaction and the cycle can continue as long as there is enough glucose-6-phosphate and α-ketoglutaric acid present. The products of these reactions build up as the reaction is allowed to continue. Several thousand cycles can occur during the 2-hr period. The reactions are finally stopped by heating the solutions to 85 to 90°C to destroy the enzymes. The 6-phosphogluconolactone produced during this time is proportional to the original NADP and NADPH concentrations, but it is much larger because of the many reaction cycles that have occurred. It is only necessary to measure the 6-phosphogluconolactone concentration to obtain the original NADP and NADPH concentrations. This measurement is performed by a third enzyme reaction, again involving the reduction of NADP and NADPH. The resulting NADPH concentration is measured. Detection limits of 10^{-10} *M* have been obtained by this procedure for a 100-μl cycling volume [121]. There were only 10^{-14} moles in the cycling volume measured. The detection limit was determined by fluctuations in the reaction rates, which were sensitive to environmental factors such as the cleanliness of the glassware used.

Site Selective Spectroscopy and Selective Laser Excitation Methods

The selectivity demonstrated in the preceding two sections has been achieved by either a physical separation or selective chemistry. It would be very desirable if a complex sample could be separated spectroscopically, that is, resolving spectroscopically the individual components in a mixture. The modern laser potentially has those capabilities because of its very narrow bandwidth. If the analyte also has narrow linewidths, the laser can selectively excite the analyte even in the presence of many other components. The laser is simply tuned to an individual sharp absorption line of the species of interest and the species is excited. Other species in the mixture can only be excited if one of their absorption lines accidentally overlaps the excitation line or if there is energy transfer from the excited species. If the single species is excited, the resulting fluorescence spectrum comes only from that species. In the same way, a monochromator can be tuned to an individual sharp absorption line of the species of interest and the dye laser can be scanned over the absorption lines. The resulting excitation spectrum comes only from the species being monitored (assuming no overlap or energy transfer). By changing the excitation or fluorescence wavelengths, the fluorescence or excitation spectra of all the other components in the mixture can be obtained. One can thus achieve a spectroscopic separation of the mixture.

In many cases, there is not a finite number of unique species in a mixture, but rather an infinite number of them that together form the homogeneous linewidth. Typically the infinite number of species corresponds to an ion or molecule in an infinite number of local environments in some condensed phase. The different local environments correspond to the strains or disorder present in the condensed phase. Even in these cases particular species can be selectively excited. The laser is tuned to a spot in the inhomogeneous profile and those species that are resonant with the laser are excited and fluoresce. Since only a small subset of species are excited, the resulting fluorescence is much sharper and more structured than a traditional fluorescence spectrum. This phenomena is called laser induced fluorescence line narrowing [161]. A similar procedure yields a sharpened excitation spectrum if a particular fluorescence wavelength is monitored within the inhomogeneous line profile. The extent of line narrowing depends completely on the amount of accidental overlap of absorption lines from different species or environments. If it is extensive, there can be no line narrowing except at the wavelength of the excitation transition itself. This transition must be narrowed because all the excited species have their transition at this wavelength. One could not call this site selective spectroscopy because many sites have been excited, but it is energy selective. The use of this concept can provide the homogeneous width of lines in a sample even if there is extensive accidental overlapping of transitions.

For site-selective spectroscopy to work, it is necessary to have sharp lines associated with the species of interest. There are many different situations where sharp line spectra can be obtained for both inorganic and organic analysis procedures. We examine each in turn.

PROBE ION METHODS

The first application of site selective and selective laser excitation spectroscopies in an analytical context was in the development of probe ion techniques [162, 163]. The crystal field splittings of ions placed in a condensed phase are uniquely determined by the immediate surroundings of the ions and consequently the spectra reflect the individual site environment. In this way, the ions act as spectroscopic probes of the condensed phase. If an analyte is present in the immediate environment of an ion that can have crystal field splittings, the environment of the ion is changed and this change is reflected in its spectrum. Several difficulties are encountered in implementing this concept in the study of a real sample. First, many ions do not have transitions that are sharp enough to distinguish them from transitions in a different environment. Second, if an ion can occupy sites with many environments, the spectra become complex and overlapped. Third, there is no method that allows one to predict the environment if only the crystal field splittings are known. Each of these problems can be overcome and it is possible to use ions effectively as probes.

The sharpness of the spectral transitions from an ion in an ordered condensed phase depends on the amount of interaction between the ionic electronic states and the lattice. The lanthanide and actinide ions are particularly well suited in this regard because the optically active electrons are in the unfilled $4f^n$ and $5f^n$ orbitals, which are well shielded from the lattice by outer orbitals [164]. The spectral lines are very sharp as a consequence, although the crystal field splittings are smaller than those in the transition metal series. This makes the rare earth ions particularly useful as spectroscopic probe ions.

If a rare earth ion enters the condensed phase in a number of possible environments, the complex spectra can be greatly simplified by using a laser to excite a specific absorption line of a specific rare earth site [165, 166]. Since the absorption lines from different rare earth sites do not overlap in general and since energy transfer between sites is not observed, site selective spectroscopy is feasible. The fluorescence spectrum is then greatly simplified and is characteristic of just the site excited. In a similar manner, if a specific fluorescence line of a specific rare earth site is monitored as the laser wavelength is tuned continuously over the possible absorption lines, a single site excitation spectrum can be obtained.

The inability to determine the kind of local environment about a rare earth ion dictates that other methods must be used to provide information about this environment. In this application, one examines the changes in the spectra that occur with and without the presence of the ion that is to be analyzed. Sites that are present only when a particular ion is present would suggest that this ion is nearby a rare earth ion and is perturbing its crystal field splitting. It is also possible that the analyte ion is producing defects (e.g., cation or anion vacancies) in the environment that perturb the rare earth ion. In general, these situations can be identified because other ions that are similar can also produce the same site.

The feasibility of this idea of using lanthanide and actinide probe ions was studied by developing a method for the analysis of trace concentrations of these ions [167–169, 172]. $Ca(NO_3)_2$ is added to a solution containing trace concentrations of the various lanthanide and actinide ions to be analyzed. Other ions such as Li^+ are added as solid-state buffers. The rare earths are then coprecipitated in a CaF_2 precipitate by adding NH_4F. The precipitate is filtered, dried, ignited under reproducible conditions, and pressed into a sample holder along with other samples or reference materials. The samples are then cooled to 10°K and excited with specific wavelengths of a dye laser. The fluorescence is measured with standard monochromators. A typical spectrum is shown in Fig. 11. The lines are very sharp and allow one to easily separate absorption or emission lines of one rare earth ion from another. The method is therefore very selective to specific lanthanide and actinide ions and has detection limits of 10 fg/ml as shown by Perry et al. [173]. At the present time, it is possible to measure Pr, Nd, Pm, Sm, Eu, Tb, Dy, Ho, Er, Tm, and U by this direct method.

To extend this method to analysis of other kinds of ions, it is necessary to provide a mechanism that will cause a one-to-one association between the analyte ion and a lanthanide probe ion. The new crystal field levels that result can be selectively excited with the laser so that fluorescence comes only from lanthanides with analyte ions nearby. The method should therefore be capable of the same selectively and low detection limits as the direct lanthanide analysis.

The feasibility of associating an analyte and a lanthanide ion was first demonstrated using the charge compensation that occurs in $BaSO_4$ precipitates when aliovalent ions (i.e., ions with different charges from the lattice) enter the lattice [163]. Na_2SO_4 and small amounts of $EuCl_3$ are added to a solution containing trace amounts of PO_4^{3-}, the analyte ion. When a

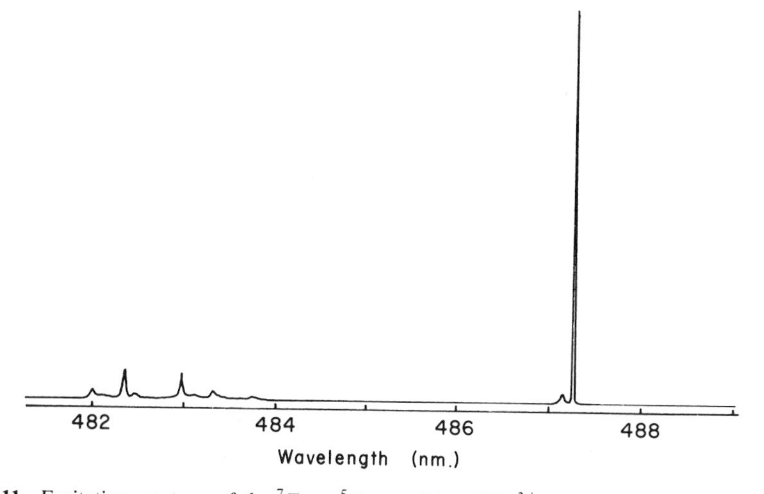

Fig. 11 Excitation spectrum of the $^7F_6 \rightarrow {}^5D_4$ transition of Tb^{3+} in CaF_2 obtained while monitoring the $^5D_4 \rightarrow {}^7F_5$ fluorescence transition at 540.42 nm.

$BaCl_2$ solution is added, the Eu^{3+} and PO_4^{3-} coprecipitate in the $BaSO_4$ precipitate and require a charge compensation because the trivalent ions are replacing divalent ions in the lattice. The charge compensation could be provided by having an oxygen vacancy created for every two PO_4^{3-} ions that enter the lattice or by having a barium vacancy for every two Eu^{3+} ions. These situations would produce the intrinsic sites. It is also possible that the Eu^{3+} and PO_4^{3-} ions could charge compensate each other. Their opposite charges would attract and cause an association to occur. The Eu^{3+} ion crystal field levels would be strongly affected by the presence of the PO_4^{3-} ion and a new site would appear in the spectrum with line positions that were characteristic of the presence of PO_4^{3-} and line intensities that were proportional to its concentration. The spectra of $Eu^{3+}:BaSO_4$ precipitates with and without the presence of PO_4^{3-} are shown in Fig. 12. An additional line appears at 578.2 nm, which is characteristic of the presence of PO_4^{3-}. Note also that the line at 574.9 nm associated with one of the intrinsic sites disappears when PO_4^{3-} is added. This disappearance indicates that PO_4^{3-} is changing the defect equilibria in the $BaSO_4$ lattice and is causing a redistribution of possible sites. This situation is not desirable from an analytical viewpoint because it means that the analyte concentration is important in determining the properties of the matrix it is in. It would be much more desirable to have another ion that has properties similar to those of the analyte ion and could function as a solid-state buffer to fix the defect equilibria independently of the analyte concentration.

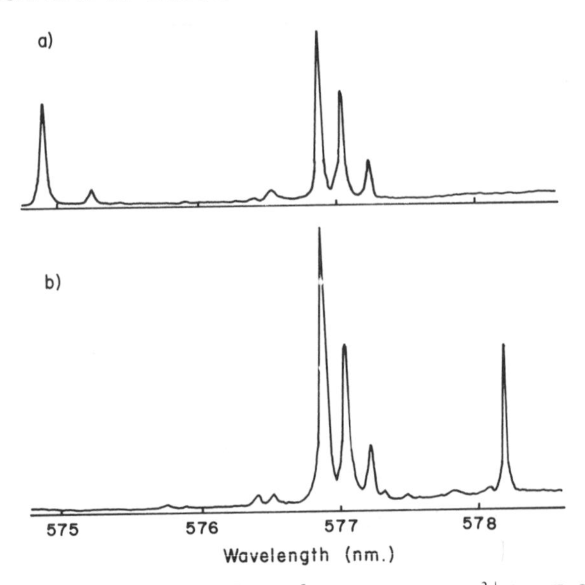

Fig. 12 (*a*) Excitation spectrum of the $^7F_0 \rightarrow {}^5D_0$ transition of Eu^{3+} in a $BaSO_4$ precipitate obtained by monitoring fluorescence at all wavelengths. Each line corresponds to one Eu^{3+} site [163]. (*b*) The same excitation spectrum as (*a*) but the precipitate was formed with a 5-μg/ml concentration of PO_4^{3-} present when the $Eu^{3+}:BaSO_4$ precipitate formed.

The spectra in Fig. 12 were obtained under experimental conditions where there was no selectivity for isolating lines from individual sites. Broad band-pass instrumentation was used that monitored all the fluorescence from the sample. One can also achieve high selectivity by using instrumentation with resolution better than about 0.1 nm and isolating the spectral transitions from the individual site of interest.

A second method of achieving a 1:1 association between an analyte and a lanthanide probe ion is to take advantage of the impurity clustering that occurs at high concentrations [170, 172]. For example, consider how one might perform an analysis for Y^{3+}, a nonfluorescent transition metal. One could add $Ca(NO_3)_2$ and a relatively high concentration of Er^{3+} to the solution containing Y^{3+} and then perform the precipitation of CaF_2 by adding NH_4F. If the concentration of Er^{3+} were sufficiently high (0.01 mole % of the Ca^{2+} concentration), the Er^{3+} in the CaF_2 lattice would begin to cluster. Dimers of Er^{3+} ions would first form in large numbers. If there were also Y^{3+} ions in the lattice, Er^{3+}–Y^{3+} dimers would be formed as well as with a concentration that depends on the relative Y^{3+} concentration. The Y^{3+} ions have a different ionic radius than Er^{3+} and would cause different crystal fields at the Er^{3+} ion than would be found for an Er^{3+}–Er^{3+} dimer. These unique crystal field levels can be selectively excited and an analysis procedure specific for Y^{3+} can be devised with a sensitivity determined by the Er^{3+}.

There is a limitation to the selectivity that can be achieved in this particular procedure because the shifts in level position are not as large as one obtains if the analyte-lanthanide probe association is achieved by charge compensation. This limitation can be overcome by judicious choice of the lanthanide probe and the fluorescence transitions examined. If there are two optically active ions near each other as there are in the dimers, energy transfer processes can occur between them where the energy provided to one ion by excitation to a specific level can be redistributed between the levels on both ions [174] as shown in Fig. 13. If the total energy of the final state is less than that of the initial state, the energy cannot be returned to the initially excited level. The energy transfer process thus quenches the fluorescence from the initially excited level. However, if the other ion in the dimer has no excited levels or has levels that do not provide appropriate matches with the initially excited level, these energy transfer processes cannot occur and the fluorescence is not quenched. Thus one can quench the large fluorescence peaks of Er^{3+}-Er^{3+} dimers by selecting specific fluorescence transitions but keep the smaller nearby peaks from the Er^{3+}-Y^{3+} dimers to achieve the high selectivity desired. This indirect method has been used to measure Ce, Gd, Yb, La, Lu, Sc, Y, Zr, Hf, and Th [170, 172].

This use of lanthanides as probe ions is still in its infancy and the applications are likely to increase. It is also possible to use these methods in disordered materials where the lanthanide ions encounter a continuum of different environments. This situation causes extensive broadening of the spectra. The fluorescence line narrowing techniques discussed earlier in this section can be easily applied to disordered systems to obtain the sharp line transitions of the individual environments. These methods have already been used to study glasses [175], but they promise to have a considerable impact on other areas, such as the study of biological systems.

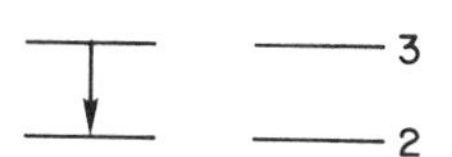

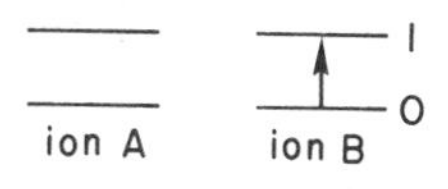

Fig. 13 When two ions are near each other, an excitation on one can be transferred to the other. Here, ion A, which was excited to level 3, transfers a portion of the initial energy to ion B.

MATRIX ISOLATION METHODS AND SHPOL'SKII SYSTEMS

Site selective spectroscopy is also possible for organic systems even though these usually have only broad spectra. The broad spectra arise because of all the conformations and environments possible in solution. A single molecule held in a unique conformation in a condensed phase has a simple spectra with sharp lines corresponding to electronic transitions (the parent transitions) and broader transitions corresponding to vibronic or phonon contributions from the condensed phase as shown schematically in Fig. 14 [176]. The relative intensity of the parent transition and the vibronic sideband depend on the relative equilibrium positions of the ground and excited states through the Franck–Condon overlap integrals and the amount of coupling between the condensed phase and the molecule. In the typical fluorescence sample one does not have molecules with unique conformations and unique environments, but all possible conformations with a number of different surroundings. The electronic and vibronic energies are slightly different for each possibility and the net result of the entire ensemble is a broad absorption or fluorescence profile.

There are special situations where a molecule can be imbedded into a matrix in either a unique conformation with unique surroundings or a small number of conformations and surroundings. The spectrum of the ensemble then becomes identical to the spectra of individual molecules in fixed conformations [177–187]. Shpol'skii discovered that when certain aromatic compounds are frozen in specific solvent matrices, the spectroscopic transitions become very sharp and structured [177–179]. The particular solvent that causes the sharp line structure depends on the aromatic compound which

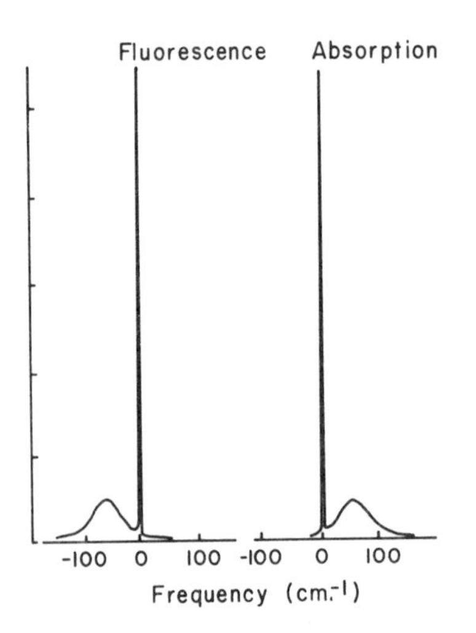

Fig. 14 This figure displays the fluorescence or absorption spectrum expected for a single isolated molecule in the region of the electronic transition [176]. In this example, the total area under the sideband is three times as large as the sharp electronic line. When the spectra of all the molecules in a sample are ensemble averaged to produce the observed spectrum, the sharp details are lost and only bands are observed.

must fit into the solvent lattice. Furthermore, not all aromatic compounds form Shpol'skii systems, that is, solvent systems with sharp line transitions. A typical fluorescence spectrum of the Shpol'skii system of perylene in *n*-heptane is shown in Fig. 15 [181]. A single parent peak appears at the low wavelength end of the spectrum and is followed by a series of peaks corresponding to the different vibrational electronic transitions at higher wavelengths. Many times there are several lines corresponding to each transition that presumably come from different possible orientations of the molecule in the solvent lattice. The sharp line structure also appears in the absorption or excitation spectra of Shpol'skii systems. The sharp lines are well-suited for the narrow bandwidths of tunable lasers and would permit selective excitation of particular molecular species in the presence of many other species. Kirkbright and DeLima [183] have demonstrated the feasibility of spectroscopically distinguishing among the different fluorescence lines from an eight component mixture of different polyaromatic hydrocarbons in a frozen *n*-octane–cyclohexane matrix when a conventional broad-band excitation source is used. Additional selectivity can be achieved by using a laser for the excitation source.

The application of selective laser excitation techniques has made Shpol'skii spectroscopy a practical possibility. Early studies indicated that the formation of the glasses was sufficiently irreproducible that quantitation was not practical [180, 183]. Recent studies by Yang, D'Silva, and Fassel [188, 189], however, show that these irreproducibility problems can be solved by working at low concentrations, using reproducible sample cooling procedures, and employing another PAH compound as an internal standard. These authors have also demonstrated the capability of selectively exciting specific sites of specific PAH compounds in complex mixtures by making use of the narrow Shpol'skii lines and the narrow laser bandwidths. Figure 16 demonstrates the procedure for 11-methylbenz(*a*)anthracene in an *n*-octane frozen

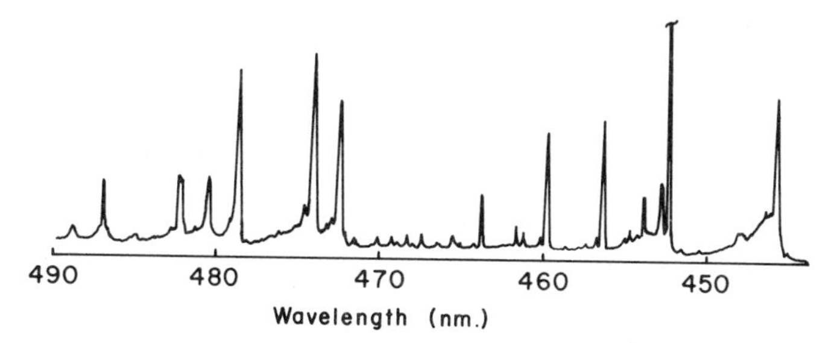

Fig. 15 Fluorescence spectrum of perylene in frozen *n*-heptane [181]. The sharp lines result because all the perylene molecules have been oriented identically by the matrix, and the observed spectrum approaches that for an isolated molecule.

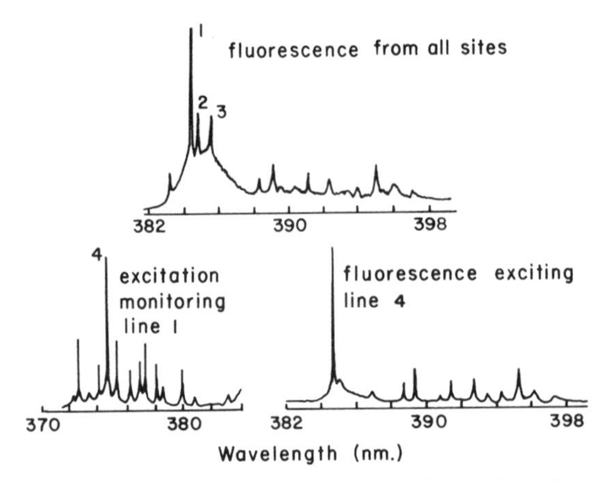

Fig. 16 The top spectrum shows the fluorescence of 11-methylbenz(*a*)anthrene in an *n*-octane frozen matrix when the laser is tuned to a wavelength where fluorescence from all sites is excited. The numbers in this spectrum label the lines from sites 1, 2 and 3. If line *1* is monitored and the laser is scanned, the excitation spectrum shown in the lower left is obtained. The lines come only from site 1. If the laser is then tuned to line *4* in this spectrum, only site 1 is excited and the fluorescence spectrum shown on the lower right is obtained. The fluorescence comes only from site 1 and it is much simpler than the top fluorescence spectrum. (After Ref. 189).

matrix. This particular PAH compound can occupy three different sites in the *n*-octane matrix. The top spectrum is the fluorescence that results when the excitation wavelength excites all three sites of 11-methylbenz(*a*)anthracene. The lines labeled *1, 2,* and *3* come from the three different sites. If the monochromator is tuned to the wavelength labeled *1* ($\lambda_{\text{fluor.}} = 384.8$ nm) and the dye laser wavelength is scanned, the excitation spectrum shown on the lower left results. The lines in this spectrum come only from site 1, since fluorescence from the other sites is absent. If line *2* or *3* were monitored, the resulting excitation spectra would be different. Thus if the laser is tuned to one of the lines in the bottom left excitation spectrum (for example, 374.7 nm), only fluorescence from site *1* is produced and the fluorescence spectrum in the lower right results. This spectrum is much simpler than the top one where all the sites contributed. By changing the laser excitation wavelength, the other sites can be selectively excited and the fluorescence spectrum of just site 2 or 3 results. The same procedure can be followed in a complex mixture of PAH compounds. Since each site and each compound has different lines, the specific line of interest can be chosen for excitation or emission and the contributions from other sites and compounds in the sample can be eliminated. The selectivity of this procedure is limited by the amount of spectral overlap between species. The procedure corresponds essentially to a spectroscopic separation. The PAH concentrations can typically be measured over the concentration range of 0.1 to 200 ng/ml without loss of linearity. These concentrations are for the final solution. Typically a dilution of 5×10^3 is required to assure reproducibility and eliminate interferences [189].

A related method of sample preparation is to use matrix isolation techniques [190–196]. A sample is sublimed onto a cold substrate along with a gas that will form the matrix. The matrix confines the environments and conformations of the analyte molecules so the spectral transitions are much sharper. For PAH compounds, the analyte/matrix molecule ratio is typically $1:10^7$ and the substrate is typically at 10°K. The entire deposition process takes less than 1 hr. To correct for any irreproducibilities in forming the matrix, internal standards are included with the sample just as with the Shpol'skii methods. The choice of the matrix is a crucial step in the analysis. Gases such as nitrogen and argon are often used. The linewidths of PAH compounds in such matrices are typically 100 to 150 cm^{-1}. Maple et al. showed that linewidths of 1 to 2 cm^{-1} could be obtained if Shpol'skii solvents such as *n*-heptane were used to form the matrix [193]. The much narrower linewidths clearly not only provide better selectivity or resolution among compounds so complex samples can be measured more easily, but they also make the peak absorption and emission strengths of transitions larger. If the laser excitation has a linewidth that is comparable, the higher peak strengths provide lower detection limits because the background fluorescence does not have sharp line structure and is discriminated against.

In addition to determining the linewidths, the choice of solvent also determines the importance of phonon sidebands on the transitions [194]. Any solvent matrix will interact with the PAH solute so that an electronic transition on the PAH chromophore can also excite a vibration in the solvent matrix. This interaction gives rise to phonon sidebands that accompany all the electronic transitions. If the interaction is large, the sidebands can become larger than the electronic transitions themselves. Interactions are particularly large if the molecules possess dipole moments. Since the sidebands are broad, selectivity is lost in the analysis. To reduce the effects of the phonon sidebands, the interaction must be made small by selecting a matrix that couples only weakly to the PAH molecule. Maple and Wehry have had good success with Ar and perfluoro-*n*-hexane matrices where the phonon sidebands are very weak even for polar PAH derivatives such as hydroxyl derivatives of naphthalene [194]. When *n*-heptane is used, on the other hand, the phonon sidebands completely obscure the electronic transitions. Neither an argon nor perfluoro-*n*-hexane matrix can be used for a conventional Shpol'skii experiment where the PAH sample is added to a liquid and frozen since argon is a gas at room temperature and perfluoro-*n*-hexane does not dissolve PAH compounds.

A nitrogen-pumped dye laser provides a good excitation source for these methods. The matrix isolation methods are linear over 5 to 6 orders of magnitude in concentration. They have detection limits that are typically 0.4 to 30 pg. The reproducibility with internal standardization is about 7%.

LASER INDUCED FLUORESCENCE LINE NARROWING AND HOLE BURNING

Although the Shpol'skii and matrix isolation methods work nicely for PAH compounds, the large majority of organic compounds do not form systems that exhibit sharp line structure. In addition, for those compounds that do form such systems, the suitable solvents or matrices are very restricted. If an analyte molecule is frozen in an arbitrary matrix, the spectroscopic transitions are usually broad because many conformations and orientations are possible for the embedded molecules. If a laser is tuned to a wavelength within the band of the parent 0,0 transition, only the molecules that have the conformation or orientation where the electronic transition is resonant with the laser wavelength are excited. If energy transfer to other molecules does not occur and if the frozen solvent does not allow the molecule to move into a different conformation, the fluorescence spectra contain only contributions from molecules in the selected conformation or orientation and consequently exhibit laser induced fluorescence line narrowing [197–199]. The fluorescence spectra are both sharp and similar in appearance to the lines in the Shpol'skii systems [200–202]. If the excitation wavelength is changed, the fluorescence spectrum also shifts, because a new set of molecules is excited. This behavior can be seen in Fig. 17 for tetracene [176].

The line narrowing technique fails when there are two or more different ways to excite a molecule with a given laser wavelength [176]. For example, if one molecule has the parent 0,0 transition resonant with the laser while a second molecule in a different conformation has the 0,1 vibrational transition also resonant, the fluorescence spectrum contains the transitions from molecules in two conformations. To prevent this occurrence, the laser wavelength must be tuned to the low-energy side of the absorption band so that only the molecules with the lowest 0,0 transitions are excited. If the excitation wavelength is at higher energy, the fluorescence spectrum can revert to a broad-band appearance. The broadening can be seen in Fig. 17 as the excitation is moved to shorter wavelengths. There are many materials, however, where the probability of the 0,0 transition is small compared with the transitions to vibrationally excited states. These are generally materials where the Stokes shift is large [176]. For example, Rhodamine 6G has a Stokes shift of 540 cm^{-1} and has the same fluorescence spectrum independently of the laser excitation wavelength. The fluorescence line narrowing techniques are presently restricted to molecules with strong 0,0 transitions and small Stokes shifts.

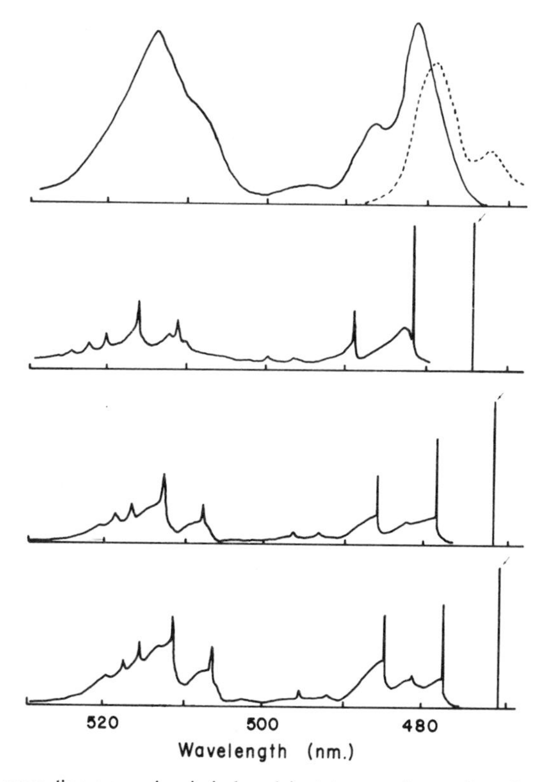

Fig. 17 Fluorescence line narrowing is induced in tetracene frozen in a 2-methyltetrahydrofuran glass by exciting with a laser (whose position is indicated by the line at the right side of the bottom three spectra) into the absorption band (the dotted spectrum on the top) [176]. The resulting fluorescence spectrum has sharp lines because only the molecules in resonance with the laser fluoresce. The broad conventional fluorescence is shown at the top. The molecules that are resonant change for the three different laser wavelengths shown in the bottom three spectra.

Brown et al. have developed procedures for the measurement of PAH samples in low-temperature glasses using the fluorescence line narrowing techniques [202]. A sample is typically diluted in a glycerol, water, DMSO mixture and is cooled 4.2°K. Dilutions of 10^4 to 10^5 are used to assure formation of a good glass. A nitrogen pumped dye laser is a convenient excitation source. The narrowed fluorescence lines are sufficiently sharp that the different components of a 14 component mixture can be resolved. Detection limits are about 20 pg/mL for perylene in the final solution after dilution.

The same techniques can be applied to matrix isolation and Shpol'skii measurements to narrow the linewidths of the transitions even more. Maple and Wehry, for example, observed a large improvement in the resolution of hydroxynaphthalene isomers when laser induced line narrowing was compared with conventional excitation sources [194]. Similar effects will occur for Shpol'skii measurements as well.

Double Resonance Excitation of Fluorescence

The Shpol'skii effect, matrix isolation methods, and laser induced fluorescence line narrowing are only effective for a limited class of molecules that have strong 0 → 0 electronic transitions. It may still be possible to achieve sharp lined spectra in a broader class of materials by using double resonance excitation techniques [203–206]. The typical broad spectra of organic materials result because the electronic energies are strongly influenced by environmental effects such as surrounding solvent and conformation. The vibrational energies are not strongly influenced, however. Thus Raman and infrared spectroscopy are capable of providing detailed structural and identifying information about a molecule. Double resonance techniques can provide a basis for combining Raman, infrared, and fluorescence spectroscopy into a common method.

A typical configuration coordinate diagram is sketched in Fig. 18. In both parts of this figure, a vibrational excitation is created either by tuning two lasers so their difference frequency excites a vibrational mode by stimulated Raman scattering (*a*) or by tuning an infrared laser to directly excite a vibrational mode (*b*). Then before the vibrational excitation can relax, another

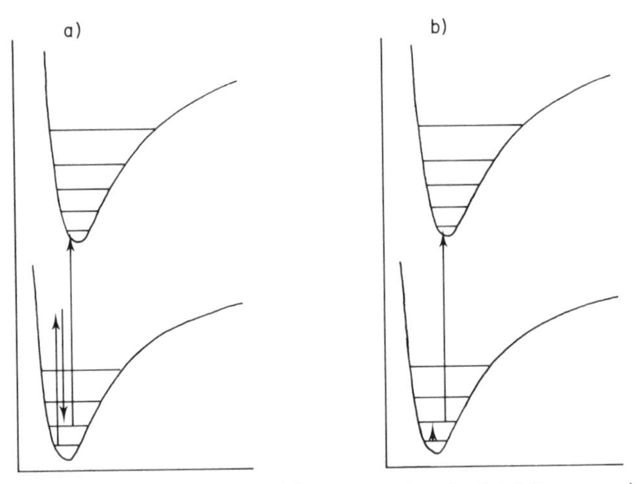

Fig. 18 (*a*) Double resonance excitation of fluorescence by stimulated Raman excitation of a vibration by two lasers tuned so their difference frequency matches the vibration followed by excitation to the excited electronic state by a third laser. (*b*) Infrared excitation of the vibration followed by the excitation to the excited electronic state.

laser excites the molecule to an upper electronic state, which can then fluoresce. By monitoring the fluorescence while scanning two lasers synchronously so their total excitation energy remains constant, one can obtain an excitation spectrum that reflects when vibrational resonances have been excited. This spectrum contains sharp lined information about the vibrational modes that are coupled with the chromophore.

The key to the successful implementation of this idea is to have sufficiently high vibrational excitation rates to overcome the rapid vibrational relaxation. A discussion of the feasibility and the possible interfering processes indicates that it should indeed be possible to overcome the rapid relaxation and that it should be possible to measure concentrations in the 10^{-8} M range [203].

A demonstration of the feasibility of this procedure was accomplished for an infrared-optical double resonance. Seilmeier and co-workers [204–206] used a mode-locked Nd:YAG laser to produce single pulses with a 6-psec width. The pulses were divided by a beam splitter into two separate beams. Tunable infrared and visible beams were then produced by nonlinear parametric processes with pulse widths of about 4 psec. The infrared and visible pulses were combined and focused into the sample where the fluorescence was generated. The fluorescence was monitored with a small spectrometer. The fluorescence background that is obtained when only the visible laser excitation is present was subtracted from the total fluorescence to obtain only the contribution from the double resonance process. A typical spectrum is shown in Fig. 19*a* for Coumarin 6 in CCl_4. The visible beam was tuned to the

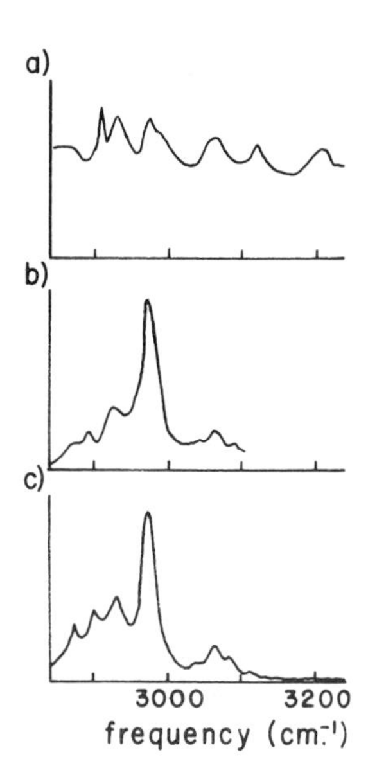

Fig. 19 (*a*) Fluorescence of coumarin 6 in CCl_4 excited by an infrared-visible double resonance. The exciting sources are tuned so the total excitation energy is always constant as the infrared source is scanned over the excitation region shown. Both excitation sources are coincident in time. (After Ref. 204.) (*b*) Same scan as in *a*, but the visible excitation sources have been delayed by 7 psec. (*c*) Infrared spectrum of coumarin 6 in CCl_4.

electronic transition and the fluorescence signal corrected for background was measured as a function of the infrared frequency. The conventional infrared spectrum for coumarin 6 is reproduced in Fig. 19*c*. The *x*-axis of the spectrum in *a* has been translated such that the total energy of the double resonance matches the *x*-axis in *c*. It is clear that considerably more structure appears in the double resonance spectrum than in a conventional excitation spectrum, but the former does not correspond to the infrared spectrum. If the two beams are delayed from each other by 7 psec and the same scan is repeated, an excitation spectrum is obtained that closely matches the infrared spectrum (Fig. 19*b*). At excitation energies around 3000 cm^{-1}, it is possible to excite both C—H stretching modes and overtones of the skeletal modes. Even though the skeletal modes are an overtone, they are still strong because they are well-coupled to the chromophore. They dominate the spectrum in Fig. 19*a*. They also relax very quickly so if the two lasers are delayed, the skeletal modes do not appear and the C—H stretching modes become dominant. Such techniques as these are exciting because they could provide a more general method for obtaining selectivity in a fluorescence measurement.

Time Resolved Spectroscopy

The preceding section illustrates some of the power of time resolved techniques. The fluorescence decay from a molecule is a characteristic of a particular molecule in a particular environment. Different molecules can have very different fluorescence decay times and it is possible to discriminate among them temporally. Knorr and Harris [207] have pointed out that the mirror relationship between many excitation and fluorescence spectra decreases the amount of information about a molecule and therefore the selectivity that might be obtainable by measuring both the excitation and fluorescence spectra for all possible combinations of fluorescence and excitation wavelengths. The lifetime of a molecule, however, is not related to its fluorescence and excitation spectrum and therefore provides a number that is characteristic of the molecule. They suggest that better selectivity can be obtained in a mixture analysis by forming the two dimensional matrix of the fluorescence intensity for all times and excitation wavelengths. The components in the mixture can be extracted by fitting procedures using standard matrix manipulations [207]. Dickinson and Wehry demonstrated that a great deal of selectivity can be achieved in mixture analysis when there are components that have very different lifetimes by using simple time discrimination procedures [196].

In addition to providing discrimination among components in a mixture, one can also provide discrimination between fluorescence and light generated coincidentally with the excitation pulse. The additional light can arise from Raman scattering or many kinds of nonlinear processes. One of the powerful spectroscopic techniques available is discrimination between two different processes on the basis of their characteristic time constants. Besides being applicable to many situations in molecular fluorescence, it is potentially applicable to atomic fluorescence, where one must discriminate between scattered laser light and atomic fluorescence.

Although time measurements and discrimination are easily accomplished when the time scales are on the order of microseconds or longer, the problem becomes much more difficult when the time scales are on the order of nanoseconds or shorter. This latter time scale, however, is the most important one because most fluorescence measurements of practical interest are made on substances with rapid decay times. Recent developments in laser technology have provided tunable laser sources that provide the capability of working in this time regime down to times of 200 fsec. Some of the most versatile lasers have proven to be the mode-locked flashlamp pumped dye lasers, mode-locked c.w. dye lasers, and the synchronously pumped dye lasers. These lasers are reviewed elsewhere [203–211]. We concentrate instead on the spectroscopic techniques that provide temporal discrimination.

PHASE MODULATION

If the excitation intensity is modulated at a frequency ν, the relative phase angle ϕ of the fluorescence from a sample with an excited state lifetime τ has the following relationship [212]:

$$\tan \phi = 2\pi\nu\tau \tag{16}$$

The error associated with this measurement is smallest when the phase angle is $\pi/4$. Many techniques exist for amplitude modulating a laser at a frequency ν. A convenient method that is able to reach the very high modulation frequencies needed to accurately measure short lifetimes is to use cavity-dumping, where a radiofrequency signal in a crystal generates a series of acoustic waves whose spacing forms a grating that can deflect the intracavity beam [213, 214]. Modulating the cavity-dumper at a particular frequency also modulates the output power of the laser at that frequency. Modulation frequencies in excess of 60 MHz can be conveniently generated and frequencies as high as 170 MHz have been reported [214], although the depth of modulation suffers at high frequencies. The modulated fluorescence signal is detected by a photomultiplier tube and is compared with a scattered signal in order to compute the phase angle difference between the two. The comparison can be performed with conventional phase sensitive electronics or a sampling oscilloscope depending on the modulation frequency [213]. The present limitation imposed on the measurement of short lifetimes is the response time of the photomultipliers. The technique is limited to measuring systems that relax as single exponential decays where (16) is valid. In addition to making lifetime measurements, it is also possible to discriminate between two processes with different relaxation rates by changing the modulation frequency. When the modulation frequency is sufficiently low, the fluorescence from both processes is able to follow the modulation. As the frequency is increased, the modulation of fluorescence from the slower relaxation process becomes smaller than that from the faster process, which is better able to follow the higher frequency excitation. The amount of discrimination depends on the difference in the two relaxation rates and the modulation frequency. An application of such a procedure would be the discrimination of Raman scattering from fluorescence. One could simply modulate the excitation at a frequency much higher than the reciprocal lifetime of the interfering fluorescence. The Raman scattering would generally have a much shorter relaxation and would be able to follow the high modulation frequency without attenuation.

FLUCTUATION ANALYSIS OF LIFETIMES

A related technique has been developed recently that utilizes the mode noise of a laser as the source modulation and measures the cross-correlation with the fluorescence to determine the lifetimes [215, 216]. In any laser, there is a series of longitudinal modes spaced evenly in frequency with an interval that depends on the distance between the mirrors in the cavity. These modes beat with each other and produce components in the power spectrum at the beat frequencies. The amplitude of each beat frequency in an argon laser with a 114.9-cm cavity is shown in Fig. 20*a*. This power spectrum was obtained by examining the scatter of the laser with a spectrum analyzer. The beat frequencies are spread evenly with a spacing of about 130 MHz and an upper limit of 2.86 GHz because of the laser linewidth. If this laser is used to excite fluorescence, the fluorescence also contains components at each of the beat frequencies, but the amplitude of each component is changed by the ability of the fluorescence to follow the beat frequency. A sample with a long fluorescence lifetime is not able to respond to a high frequency for the same reason that a charging capacitor does not follow a voltage oscillation at frequencies much higher than $1/RC$. The power spectrum for Rhodamine B is shown in Fig. 20*b*. The lifetime can be obtained from these data by first dividing the two power spectra for scattered light and fluorescence. The attenuation that occurs as a function of frequency is then obtained. For an exponential decay, the attenuation should have a Lorentzian behavior as a function of frequency with a width at half-maximum equal to the reciprocal of the fluorescence lifetime. For Rhodamine B, a lifetime of 3 nsec was measured [215]. Thus frequencies higher than about 300 MHz are strongly attenuated. Since the beat frequencies are spaced by 130 MHz, the number of beat frequencies that can be used for the measurement is smaller for

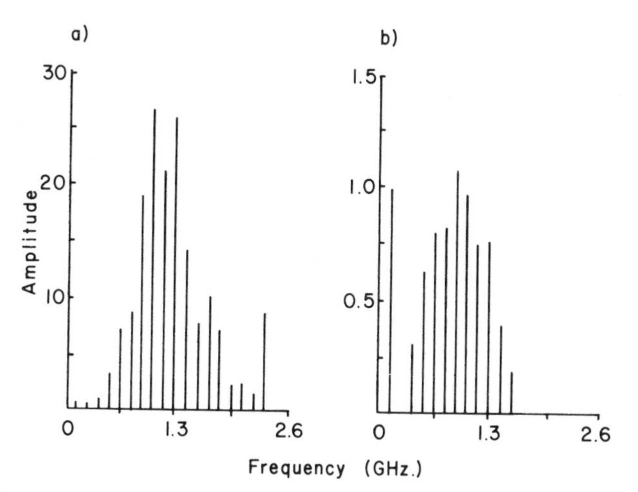

Fig. 20 (*a*) The amplitude of the beat modes in an argon laser that is used to excite fluorescence [166]. (*b*) The amplitude of the fluorescence from Rhodamine B at the modulation frequencies of the argon laser modes. Note the twentyfold expansion of the y-axis that was required to normalize the lowest mode at 130 MHz for a and b. The amplitudes have been reduced by the long response time associated with the Rhodamine B lifetime.

Rhodamine B than for samples having shorter lifetimes. Consequently, the measurement accuracy is poorer. For samples with much longer lifetimes, other lasers or techniques would have to be used to provide frequencies lower than 130 MHz. The limitation on samples with faster lifetimes is determined by the response time of the photomultipliers.

The technique has some very important advantages. The high modulation frequencies are produced without the complications and expense of mode-locking or cavity-dumping. The technique can also be used to temporally discriminate among processes that have different relaxation rates. One can simply monitor the fluorescence at a frequency where the fluorescence from the slower process is attenuated over that of the faster process. This technique would also be useful for attenuating the fluorescence interference in Raman scattering measurements. With simply a modification to the electronics, it could be implemented on any of the commercial Raman instruments, which of course have lasers with suitable mode noise. This modification would provide a powerful extension of the present capabilities of Raman spectroscopy.

PULSE EXCITATION METHODS

A widely used method of measuring fluorescence relaxation times is to excite the sample with a very short pulse and monitor the exponential decay that follows. Lasers have made a significant contribution to this method because pulses of several nanoseconds are easily produced by simple pulsed lasers, Q-switching pulsed lasers, or mode-locking continuous lasers, and pulses of several picoseconds can be produced with mode-locking techniques. These pulse widths are a great improvement over those of conventional sources, which have low intensities and nanosecond pulsewidths. If the excitation pulses are much shorter than the fluorescence relaxation, it is not necessary to deconvolute the temporal profile of the excitation from the observed data. Additionally, faster fluorescence relaxations can be measured with short excitation pulses.

The decay measurement technique is determined by the decay time. When the relaxation time is longer than the average photomultiplier pulse, there are several instruments available for making the measurement. A boxcar integrator or a gated photomultiplier or vidicon tube can be used to sample the fluorescence transient during a selectable gate period after the excitation. By moving the gate to different delay times after the excitation, one can determine the entire transient profile. This kind of method requires pulse-to-pulse stability of the excitation source intensity over the time period of measurement. Fast transient recorders can also record the entire fluorescence decay on a single excitation pulse. A computer or multichannel analyzer can then signal average the individual transients over a number of pulses. This method does not require stability of the excitation source.

Many fluorescence decays are much shorter than the average photomultiplier pulse and these simple methods cannot be used. Instead, methods that measure the time delay between the excitation and a particular point on the photomultiplier pulse are used. A number of different systems have been described for performing this measurement [217–226]. Typically, a fast photodiode is used to detect the excitation pulse and supply a pulse that starts a time-to-amplitude converter (TAC). The TAC is a fast ramp generator whose ramp is started by a start pulse and stopped by a stop pulse. The final voltage of the ramp is directly proportional to the time between the start and stop pulses. The stop pulse in the experiment is derived from a single fluorescence photon. If the fluorescence intensity is made sufficiently low that only one photon is likely to be detected per excitation pulse, the time delay between the detection of the photon and the excitation is exponential for simple fluorescence decays. The arrival of the first fluorescence photon would not be exponentially distributed in time if more than one photon were detected after the excitation because of the much higher probability for being emitted shortly after the excitation. It is thus very important to make sure the intensity is low enough to prevent distortion of the data.

To provide the best time resolution, the stop pulse is referenced to a particular *part* of the rapidly rising edge of the photomultiplier pulse so that the actual time resolution can be much better than the pulse width. A discriminator that supplies the stop pulse is set to trigger at a particular voltage on the photomultiplier pulse. The time resolution that results from this arrangement depends on the reproducibility of the pulse amplitudes for a single photoelectron event and on the rate of rise of the leading edge of the photomultiplier pulse. A method for improving the reproducibility of pulse amplitudes is to operate the photomultiplier at a very high potential difference between the photocathode and the first dynode [223]. Thus a single photoelectron crashing into the first dynode would produce many more secondary electrons and consequently the statistical uncertainty would be reduced over the situation where only a few photoelectrons are produced. Harris and Lytle [223], for example, have operated an RCA 8850 photomultiplier at 980 V between the photocathode and the first dynode and achieved a single state gain of 55. This operation caused a very sharply defined pulse height distribution for single photoelectrons. The timing errors were substantially reduced as a result.

Wild and co-workers [224] employed a double discrimination to achieve the same goal. A differential discriminator was used to provide an upper and lower discriminator level that determined the window where the photomultiplier pulse amplitude must fall to produce an output pulse. The timing of the output pulse was derived from the first discriminator level. A narrow distribution of pulse amplitudes was thus achieved by rejecting any pulse that had an amplitude other than that selected by the window. This method does not provide the same sensitivity as the previous method because many of the photon detection events that do not have amplitudes within the window are rejected and lost.

Harris and Lytle [223] have developed a clever method of achieving better time resolution by using a sampling oscilloscope. The sweep on the oscilloscope is started with a pulse from a fast photodiode that monitors the excitation source. A photomultiplier pulse is divided into two identical pulses that are brought into the two input channels of a sampling oscilloscope. One of the two pulses is delayed 2 nsec by having it travel through a longer cable. The sampling oscilloscope is set to take a sample of both channels a short and adjustable time after it is triggered. Since the two channels contain the same but delayed information, one is essentially measuring the instantaneous voltage at two points on the pulse separated by 2 nsec. If this sampling occurs symmetrically on either side of the peak of the pulse, the two voltages measured at the two channels are equal. Pulses that occur at other times relative to the trigger are not sampled symmetrically and the two voltages are not equal. This method locates the pulse precisely in time and permits a much higher time resolution for a sampling procedure. To count the number of pulses that have met this condition, the output from the oscilloscope channel with the delayed photomultiplier pulse is sent to a discriminator that is triggered at a set voltage. The output from both channels is also subtracted electronically and if the result is sufficiently low, a logical pulse is produced, indicating the pulse is symmetrically placed. This logical pulse is combined with the discriminator pulse and a pulse indicating an excitation pulse has occurred in a logical AND gate. The output of this gate is sent to a linear ratemeter that indicates the count rate of events within the narrow time window set. If this rate is plotted as a function of the delay time set on the sampling oscilloscope, the fluorescence relaxation is obtained. This method gave a transient response of 0.49 nsec for the full width at half-maximum for the photomultiplier pulse timing. Lifetimes down to 50 psec were measured with a 6% error by deconvoluting the instrumental profile from the observed time response.

It is mentioned earlier that the fluorescence intensity in these kinds of experiments must be low enough that at most, only one photon will occur per excitation pulse in order to have that one pulse representative of the entire decay curve. This requirement limits the signals that can be observed. Meltzer and Wood [221] have described a system that uses a pile-up detector to determine when two photons have been emitted during an excitation. If this occurs, a pulse is produced that rejects the data obtained during that excitation pulse. The result is that one can allow high fluorescence intensities to reach the photomultiplier without distorting the measurement. Clearly, one could not go to much higher intensities where one is likely to obtain more than one photon because most of the data would then be rejected and the measurement time would be increased.

Ramsey et al. have used a double balance mixer to perform a cross-correlation of the fluorescence decay that results after excitation with a mode-locked laser and the pulses in the mode-locked laser [216, 227]. The time between the two signals in the cross-correlation is varied by changing an optical delay line in the excitation beam. In this approach, the mode-locked laser pulses serve as a gate that can be scanned with the delay line over the fluorescence transient. The cross-correlation then traces out the fluorescence transient convoluted with the instrumental response function. Decay times of 80 psec have been measured and decay times of 50 psec should be possible.

The sampling of photon pulses after the initial pulsed excitation has been used effectively to reject fluorescence interference from Raman scattering measurements [220, 222, 226]. Raman scattering is a very rapid process compared with fluorescence. Therefore if an excitation pulse that is much shorter than a fluorescence lifetime is used to excite Raman scattering and if only photons that occur during the excitation are counted, one can easily discriminate between the Raman and fluorescence contributions to the signal since most fluorescence occurs later in time. The rejection obtained depends on the fluorescence lifetime relative to the time required for sampling. Long-lived fluorescence can be more effectively rejected than short-lived fluorescence. A fluorescence rejection of 34 was obtained for acridine orange (4.5-nsec lifetime) and 115 was obtained for rubrene (15.6-nsec lifetime), and it was anticipated that this rejection could be improved. One can also operate with a sampling time that is delayed from the excitation pulse. In this way, the Raman scattering can be rejected. This operation provides a greater rejection factor because one is sampling at times where nonresonant Raman scattering cannot exist. It is a useful technique at low concentrations, where Raman scattering from the solvent can contribute to the blank measurement.

FLUORESCENCE GAIN

All the preceding methods are limited in resolving short lifetimes by the electronics, in particular the photomultiplier tube. Although there remain ways of optimizing the photomultiplier that might lower the time response by a factor of 2, it is not a component that can be used over the range of times characteristic of molecular relaxation. A number of methods have been invented since the introduction of picosecond lasers that use light itself to access this time regime [228–232]. One technique for accomplishing this goal is fluorescence gain spectroscopy [233–235]. Instead of detecting the excited electronic state by measuring its spontaneous fluorescence, one can pass a coherent beam through the excited medium at a wavelength that can stimulate the emission from the excited state and detect the increase in the beam intensity. The instrumental block diagram is sketched in Fig. 21 [235]. A c.w. argon laser is mode-locked and part of the beam is split and used to synchronously pump a dye laser. The remainder of the argon laser beam is chopped at 10 kHz, combined with the beam from the dye laser, and focused into the sample, which absorbs at the argon laser wavelength. The combined beams then pass through a filter that removes the argon laser light, and the dye laser light is detected with a photodiode. A lock-in amplifier is used to measure the 10-kHz component of the dye laser beam. This component can arise if the sample molecules have been excited by the argon laser during the on time of the chopped signal and then stimulated to emit by the dye laser. In this experiment, the sample solution is flowed to eliminate thermal blooming effects in the focal region. The ability of this method to detect an excited state is primarily determined by the shot noise and fluctuations in the dye laser beam at the chopping frequency. A minimum detectable concentration of 10^{-6} M was measurable with the instrument in Fig. 21. The time resolution is determined by the width of the argon pulse—about 250 psec. It is possible to measure much shorter lifetimes by using a different arrangement of lasers. Shank and co-workers [234] used a passively mode-locked dye laser

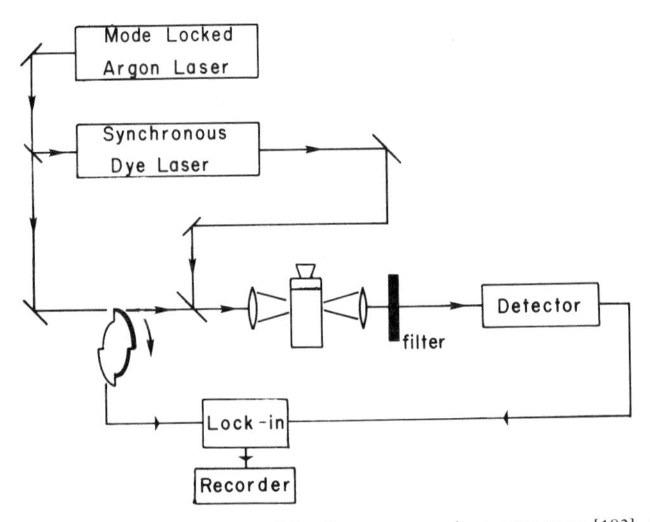

Fig. 21 Block diagram of apparatus used for fluorescence gain spectroscopy [183]. The beam from a mode-locked argon laser is split. Part is chopped and excites analyte molecules, while the second portion pumps a synchronous dye laser. The resulting picosecond pulses are recombined with the argon laser beam and focused into the sample. The dye laser beam stimulates emission from the sample, which has been excited by the argon laser. The modulation thus introduced in the dye laser beam is then detected.

with about 2-psec pulse widths to stimulate emission. The dye laser was frequency doubled to provide an ultraviolet excitation pulse that was modulated. Time resolutions of 0.2 psec were achieved with this sytem.

The fluorescence gain measurement is an example of a more general method where light replaces electronics in acquiring the data on a rapid time scale. A number of clever methods have made it possible to obtain temporal information on the picosecond time scale. These methods have been reviewed elsewhere [228, 229].

4 PHOTOIONIZATION AND PHOTODISSOCIATION METHODS

Several innovative techniques recently have been developed for detecting excited atomic and molecular states that do not depend on light emission from the excited state but rather on either detecting photoelectrons that result from ionizing the excited state or measuring the change in the plasma impedance of a discharge when excited states are present. These new techniques offer exciting new possibilities for analytical spectroscopy.

Single Atom Detection by Resonance Ionization

Hurst and co-workers developed a method of detecting an excited state that is capable of measuring a single atom in the laser beam [236–238]. The method is perhaps best illustrated in an experiment where the single Cs atoms are detected after being formed and ejected into the detection volume with a 14% probability from the fission of ^{252}Cf [238]. The apparatus is sketched in block diagram form in Fig. 22. The occurrence of a fission event is detected by a surface barrier detector mounted directly behind the ^{252}Cf source, which starts a sequence of timing pulses. The first pulse controls the timing on a collector plate that is biased to remove the ions formed by the fission fragments from the experiment chamber. After 40 μsec, the collector plate is pulsed negative in order to deflect the photoelectrons that will be generated into the proportional counter. After 50 μsec, a flashlamp pumped dye laser is fired at a wavelength corresponding to one of the $6^2S_{1/2} \rightarrow 7^2P_{1/2,3/2}$ transitions of Cs (455.5 and 459.3 nm). The dye laser output is focused into the path of the fission fragment. If the fission fragment is Cs, the laser causes excitation to the excited ^2P state, from which a second transition can occur to the ionization continuum (this double resonance excitation to a continuum can be used for any system where the excited state is more than halfway to the continuum). The photoelectron from the ionized Cs is directed through the aperture in the field plate with 65% collection efficiency. The electron enters a proportional counter where gas amplification occurs, producing an output pulse that is proportional to the number of elec-

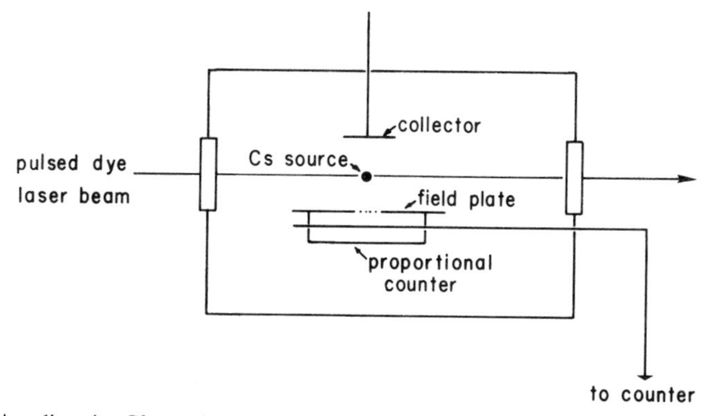

Fig. 22 A radioactive Cf sample located below the laser beam generates single Cs atoms by fission. A surface barrier detector mounted below the source detects the fission fragment ejected in the opposite direction and provides the triggering pulse for a dye laser. Before the laser is triggered, the collector plate sweeps out the free electrons produced by the fission fragment and is then pulsed negatively. The laser is fixed and ionizes any Cs atom that is in its path. The resulting electron drifts into the proportional counter and is counted.

trons that generated it. Pulse height discrimination can be used to accept only events caused by single electrons and to reject signals that come from background noise. In these experiments, the background noise was produced primarily by radiofrequency noise from the pulsed laser. Because of this discrimination, 20% of the single electron events were also rejected. The counters were also gated to reject spurious pulses from natural radioactivity that was not synchronous with the laser. Other background contributions were synchronous with the laser and contributed an intensity twice as large as the intensity from the resonant ionization of single Cs atoms. It is believed that this background came from photodetachment from the negative ions created by the passagae of the fission fragment. This background could be overcome by signal averaging over many events and comparing the results at wavelengths on and off resonance with those of Cs. The number of single photoelectron events that occurred per 100 fissions that produced a heavy mass fragment is shown in Fig. 23. Each point in this figure represents 800 laser shots. The background contribution has been subtracted from each point by measuring the background at 450.5 nm. The two points at the wavelengths of the Cs transitions are significantly different from the neighboring nonresonant wavelengths and represent the ultimate in analytical detection limits—single atom detection. The eight single photoelectron events per 100 heavy-mass fission fragments is within experimental error of the number expected (14% Cs fragments × 80% counting efficiency × 65% collection efficiency).

Several factors contributed to the success of these experiments. The pulsed dye laser is focused sufficiently to saturate the atomic transitions, making

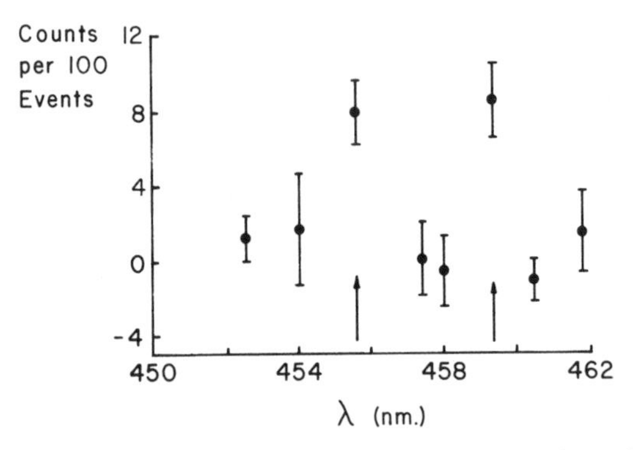

Fig. 23 The count rate produced by 100 heavy atom fission events is shown for different laser wavelengths. The corrected counts (background was subtracted using the first point at 450.5 nm) are equal to the background at all wavelengths except the two at the transitions to $7^2P_{3/2}$ and $7^2P_{1/2}$ levels (indicated by arrows in the figure).

photoionization a very probable event once a Cs atom enters the beam. The proportional counter detection method is particularly well-suited to this application because of the high collection efficiency. A single photoelectron can be observed with a 95% probability. Optical methods of detection collect only a fraction of the emission events because of the finite solid angle and the quantum efficiencies of the available detectors. Finally, this method has not been sensitive to the same background problems that have limited optical methods. The background that was observed from Cs fission fragments of ^{252}Cf was high compared with the background found in earlier experiments where Cs atoms were generated by a different method [236, 237]. The low background permits the single photoelectron events to be easily observed.

A number of implications arise from these experiments. First, the experiments clearly establish that lasers are capable of measuring single atoms with high selectivity if only we are clever enough in devising techniques. There are a number of scientific applications for such methods, particularly in nuclear physics, where one is interested in measuring small atomic concentrations, including experiments where solar neutrinos are detected by nuclear transmutations of elements or where new heavy elements are created in collisions between heavy atoms. The technique is particularly interesting because of the time resolution available. Experiments similar to that described for ^{252}Cf can be performed because the detection can be linked directly in time with an event creating the atom.

As is pointed out earlier, optical techniques of detecting excited atomic states have also reached the limit of measuring single atom concentrations in the volume of the laser excitation region [88, 111]. This is indeed true, although there is a subtle difference. Optical detection has not yet reached the same level of detection efficiency that has been achieved with the proportional counter and consequently the single atom must be recycled to the excited state repeatedly in order to generate a detectable flux of photons. This recycling causes a loss in the time resolution compared with the method of Hurst et al. [236]. An experiment by Balykin et al. has come the closest to providing a true optical detection of single atoms [88]. The cell they used in their experiments is sketched in Fig. 24. A heated Na oven ejects Na atoms at an adjustable rate across a c.w. dye laser beam. The beam is directed down a series of baffles on the inside of the blackened cell and is captured by a light trap to keep scattered light at a minimum. Two photomultipliers constantly observe the interaction region where the Na beam intersects the laser beam. If a Na atom is in the beam, it will be successively excited during its 20-μsec time in the beam emitting about 10^3 photons and cause a coincidence signal between the two detectors. Scattered photons, however, are random and tend to be rejected by the coincidence circuit. This experiment succeeded in measuring a flux of 10 atoms/sec, although individual pulses from individual

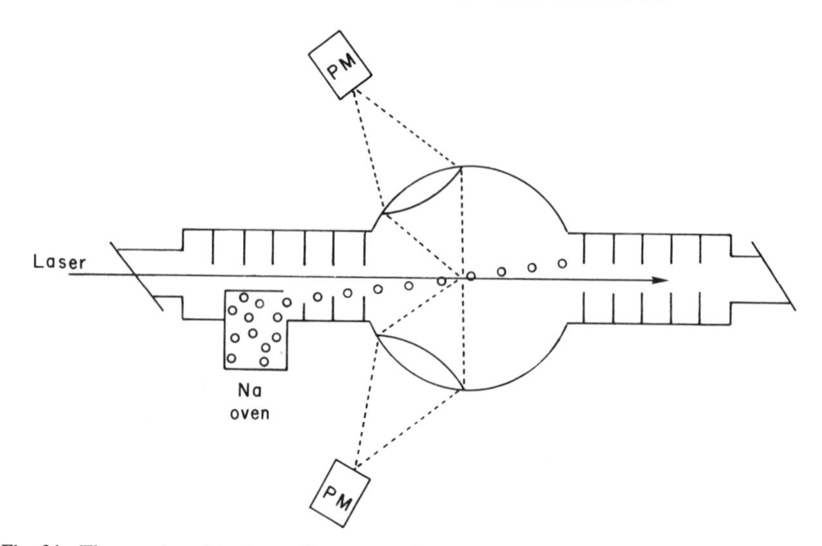

Fig. 24 The vessel used to detect Na atoms optically. A continuous beam of Na atoms is generated in an oven. A laser intersects that beam and the resonance fluorescence is detected by two photomultiplier tubes operated in ccincidence. Extensive baffling reduces scattered light.

atoms crossing the beam still could not be detected above background. The lower efficiency of optical detection requires that an atom remain in the laser beam long enough to emit a detectably large number of photons, which causes an inherently poorer time resolution than methods capable of detecting a single emission event.

Optogalvanic Spectroscopy

A very different method of detecting excited states was introduced by Green and co-workers [239] for analytical purposes after its discovery as an interesting phenomena many years before. It was found that if a laser is tuned to many of the lines of an atom, ion, or molecule in a plasma or flame, the electrical impedance of the medium is changed. The mechanisms behind the impedance changes have been studied by several workers [240–245]. The excited states of an atom or ion are closer to the ionization continuum than the ground state and ionization can occur more readily by thermal excitation. The ionization changes the impedance of the flame, which can then be detected by measuring the current through the flame with an applied potential. The sensitivity for a particular transition ν_0 that terminates at a level that is at an energy E_i^* below the ionization continuum can be approximated by the equation [244]

$$S(\nu_0) = C_1 X_B \beta [B_{12}(\nu_0) E_{\nu_0} e^{-E_i^*/kT}]^{C_2} \tag{17}$$

where C_1 and C_2 are constants, X_B is the Boltzmann factor that describes the population of the initial state of the transition ν_0, β is the atomization efficiency for the particular flame, $B_{12}(\nu_0)$ is the Einstein B coefficient for the absorption transition, $E_{\nu 0}$ is the laser peak spectral irradiance, kT is thermal energy of the flame, and $S(\nu_0)$ is the sensitivity expressed in nA ml/ng. Both C_1 and C_2 are constants that depend on the particular equipment and lasers used. For a slot burner with an air–acetylene flame and a flashlamp pumped dye laser, $C_1 = 5 \times 10^{-4}$ and $C_2 = 0.65$ [244]. This formula can then reproduce experimental sensitivities within an order of magnitude [246]. It provides a convenient guide for the selection and optimization of experimental parameters.

A typical configuration for analytical work with the optogalvanic effect is sketched in Fig. 25 [247, 248]. A commercial premixed, slot burner with a fuel-rich air–acetylene flame is used to atomize a sample. A pair of tungsten rods are suspended on either side of and close to the flame. The rods are established at a negative 500 V relative to the burner head and the impedance of the flame is measured by the current flow. A pulsed laser is directed down the axis of the flame. The current pulse that results when the laser is tuned to a flame resonance is converted to a voltage pulse by a simple preamplifier, amplified, and measured by a sample-and-hold and A-D converter with a combined aperture time of 10 nsec.

The performance of this system has been evaluated for 18 elements that have detection limits between 1 pg/ml and 100 mg/ml. These limits are tabulated in Table 2. They are somewhat better than limits of other flame methods for the majority of cases and are comparable to laser excited atomic fluorescence and ICP methods. They are determined by the fluctuations in the DC background currents from the flame. Light scattering that plagues many atomic fluorescence measurements does not hinder optogalvanic

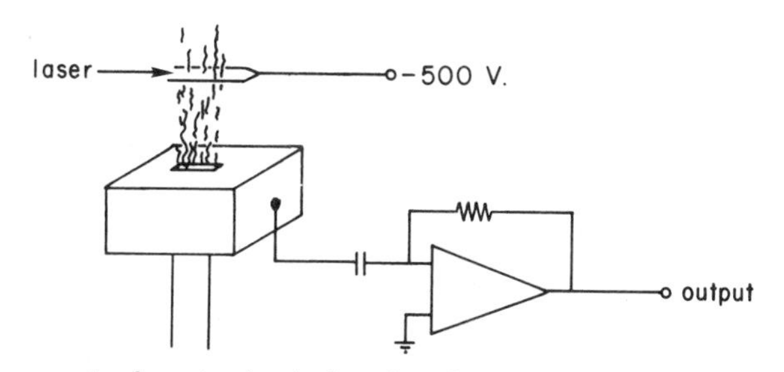

Fig. 25 The apparatus for optogalvanic detection of laser excited atoms in a flame consists of two tungsten rods held at -500 V on either side of a flame. The flame resistance changes that result from the changes in state population after laser excitation are detected by the resulting current changes.

measurements. At high concentrations, there is no postfilter effect, but there is an influence from changes in the flame impedance from the high concentration of analyte. Prefilter effects are similar to atomic fluorescence. The linear dynamic range extends over 4 to 5 orders of magnitude.

Since the method is based on impedance changes in the flame, it is sensitive to other ions that can be easily ionized [246]. For example, if alkali atoms enter the flame and are thermally ionized, the impedance of the flame decreases markedly and masks impedance changes caused by the optogalvanic effect. Sodium was observed to interfere at the 1- to 10-μg/ml level. These effects were diminished by increasing the potential on the electrodes to higher voltages and by using flat electrodes with a large surface area. The problem could be almost eliminated by sending the laser beam very close to the electrodes. To get it close enough, the electrode was actually placed within the flame [249]. It was necessary to water cool the electrode. This approach has increased the tolerance of optogalvanic spectroscopy to over 3000 μg/ml of Na.

The efficiency of the optogalvanic effect depends on the energy required to reach the ionization continuum. If an atom can be excited closer to the continuum, it can be ionized more easily and will provide a larger signal. Turk and co-workers used two lasers to sequentially excite atoms to higher states using real intermediate levels [250]. Although the total probability for making a double transition is necessarily less than that of a single transition, the exponential dependence on the energy gap to the ionization continuum makes the double resonance efficient. A sensitivity enhancement of 7 to more than 1600 was reported over the single excitation. This enhancement is somewhat misleading because the single excitation used in the comparison represented the first excitation used in the double resonance and not the best single excitation that might have been used. The important aspect of this work is the illustration of the increased sensitivity that is available when the sample is excited nearer to the continuum.

The linewidths of the transitions measured using the optogalvanic effect are often determined by Doppler broadening. Lawler and co-workers have developed a method to eliminate the Doppler broadening and achieve a linewidth determined only by homogeneous broadening [251]. The beam from a dye laser is split in two and each beam is chopped at a different frequency, one at f_1 and one at f_2. The two beams are directed through the sample from opposite directions so the beam profiles overlap. A particular velocity subset of atoms is resonant with each beam, but since the beams are counterpropagating, they are not the same subset of atoms. If the velocity subset corresponds to those atoms that are stationary for the velocity component along the beams, then the beams in fact are resonant with the same subset of atoms. The optogalvanic effect is then modulated at both f_1 and f_2, but if the beam intensities are near to the saturation regime the nonlinear response causes additional frequencies at $(f_1 + f_2)$ and $(f_1 - f_2)$. If $(f_1 + f_2)$ is monitored with a lock-in detector, a spectral scan of the dye laser produces a Doppler-free linewidth. Lawler and co-workers used a c.w. dye laser to record the linewidth of He in a low-pressure discharge. They measured a width of 130 MHz, a value limited by pressure broadening. The narrow linewidths that are available with this method could potentially be useful for measuring specific isotopes of atoms.

Another influence of the optogalvanic effect on analytical chemistry arises in a different context. In all applications where narrow- bandwidth lasers are used, it is necessary to tune the laser to the analytical wavelength and keep it there. If the analyte has narrow linewidths, this step becomes both important and difficult—difficult because many commercial lasers do not have long-term wavelength stability and drift off the line. Many methods have been used to establish the the correct wavelength:

1. Use of any auxiliary flame or furnace with a high concentration of analyte as a monitor.
2. Use of a counter, encoder, or other readout to measure the wavelength setting.
3. Use of a separate monochromator to measure the wavelength output.
4. Use of an etalon to count fringes in keeping track of the wavelength.
5. Use of the optogalvanic effect in a hollow cathode discharge [252].

The first four methods involve more instrumental complexity, expense, or inaccuracy than the fifth method, which we consider in more depth.

The most convenient lamp for generating a wavelength standard is the hollow cathode lamp. The laser output is directed down the axis of a hollow cathode discharge while the current through the lamp is monitored for changes across a ballast resistor [252]. Whenever the laser wavelength matches transitions with the element in the hollow cathode or the fill gas, the current changes because of the optogalvanic effect. The current changes generate a potential change across the ballast resistor, which can be AC coupled to detection electronics. By scanning the laser wavelength, the spectrum of the hollow cathode lamp can be swept out in the same way that a monochromator can be calibrated with a hollow cathode lamp. Since the hollow cathode transitions are often sharper than the laser bandwidth, a scan of the laser wavelength across a line also provides the laser bandwidth.

In addition to calibration of laser wavelength, optogalvanic detection with a hollow cathode lamp can be used to stabilize the laser wavelength at the transition of interest [108]. Once the laser wavelength is set to the correct transition using a hollow cathode lamp appropriate for the analyte of interest, the optogalvanic effect will provide a continuous monitor of the wavelength. If that signal is compared against a reference signal to generate an error signal, that error signal can drive the wavelength controls of the laser to correct for any wavelength error.

Molecular Spectroscopy by Photoionization and Photodissociation

Photoionization of molecules by single photon absorption has been studied extensively, especially in connection with ion sources for mass spectroscopy. This method has been hindered by the lack of good light sources and the difficulties in obtaining suitable optics in the vacuum UV region of the spectrum. Synchrotron radiation sources have been excellent, but such facilities are sharply limited to a few large installations. It has been demonstrated that present lasers can efficiently excite multiple photon transitions that result in the ionization and/or dissociation of the molecule [253-255]. A number of different schemes are possible [255]. One can tune a very intense infrared laser to a vibrational transition and excite a molecule up the vibrational ladder until it dissociates or ionizes. One can also tune a laser to an electronic transition and cause a molecule to dissociate and/or ionize after a series of transitions. There are also the various combinations of these ideas where one excites a vibrational mode and then causes further electronic excitation until ionization and/or dissociation occurs. A great deal of research has been done to determine the feasibility of performing selective photochemistry and isotope separations with these techniques [255].

The most extensive studies aimed at developing these multiphoton methods for analytical measurements use electronic excitations [256-258]. A multiphonon induced ionization and/or dissociation is resonantly enhanced by an intermediate electronic state. It is generally believed that one or more photons cause an excitation of the intermediate state and subsequent photons further excite the molecule to the ionization continuum. A wide variety of molecules have been studied including acetone, biacetyl, glyoxal, benzophenone [256], benzene [254, 256-258], naphthalene, azulene [256], aniline [259-261], acetaldehyde [262], pyridine, pyrazine [263], hexatriene [264, 265], Na_2, K_2 [266, 267], BaCl [267], I_2 [268-270], NO [257], butadiene [257], triethylenediamine [271], methyl ethyl, propyl, and butyl iodide [272], toluene [273, 274], anthracene, phenanthrene, benzanthracene, acenaphthene [275], pyrrole [276], cyclopropane [277], ethyleninine, and 1,4-diazabicyclo[2,2,2]octane (DABCO) [278]. These studies have provided a reasonably well-defined picture of the multiphoton ionization process and its capabilities.

The typical instrumental arrangement is sketched in Fig. 26. A relatively high power laser is usually used as the excitation source. The N_2 pumped dye laser is a convenient source, but it generally must be focused into the sample

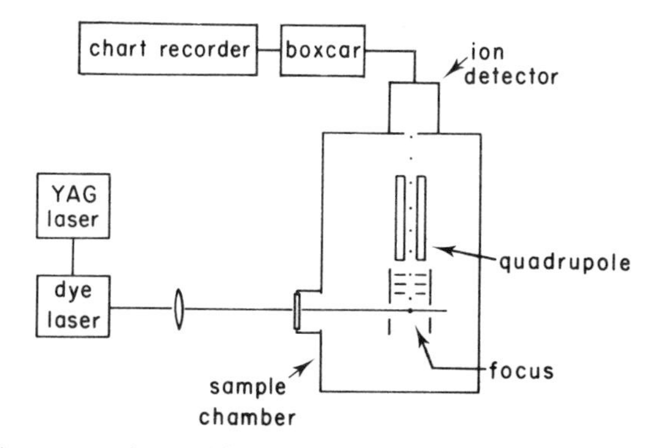

Fig. 26 Typical apparatus for studying multiphoton-induced dissociation. A dye laser is focused into a sample chamber and the ions created are measured. The sample chamber contains a molecular beam at the focus for this illustration. The ions are detected by a quadrupole mass spectrometer. (After Ref. 257.)

to achieve power densities high enough for multiple photon transitions [257]. The ionization region is more restricted and the sensitivity of the apparatus is lower as a result. A Nd:YAG pumped dye laser has much more power, and efficient ionization can be achieved even in an unfocused beam [256]. A second laser can provide more information and selectivity by separating the excitation and ionization steps [68]. The first laser excites the sample to an intermediate state while the second laser excites it further to the continuum.

The samples used have been a simple gas cell [270], the effluent from a gas chromatograph [275], or a molecular beam [259, 260, 266, 267, 269]. If the sample is at room temperature, the thermal population of rotational and vibrational modes decreases the fraction of molecules that can be resonant with the excitation beams. The large cooling effects in a supersonic molecular beam depopulate the vibrational and rotational modes so the ionization efficiency is enhanced [270]. The individual spectral transitions are also sharpened so the excitation selectivity is higher.

The ionization in the sample can be measured directly with a pair of collector plates or a channeltron ion multiplier [270]. This approach provides the total ion current as a function of the laser excitation wavelength. To measure a particular ion fragment, a mass spectrometer is often used as the ionization detector. Both time of flight and quadupole spectrometers have been used. This approach provides the ionization efficiency as a function of both the laser wavelength and the fragment mass. A typical two dimensional display is shown in Fig. 27 for the photoionization and dissociation of benzene vapor [257]. Benzene is excited from the ^1E_g ground state to the $^1A_{1g}$ excited state by a two-photon transition. This is followed by a one-photon transition to the ionization continuum.

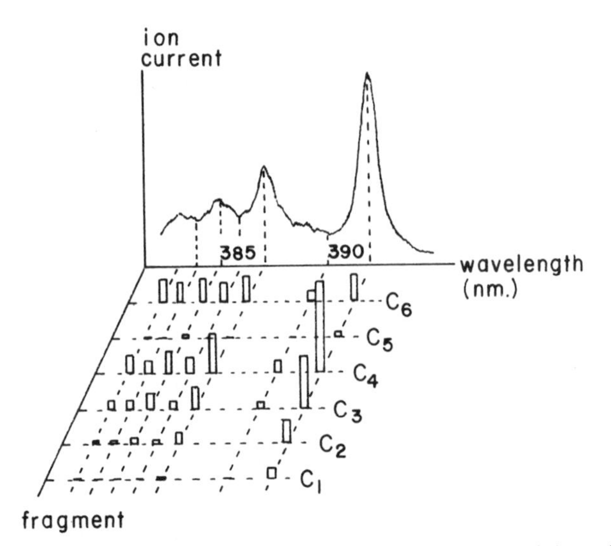

Fig. 27 Three dimensional plot of the ion current detected as a function of the excitation wavelength for different fragments $C_1 - C_6$. The quadrupole mass spectrometer did not resolve the number of hydrogens attached to each carbon fragment. (After Ref. 257.)

The benzene spectrum in Fig. 27 is representative of the molecules that have been studied. In general the multiphoton ionization spectrum is quite similar to the fluorescence excitation spectrum of the intermediate state. The same transitions are observed, although some lines may have some changes in relative intensities since the multiphoton ionization spectrum reflects additional transitions required to reach the continuum. Occasionally there is some background due to a direct multiphoton transition to the continuum without any enhancement from an intermediate state [257, 270]. In addition, it is possible to have other resonances in the same wavelength region when it is possible to reach a different intermediate state by a different number of photons. Nevertheless, the excitation spectra monitored by detecting ionization versus fluorescence remain very similar. The similarity persists regardless of which fragment ion is monitored.

The fragmentation pattern changes as a function of the laser power density [256–259, 268–270]. The nature of the variation is a subject of much interest because it provides information about the mechanisms behind multiphoton ionization. Many effects combine to produce the final curve of ionization efficiency for a particular fragment as a function of the laser power [256, 257, 262]. For example, if a particular fragment is the result of a four-photon total absorption, one would expect a quartic dependence on laser power. Such a dependence is seldom seen. Instead the power law depends on the number of photons involved in the slowest step in the excitation process, such as the excitation to the intermediate state [257]. Even here the expected dependence is not always seen for a number of reasons. The focusing of a laser results in such high power densities that the transitions can become saturated. As the power is changed, the part of the focal volume where saturation is present changes [262]. Relaxation effects such as autoionization can depopulate states on the way to the ionization continuum. Finally the ions and fragments themselves are in the beam long enough that they also can be excited further. The complexities in obtaining definitive data can be simplified by using two lasers—one for exciting the intermediate state and one for exciting into the ionization continuum [268]. The power of each laser can be varied separately to isolate the power dependence for the two important steps in the ionization process.

The efficiency of producing each fragment ion varies with the laser power, so the fragmentation pattern is dependent on the laser power and the focusing parameters. As might be expected, at higher laser powers the fragmentation becomes more extensive. It is also striking that large amounts of energy can be deposited in a molecule without much difficulty. For example, the production of C_1 fragments in benzene requires nine or more photons and it is still an efficient channel [257, 258, 270]. More fragmentation is produced in benzene than can be achieved with electron impact ionization. In other molecules the multiphoton ionization is quite soft and produces little fragmentation in comparison with electron impact ionization [256].

The excitation spectra can contain a great deal of information about the molecular states since the intermediate state controls the spectroscopy that is observed. If the excitation of the intermediate state requires a two-photon absorption, that absorption reflects the same polarization characteristics as a traditional two-photon experiment [263, 265, 271–273, 277]. By comparing the ionization efficiency for two different polarizations, one can get information about the symmetry of the intermediate state. This technique has been used by several groups to identify the nature of the intermediate state [263, 265, 271–273, 277]. One can also get the relaxation rate of the intermediate state by using two lasers again to isolate the two steps in the ionization process. By measuring the ionization as a fraction of the time delay between the two laser pulses, one obtains the time dependence of the intermediate state decay [278].

The efficiency and detection limits for multiphoton ionization methods have been studied by several groups [256, 257, 270, 275, 279]. In the regions of and during the times of the laser excitation of the sample, the ionization efficiency exceeds that of electron impact ionization, perhaps by a large amount. The problem is that the laser is generally exciting the sample for a very short time [256, 257]. A N_2 or a Nd:YAG laser has a pulse width of about 10 nsec and a repetition rate of 10 to 100 Hz. The duty cycle is only 10^{-6} to 10^{-7}. If the comparison is with fluorescence methods, multiphoton ionization is more efficient because of the near unit quantum efficiency of detecting ionization [279]. Frueholz and co-workers have found detection limits of 10^7 molecules/cm^3 for naphthalene vapor [279]. The ionization was produced by a frequency doubled flashlamp-pumped dye laser and was detected by a cylindrical proportional counter. No mass discrimination was used. This detection limit was 3 or 4 orders of magnitude higher than one would estimate should be achievable and there is optimism that it can be lowered substantially. Brophy and Rettner reported similar sensitivities [259]. Although they did not report quantitative values, they were able to see contaminants estimated at less than 3×10^8 molecules/cm^3.

Perhaps more importantly than low detection limits, multiphoton ionization techniques promise a good selectivity for specific molecules. Gas-phase molecular spectra are structured and are characteristic of the specific molecule. They are particularly well defined in supersonic molecular beams. Leutwyler and Even succeeded in resolving the isotope shift in aniline, suggesting the possibility of obtaining isotope-specific molecular ionization [260]. The spectra can be complicated by other resonances involving a different number of photons. Thus an electronic state at 30,000 cm^{-1} can resonantly enhance ionization from two photons of 15,000 cm^{-1}, but this transition can overlap when there is an electronic state at 45,000 cm^{-1}, which will resonantly enhance a three-photon process. The nonresonant multiphoton ionization directly to the continuum will always provide a background ionization against which the resonant signals must be discriminated. The background can become a problem when there is a sample component at concentrations much higher than that of the component of interest, as might be the case in chromatographic applications where a solvent is present.

The multiphoton ionization methods promise important applications for chemical measurements, especially when teamed with other techniques. They provide one more technique for the variety of hypenated methods [280]. Mass spectroscopy has been the most studied partner for multiphoton ionization. A mass spectrum would be available for each laser wavelength to provide a two dimensional fingerprint that would be specific for a molecule. Some specificity is lost because the mass spectrum of a molecule does not change markedly with laser wavelength. The fact that the fragmentation depends on laser intensity can be viewed as a complication to the interpretation of the data or as a third dimension that is available for further specifying the molecule.

Gas chromatography also has been studied as a potential partner to multiphoton ionization methods by Klimcak and Wessel [275]. They used either a flashlamp-pumped or nitrogen-laser-pumped dye laser to excite the effluent from a gas chromatograph as it passed through a low-volume proportional cell. They demonstrated the capability for resolving several five-component mixtures of PAH compounds with detection limits between 0.3 and 50 pg. Some of the PAH compounds included halogenated naphthalenes, which have no appreciable fluorescence. They could nevertheless be measured easily by the multiphoton ionization method. The detection limits were determined by fluctuations in the background signal that came from other materials bleeding from the column. The selectivity of this combination could be improved further with the addition of mass discrimination. A simple time-of-flight method would be readily adapted to this situation.

Siomos and co-workers have developed a liquid phase analogy to the gas phase multiphoton ionization process [281]. They detect the ionization signal by a parallel plate charge sensitive detector with a 3-kV potential drop giving 15-kV/cm electric field immersed directly in the solution. A N_2 pumped dye laser is used as the excitation source to achieve an excellent signal level. Figure 28 shows the spectrum they obtained for a 5×10^{-6} M solution of fluoranthene in n-pentane. The transitions arise from two-photon excitation to a state that is embedded in the ionization continuum and can therefore autoionize.

A number of other methods will be developed that use some of the unique capabilities of multiphoton ionization. The polarization selection rules that are important for multiphoton processes have not been utilized for analytical measurements in this area, but they will be able to provide more information

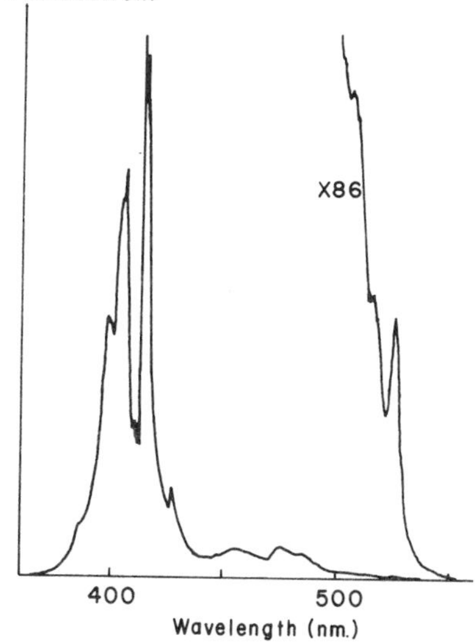

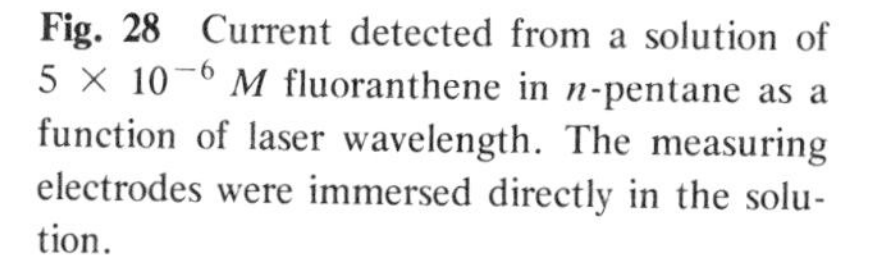
Fig. 28 Current detected from a solution of 5×10^{-6} M fluoranthene in n-pentane as a function of laser wavelength. The measuring electrodes were immersed directly in the solution.

about molecular symmetry and identity. One can also expect that the use of two or more lasers will provide important new advances for selectivity in measurements.

5 COHERENT RAMAN SPECTROSCOPIES

Raman Scattering

The major analytical impact of the laser has historically been on the field of light scattering, particularly Raman spectroscopy [282, 283]. Not only has the laser's intensity permitted Raman spectra to be obtained quickly and conveniently from low concentration samples, but it has also spawned the development of a broad series of coherent Raman spectroscopies that have become very exciting to analytical spectroscopists [284–293]. Unfortunately, the fundamental physical phenomena underlying these new developments can be somewhat obscured by the mathematics used in the description. One of the purposes of this part of the Chapter is to use the mathematics in an intuitive manner to both describe and unify the many different coherent Raman techniques.

Light is scattered by matter because the oscillatory electric field of an incoming light wave induces an oscillatory polarization in matter. One can ignore the magnetic component of a light wave as long as the electron motion induced in the matter does not approach relativistic velocities. This assumption limits photon energies to being lower than those of hard X-rays [294]. Since the oscillatory polarization is a periodic movement of electrical charges, it may also generate a light wave itself and result in light scattering. A number of approaches, including both quantum mechanical and classical models, can describe the scattering quantitatively [294–296]. We use the classical approach here because of its strong connection with intuition, although the connection to the quantum mechanical models is also made.

When an electric field is applied to a material, a polarization is induced. For a single molecule, the induced dipole moment $\vec{\mu}$ is proportional to the electric field $\vec{E}$ through the polarizability α [284, 293].

$$\vec{\mu} = \vec{\vec{\alpha}}\,\vec{E} \tag{18}$$

For an ensemble of molecules, the net polarizability $\vec{P}$ is proportional to the electric field through the susceptibility $\vec{\vec{\chi}}$.

$$\vec{P} = \vec{\vec{\chi}}\,\vec{E} \tag{19}$$

Both the susceptibility and the polarizability are tensors, a fact that we largely ignore in later discussions for notation simplicity, although their inclusion is a simple matter that is described in many other places [284–286]. We assume that the electric and dipole vectors are colinear and are directed along a coordinate axis. We also ignore any permanent dipole moments in this section. Both (18) and (19) assume the molecular distortions caused by the external field are sufficiently small that the bond energies are not changed. However, at higher fields, this approximation breaks down and there is no longer a linear relationship between the induced polarization and the electric field. A saturation of the polarization must be reached and eventually the material must undergo dielectric breakdown. To take account of the nonlinear behavior, the polarization is written as a Taylor series in the electric field.

$$P = \chi^{(1)}E + \chi^{(2)}E^2 + \chi^{(3)}E^3 + \cdots \tag{20}$$

In this expression, $\chi^{(i)}$ represents the ith order susceptibility. An analogous expression could be written for the single molecule in terms of the polarizability. It is important to recognize that (20) is simply a convenient method for expressing the distorted polarization since there is not an analytical solution for the relationship between P and E.

The various order susceptibilities in (20) are functions of the state of the molecule, the frequencies of oscillation that are involved, and the vibrational motion of the molecule. We concern ourselves with the form of all these dependences in the course of this chapter, but let us first examine the dependence on vibrational motion. The polarizability of a single molecule depends on how easily one can cause distortions in the electron clouds. If the molecule is vibrating, the bonding energies and shapes of the electron clouds is slightly modified and therefore so is the polarizability. Again, the changes are very small and can be conveniently expressed as a Taylor's series expansion in terms of a vibrational coordinate X that measures the amplitude of vibrational motion [282]:

$$\alpha = \alpha_0 + \left(\frac{\partial \alpha}{\partial X}\right)X + \cdots \tag{21}$$

The $\partial\alpha/\partial X$ describes the polarizability changes for small vibrational amplitudes. An analogous expression can be written for susceptibility. If (21) is substituted into (18),

$$\mu = \left[\alpha_0 + \left(\frac{\partial \alpha}{\partial X}\right)_0 X\right]E \tag{22}$$

one gets two terms that can produce an oscillatory dipole moment (or analogously an oscillatory polarization)—one that is independent and one that is dependent on vibrational motion. The first term produces Rayleigh light scattering in a disordered material that has fluctuations in its density and it also produces the index of refraction for the forward propagating wave. An oscillatory dipole results with the same frequency dependence as the incoming electric field. The second term represents the mechanism by which vibration that has no oscillatory dipole of its own can cause oscillatory changes in the polarizability. The effect is very small, but it produces Raman scattering. The second term is therefore called the Raman term and represents the physical mechanism for coupling a molecular vibration with an electromagnetic wave.

Let us assume that E and X have the following functional forms:

$$E = E_0 \cos \omega t \tag{23}$$

$$X = X_0 \cos \omega_V t \tag{24}$$

If these expressions are substituted into (22) and we examine only the second term

$$\mu = \left(\frac{\partial \alpha}{\partial X}\right)_0 X_0 E_0 \cos (\omega_V t) \cos \omega t \tag{25}$$

Using elementary trigonometric relationships, we obtain the product of two different oscillatory functions in terms of their sum and difference frequencies:

$$\mu = \frac{1}{2}\left(\frac{\partial \alpha}{\partial X}\right)_0 X_0 E_0[\cos (\omega + \omega_V)t + \cos (\omega - \omega_V)t] \tag{26}$$

This expression shows that the induced molecular dipole (or equivalently the polarization) oscillates at the sum and difference of the incoming light frequency and the molecular vibration frequency. The oscillating dipole in turn launches light waves corresponding to these frequencies. The first term corresponds to anti-Stokes Raman scattering, while the second term corresponds to Stokes Raman scattering.

The procedure used to obtain the sum and difference frequencies in (26) is repeated many times in this section and it is a result that is obtained in many contexts. Whenever two waves are mixed, the sum and difference frequencies are produced. This idea of course forms the basis of the communications industry where a signal and a carrier frequency are mixed for transmission. It is more convenient mathematically to work with complex notation instead of relations such as (23) and (24). The oscillatory electric field is then written as

$$E = \tfrac{1}{2}[E_0 e^{i\omega t} + E_0^* e^{-i\omega t}] \tag{27}$$

where we have allowed the possibility that the amplitude E_0 might be a complex number. Since E must be a real physical quantity, the second term must be the complex conjugate of the first term. Equations (23) and (27) are mathematically equivalent except for a possible phase difference. By substituting equations such as (27) into (22), one can get (26) more conveniently.

Before we proceed further, it is of interest to examine the quantum mechanical model for Raman scattering. In quantum mechanics, a molecule is described by a series of stationary electronic states with the lowest energy state normally populated. One usually considers an incoming light wave as a very small perturbation on the states that does not distort any of the orbitals. Of course, one must distort the electron clouds to induce a polarization. To describe the distortions, one must mix in contributions from all the other possible states of the system weighted according to their difference in energy from the exciting photon. If the exciting frequency approaches a natural resonant frequency of one state of the molecule, the distortion associated with that state becomes large and provides a resonant enhancement of the Raman scattering. The picture usually drawn of the Raman scattering involves a transition to a virtual state and a return transition from that state producing the scattered photon. The virtual state corresponds to the distorted molecule. In the quantum mechanical picture, an incoming photon produces a distortion in the molecular electron cloud that can relax back to the ground state and produce a photon at the same energy as the incoming photon (Rayleigh scattering) or it can relax to a vibrationally excited state and produce a Raman scattered photon.

The classical picture of Raman scattering described by (26) predicts incorrectly that the Stokes and anti-Stokes scattering intensities are equal. This prediction arises because we assumed there was an initial vibrational excitation. This assumption is not true because in the quantum mechanical picture, the relative intensities of Stokes and anti-Stokes scattering depend on the population of the vibrationally excited state.

Coherent Raman Techniques

Let us now return to (20) and consider the consequences of having the nonlinear terms in the expression for the polarization. The term involving the second-order susceptibility is not discussed in this chapter. Since this term involves a product of two electric vectors, it is clear that it can lead to second harmonic and sum and difference frequency generation [285]. This term also leads to the hyper-Raman effect where incoherent scattering at $(2\omega \pm \omega_v)$ occurs [297]. Instead we consider the third-order term that couples three electric waves together to produce a fourth wave (four wave mixing). Let us assume there are two incident electromagnetic waves (or two lasers) whose frequencies are ω_L and ω_S. Using the same logic employed in getting the sum and difference frequencies for (26), for the third order term we obtain the polarization oscillating at frequencies of $3\omega_L$, $3\omega_S$, $(2\omega_L + \omega_S)$, $(2\omega_S + \omega_L)$, $(2\omega_L - \omega_S)$, $(2\omega_S - \omega_L)$, ω_L and ω_S. The first four frequencies represent very promising methods that are currently being researched for generating coherent sources at high energies in the UV, far UV, and even X-ray spectra regions from lower energy coherent sources [285]. We are concerned with the latter four frequencies, which give coherent anti-Stokes Raman scattering (CARS) [288–291], coherent Stokes Raman scattering (CSRS, pronounced scissors) [298], the inverse Raman effect or stimulated Raman loss spectroscopy [292, 299], and stimulated Raman scattering or stimulated Raman gain spectroscopy [292, 293]. An equivalent way of understanding the generation of these frequencies is to realize that ω_L and ω_S beat with each other to produce the difference frequency $(\omega_L - \omega_S)$, which in turn interacts with the third wave at either ω_L or ω_S to produce the sum and difference frequencies. This process is greatly enhanced if there are natural resonances in the material, in particular if $(\omega_L - \omega_S)$ matches a vibrational frequency ω_v. The generation of frequencies at $(2\omega_L - \omega_S)$ and $(2\omega_S - \omega_L)$ can still occur, however, if $(\omega_L - \omega_S)$ is nonresonant with a vibrational frequency since a polarization can be induced by a time-varying electric field at any frequency. As we see later, this nonresonant contribution to the scattering is the major limitation at present on many of the coherent Raman techniques. The different techniques are summarized in Fig. 29.

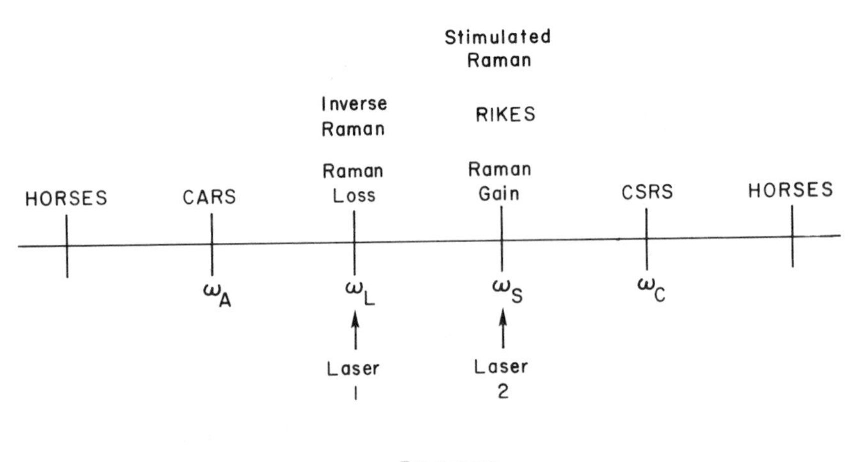

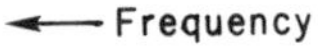

Fig. 29 When two laser beams at a frequency of ω_L and ω_S, where $\omega_L > \omega_S$, are brought together in a medium where there is a vibrational resonance at $(\omega_L - \omega_S)$, the two beams interact to cause generation of new beams at ω_A and ω_C, and additionally light is scattered out of the beam at ω_L and into the beam at ω_S. The beams at ω_A and ω_C give rise to coherent anti-Stokes Raman spectroscopy (CARS) and coherent Stokes Raman spectroscopy (CSRS). Interactions between the new beams generated and any of the other beams present produce higher order scattering processes called HORSES for higher order Raman spectral excitation studies.

Our intuitive reasoning can be easily supported mathematically. The incoming electric field $E(z, t)$ consists of the two excitation lasers at ω_L and ω_S and can be written as [293]

$$\begin{aligned} E(z, t) = \tfrac{1}{2}\{&E_L(z, t) \exp[i(k_L z - \omega_L t)] + E_L^*(z, t) \exp[-i(k_L z - \omega_L t)] \\ &+ E_S(z, t) \exp[i(k_S z - \omega_S t)] \\ &+ E_S^*(z, t) \exp -i(k_S z - \omega_S t)\} \end{aligned} \tag{28}$$

Note we are not admitting the possibility of any new waves being generated that add to $E(z, t)$. Thus we are restricted to the low signal limit where appreciable CARS, CSRS, etc. amplitudes are not created. If this expression is substituted into (20) and we restrict our consideration to the third order term only, we obtain the expression

$$\begin{aligned} P^{(3)}(z, t) = \tfrac{1}{8}\{&\chi_1^{(3)} E_L^3 \exp[i(3k_L z - 3\omega_L t)] + \chi_2^{(3)} E_S^3 \exp[i(3k_S z - 3\omega_S t)] \\ &+ 3\chi_3^{(3)} E_L^2 E_S \exp\{i[(2k_L + k_S)z - (2\omega_L + \omega_S)t]\} \\ &+ 3\chi_4^{(3)} E_L E_S^2 \exp\{i[(2k_S + k_L)z - (2\omega_S + \omega_L)t]\} \\ &+ 3\chi_5^{(3)} E_L^2 E_S^* \exp\{i[(2k_L - k_S)z - (2\omega_L - \omega_S)t]\} \\ &+ 3\chi_6^{(3)} E_L^* E_S^2 \exp\{i[(2k_S - k_L)z - (2\omega_S - \omega_L)t]\} \\ &+ 3\chi_7^{(3)} |E_L|^2 E_L \exp[i(k_L z - \omega_L t)] \\ &+ 6\chi_8^{(3)} |E_L|^2 E_S \exp[i(k_S z - \omega_S t)] \\ &+ 3\chi_9^{(3)} |E_S|^2 E_S \exp[i(k_S z - \omega_S t)] \\ &+ 6\chi_{10}^{(3)} |E_S|^2 E_L \exp[i(k_L z - \omega_L t)] \\ &+ \text{complex conjugate}\} \end{aligned} \tag{29}$$

Since the third-order susceptibility is frequency dependent, each term has been written with a different susceptibility [284]. The first and second terms describe frequency tripling of ω_L and ω_S, respectively. The third and fourth terms describe frequency summation. The fifth and sixth terms describe CARS and CSRS, respectively, while the eighth and tenth terms describe Raman gain and Raman loss. The seventh and ninth terms describe waves propagating at ω_L and ω_S, respectively, with amplitudes that do not depend on the second frequency. These terms cause self-induced refractive index changes and can lead to self-focusing, self-steepening, self-phase modulation, and self-induced polarization changes [285, 286]. We are not concerned here with these effects and ignore these two terms.

It should be noted at this point that there are several conflicting conventions for expressing the third-order susceptibility. Our expression in (29) rests on the definition of $\chi^{(3)}$ given by (20). Maker and Terhune introduced a different convention where the coefficients of each term in (29) are combined into a new definition of susceptibility so that the generation of a third harmonic signal from a single laser does not have a numerical argument [284]. Thus each of the $\chi_i^{(3)}$ in (29) is four times as large as the value of $\chi^{(3)}$ of the Maker and Terhune convention. It becomes important in reading the literature of this field to determine which convention the authors have adopted.

Thus far we have seen that an incoming set of light waves will drive the polarization. An oscillating polarization defined by (29) in turn launches an electromagnetic wave itself. One can derive an expression from Maxwell's equations that relates the oscillating polarization to the electric fields produced:

$$\nabla^2\vec{E} - \frac{n^2}{c^2}\frac{\partial^2\vec{E}}{\partial t^2} = \frac{4\pi}{c^2}\frac{\partial^2\vec{P}}{\partial t^2} \tag{30}$$

where n is the index of refraction at the frequency of the resultant electric field oscillation [293, 296]. If any of the terms in (29) are inserted for the polarization, the resulting electric field can be easily obtained. Any of the polarization terms can be written in the form

$$\vec{P} = \tfrac{1}{2}[Ae^{i(kz-\omega t)} + A^*e^{-i(kz-\omega t)}] \tag{31}$$

and we assume that the electric wave that is launched has the form

$$\vec{E} = \tfrac{1}{2}[E_0e^{i(k'z-\omega t)} + E_0^*e^{-i(k'z-\omega t)}] \tag{32}$$

because the frequency of the electric field must be the same as the polarization that produces it. However, it is very important to note that the wave vector k' of the electric field is different from the wave vector k of the polarization wave that launches it. For example, the fifth term in (29) oscillates at a frequency of $(2\omega_L - \omega_S)$ and possesses a wave vector of $(2k_L - k_S)$. For colinear waves, we could use the relationship $k = n\omega/c$ where n is the index of refraction in the medium, to rewrite the wave vector amplitude in terms of the frequency. Then we would find the polarization wave vector amplitude to be $(1/c)(2n_L\omega_L - n_S\omega_S)$. The electric field wave that is launched by the polarization has a frequency $\omega_A = 2\omega_L - \omega_S$ and a wave vector $(n_A\omega_A/c)$. This wave vector is in general different from the polarization and becomes equal only if n_L, n_S, and n_A are equal or if proper precautions are taken for phase matching. The wave vector of the polarization is determined completely by the wave vectors of the lower frequency electromagnetic waves driving it since they are anchored to each other, but the new electromagnetic wave propagates independently. The dispersion of the medium therefore causes a fundamental difference between the wavelengths of the electromagnetic wave and the polarization that launches it. If you substitute (31) and (32) into (30), consider only the components in the z direction, and neglect terms containing $\partial^2 E_0/\partial z^2$ and $\partial^2 E_0/\partial t^2$ (i.e., assume the amplitude does not vary rapidly in time or space), the following expression results:

$$\frac{\partial E_0}{\partial z} + \frac{n}{c}\frac{\partial E_0}{\partial t} = \frac{2\pi i A\omega}{nc} e^{i(k-k')z} \tag{33}$$

and a comparable equation results for the complex conjugate. This expression can then be readily used to find the effect of any of the terms in the oscillating polarization.

Let us first obtain the expression for generating a CARS signal. The CARS signal arises from the fifth term in (29). If (33) is applied to this term, we obtain the expression

$$\frac{\partial E_A}{\partial z} + \frac{n_A}{c}\frac{\partial E_A}{\partial t} = \frac{2\pi i\omega_A}{n_A c}\left(\frac{3}{4}\chi_5^{(3)}E_L^2 E_S^*\right)\exp\left[i(2k_L - k_S - k_A)z\right] \tag{34}$$

where E_A, ω_A, n_A, and k_A refer to the wave at the anti-Stokes frequency at $\omega_A = (2\omega_L - \omega_S)$ and $\frac{1}{2}A = \frac{3}{8}\chi_5^{(3)}E_L^2 E_S^*$. We assume steady-state conditions $(\partial E_A/\partial t = 0)$ and integrate (34) over the length of the material l.

$$E_A = \frac{3\pi\omega_A\,\chi_5^{(3)}E_L^2E_S^*}{2cn_A(2k_L - k_S - k_A)}\{\exp\left[i(2k_L - k_S - k_A)l\right] - 1\} \tag{35}$$

A similar expression is obtained for the complex conjugate. The intensity of an electromagnetic wave is related to its electric field by [294, 294]

$$I = \frac{cn}{8\pi}EE^* \tag{36}$$

so that multiplying (35) by its complex conjugate and simplifying, we obtain the expression for the intensity of the CARS wave:

$$I_A(z = l) = \frac{144\pi^4\omega_A^2|\chi_5^{(3)}|^2 I_L^2 I_S l^2}{c^4 n_A n_L^2 n_S}\cdot\frac{\sin^2\left[(2k_L - k_S - k_A)l/2\right]}{\left[(2k_L - k_S - k_A)l/2\right]^2} \tag{37}$$

The coherent Stokes Raman scattering arises from the sixth term in (29) and is very similar to the term for CARS. The expression for the CSRS intensity follows the treatment for CARS and results in

$$I_C = \frac{144\pi^4\omega_C^2|\chi_6^{(3)}|^2 I_S^2 I_L l^2}{c^4 n_C n_S^2 n_L}\cdot\frac{\sin^2\left[(2k_S - k_L - k_C)l/2\right]}{\left[(2k_S - k_L - k_C)l/2\right]^2} \tag{38}$$

where I_C represents the Stokes shifted scattering relative to the excitation frequency ω_S.

The intensity of the excitation beam at ω_S can also be affected as it propagates through the medium. If we ignore self-induced effects, we must consider the effect of the eighth term in (29). Using (33) and assuming steady-state conditions, we obtain

$$\frac{\partial E_S}{\partial z} = \frac{2\pi i \omega_S}{n_S c} \left[\frac{6}{4} \chi_8^{(3)} |E_L|^2 E_S \right] \tag{39}$$

This expression can be readily intergrated to give

$$\frac{E_S(z = l)}{E_S(z = 0)} = \exp \left[\frac{3\pi i \omega_S \chi_8^{(3)}}{n_S c} |E_L|^2 l \right] \tag{40}$$

The intensity can be found using (36) remembering that $\chi^{(3)}$ can be complex.

$$\frac{I_S(z = l)}{I_S(z = 0)} = \exp \left[- \frac{48\pi^2 \omega_S I_L l}{n_L n_S c^2} \operatorname{Im} \chi_8^{(3)} \right] \tag{41}$$

This expression states that the intensity of the beam will propagate unchanged if the $\chi_8^{(3)}$ is a real quantity. We see later that $\chi_8^{(3)}$ can have imaginary components when $(\omega_L - \omega_S)$ approaches a vibrational frequency because of the energy dissipation introduced. We also see that the form of $\chi_8^{(3)}$ causes a positive exponent and results in amplification of the beam as it propagates through the material.

The intensity of the excitation beam at ω_L is affected in the same way through the tenth term in (29). Proceeding in exactly the same manner, we obtain

$$\frac{I_L(z = l)}{I_L(z = 0)} = \exp \left[- \frac{48\pi^2 \omega_L I_S l}{n_L n_S c^2} \operatorname{Im} \chi_{10}^{(3)} \right] \tag{42}$$

The form of $\chi_{10}^{(3)}$ is such that when a vibrational resonance is approached, the exponent is negative and an attenuation of the beam occurs as it propagates through the material.

In summary, the third-order susceptibility can mix the frequencies of our input beams to produce a polarization that oscillates at new output frequencies (the CARS and CSRS beams) or at the input frequencies. The oscillating polarization produces new electric waves that can be related to the intensities of new electromagnetic waves. The new electromagnetic wave frequencies that represent the CARS and CSRS outputs vary according to the phase matching of the different waves. Since they are generated by either the real or imaginary components of $\chi^{(3)}$, they are present in any medium or frequency. The electromagnetic waves generated at the input frequencies exhibit exponential intensity increases or decreases as the waves propagate through the material when a vibrational resonance is approached. This behavior only occurs when $\chi^{(3)}$ is imaginary, that is, when there is dissipation in the medium.

Classical Treatment of Nonlinear Effects

MECHANICAL MODEL FOR $\chi^{(3)}$

To have an understanding of how the waves at $(2\omega_L - \omega_S)$, $(2\omega_L - \omega_L)$, ω_S, and ω_L behave in detail, it is necessary to obtain information about the functional form of the third order susceptibility. A number of models can be used to generate $\chi^{(3)}$ [286]. We first examine a model for the polarization that is based on a simple harmonic oscillator picture because it provides both a correct functional form and insight into the fundamental phenomena [288, 293]. Later we examine a more detailed model that describes the resonant enhancements that occur when real molecular levels are at the appropriate energies.

If an oscillatory electric field is applied to a molecule, a small oscillatory displacement is induced that results in a polarization. If X_{av} represents the average displacement of the molecules and N represents the density of molecules, the part of the polarization that is correlated with vibrational motion is [see (22)]

$$P = N \frac{\partial \alpha}{\partial X} X_{av} E \tag{43}$$

assuming a one dimensional problem [293]. We must obtain an expression that relates the oscillatory displacement X_{av} to the driving electric fields.

From Newton's laws of motion, we can write the following expression:

$$\frac{\partial^2 X_{av}}{\partial t^2} + \Gamma \frac{\partial X_{av}}{\partial t} + \omega_V^2 X_{av} = \frac{1}{m} F \Delta \tag{44}$$

where Γ is a constant describing the losses in the molecule (the damping term for a harmonic oscillator), ω_V is the undamped vibrational frequency, m is the effective mass of the oscillator, F is the driving force, and Δ is a fraction that is included to account for depopulation of the ground state if large vibrational excitation is obtained. This fraction is equal to unity for small driving fields and low temperatures. The driving force is [293]

$$F = \frac{1}{2} \left(\frac{\partial \alpha}{\partial X} \right)_{X=0} E^2 \tag{45}$$

where we choose E to be given by (28). If this expression is substituted into (44), we can solve for X_{av} using standard techniques [293, 294]. If the result is substituted into (43), the following equation results:

$$P = \frac{N\Delta}{8m} \left| \frac{\partial \alpha}{\partial X} \right|^2 \left\{ \frac{E_L E_S^* \exp\{i[(k_L - k_S)z - (\omega_L - \omega_S)t]\}}{\omega_V^2 - (\omega_L - \omega_S)^2 - i\Gamma(\omega_L - \omega_S)} \right.$$

$$+ \left. \frac{E_S E_L^* \exp\{-i[(k_L - k_S)z - (\omega_L - \omega_S)t]\}}{\omega_V^2 - (\omega_L - \omega_S)^2 + i\Gamma(\omega_L - \omega_S)} \right\}$$

$$\cdot\{E_L \exp[i(k_L z - \omega_L t)] + E_L^* \exp[-i(k_L z - \omega_L t)]$$
$$+ E_S \exp[i(k_S z - \omega_S t)] + E_S^* \exp[-i(k_S z - \omega_S t)]\} \qquad (46)$$

The important part of the expression for X_{av} is contained in the large bracket. It shows that the displacement is being driven by the difference frequency $(\omega_L - \omega_S)$ and will become resonantly enhanced when this frequency matches a natural vibrational frequency ω_V. The amplitude is limited by the damping Γ, which occurs as an imaginary term representing energy loss. If (46) is expanded, we obtain four different frequencies that correspond to the ones of interest in (29). By comparing (29) and (46), we can obtain explicit expressions for $\chi^{(3)}$.

$$\chi_5^{(3)} = 2\chi_{10}^{(3)} = \frac{N\Delta}{3m}\left(\frac{\partial\alpha}{\partial X}\right)^2 \frac{1}{\omega_V^2 - (\omega_L - \omega_S)^2 - i\Gamma(\omega_L - \omega_S)} \qquad (47)$$

$$\chi_6^{(3)} = 2\chi_8^{(3)} = \frac{N\Delta}{3m}\left(\frac{\partial\alpha}{\partial X}\right)^2 \frac{1}{\omega_V^2 - (\omega_L - \omega_S)^2 + i\Gamma(\omega_L - \omega_S)} \qquad (48)$$

The susceptibilities reach a maximum when $(\omega_L - \omega_S)$ becomes resonant with a vibrational frequency. They are prevented from reaching infinite values by the damping coefficient that appears as the imaginary term. If there were no damping, the susceptibility would be real and there would be no energy dissipated.

There are contributions to the net susceptibility of a material other than vibrations [285, 286]. Since any material exhibits a nonlinear polarization at high electric fields, there is always a nonvanishing $\chi^{(3)}$ that does not depend on nearby resonances and is consequently labeled $\chi_{NR}^{(3)}$ for the nonresonant susceptibility. This contribution generally does not involve energy dissipative terms and $\chi_{NR}^{(3)}$ is consequently real. It does have a slight frequency dependence [286]. It is also possible to excite electronic resonances in a material. In analogy with $(\omega_L - \omega_S)$ driving vibrational resonances, $(\omega_L + \omega_S)$ can drive electronic resonances and cause a two photon contribution to $\chi_{TP}^{(3)}$ that can have both real and imaginary parts. The response of the system depends on all possible contributions [293].

$$\chi_{Net}^{(3)} = \chi_5^{(3)} + \chi_{NR}^{(3)} + \chi_{TP}^{(3)} \qquad (49)$$

It should again be pointed out that the convention of Maker and Terhune includes an additional factor of ¼ in the definition of the susceptibility [284]. One must therefore divide the susceptibilities in this chapter by 4 to reproduce expressions of some other authors.

COHERENT RAMAN INTENSITIES AT RESONANCE

We are now in a position to look in detail at the different types of coherent Raman spectroscopy. We are particularly interested in the situation where the difference between the two laser frequencies $(\omega_L - \omega_S)$ matches a vibrational frequency. Other resonances with electronic transitions are not considered. Our discussion predicts four frequencies that must be considered, each differing from the previous one by $\omega_V = (\omega_L - \omega_S)$; $\omega_A = 2\omega_L - \omega_S$, ω_L, ω_S, and $\omega_c = 2\omega_S - \omega_L$. The intensity at resonance of these four waves can be obtained by substituting either (47) or (48) into (37), (38), (41), and (42) and setting $(\omega_L - \omega_S) = \omega_V$.

$$I_A(z = l) = \frac{16\omega_A^2 N^2 \Delta^2 I_L^2 I_S l^2 B^2}{n_A n_L} \left\{ \frac{\sin[(2k_L - k_S - k_A)l/2]}{[(2k_L - k_S - k_A)l/2} \right\}^2 \tag{50}$$

$$I_C(z = l) = \frac{16\omega_C^2 N^2 \Delta^2 I_S^2 I_L l^2 B^2}{n_C n_S} \left\{ \frac{\sin[(2k_S - k_L - k_C)l/2]}{[(2k_S - k_L - k_C)l/2} \right\}^2 \tag{51}$$

$$I_L(z = l) = I_L(z = 0) \exp - \left(\frac{8\omega_L N \Delta I_S l B}{n_S} \right) \tag{52}$$

$$I_S(z = l) = I_S(z = 0) \exp \left(\frac{N \Delta I_L l B}{n_S} \right) \tag{53}$$

where

$$B = \frac{\pi^2 |\partial\alpha/\partial X|^2}{n_L n_S c^2 m \Gamma(\omega_L - \omega_S)}$$

The following picture now emerges. Two beams at ω_L and ω_S $(\omega_L > \omega_S)$ propagating through a material can resonantly and coherently excite molecular vibrations, which in turn can mix either with the wave at ω_L and scatter into a wave at ω_A (CARS) or with the wave at ω_S and scatter into a wave at ω_C (CSRS). In addition, the resonant production of a molecular vibration furnishes a mechanism for coupling the waves at ω_L and ω_S. Molecular vibrations are created by $(\omega_L - \omega_S)$ and the excess energy appears as a scattered photon at ω_S causing exponential gain at ω_S (Raman gain spectroscopy or stimulated Raman scattering) and exponential loss at ω_L (Raman loss spectroscopy or the inverse Raman effect).

It is very important to point out that all our discussions are valid only in the limit of small intensities at ω_A and ω_C because we neglected to include the electric fields associated with such waves in the beginning of the problem (28) [288]. We also neglect any change in the two excitation beam intensities when the intensity at ω_A and ω_C is calculated. We are therefore unable to describe more complex interactions among the four propagating waves or the generation of higher order scattering effects [e.g., generation of waves at $(3\omega_L - 2\omega_S)$, and so on] by further mixing processes. These higher order scattering processes have been named HORSES for higher order Raman spectral excitation studies [298].

Quantum Mechanical Form of $\chi^{(3)}$

To obtain a deeper understanding of the nonlinear techniques and their potentialities, particularly for resonantly enhanced CARS, it is necessary to discuss the quantum mechanical description of $\chi^{(3)}$. We see above how the application of three oscillatory electric fields associated with three lasers can induce nonlinear distortions in the electron clouds surrounding a molecule. Quantum mechanically, these distortions must be described by a linear combination of all the stationary states of the molecule. One says that there is a mixing of states induced by the oscillatory fields. For four wave mixing processes such as CARS, this mixing becomes complex and involves third order perturbation theory. Since there can be three fields present simultaneously from the excitation lasers (we neglect the field that is generated by the nonlinear process), there is no preferred way of accomplishing the mixing and all possible orderings are equally weighted. The second laser, for example, can mix some state with the ground state, the third laser can bring a contribution from another state into that mixture, and the first laser can add still another contribution from a different state into the now complex mixture of states. To correctly describe the nonlinear process, it is necessary to consider both the mixing of the bra and ket parts of the wavefunction with all possible other states [300–306]. Although the procedure is straightforward, it is tedious. Diagrammatic techniques have been developed that allow for a systematic approach to the problem (the tedium remains though). The complete set of diagrams associated with production of a polarization at $2\omega_1 - \omega_2$ is shown in Fig. 30. Two vertical lines are drawn, the left representing the time evolution of the bra and the right representing the time evolution of the ket [300–306]. Time begins at the bottom of each line and evolves vertically. Absorption of photons is indicated by positively sloped lines, while emission is indicated by negatively sloped lines. Each line is labeled by the photon frequency it represents. Since we are interested in describing production of a wave at $2\omega_1 - \omega_2$, there are two absorptions at ω_1 and emission at ω_2. Since we are neglecting any further mixings that might be caused by the new wave at $2\omega_1 - \omega_2$, it does not enter into the mixing shown in Fig. 30. The state that results after each interaction is indicated beside each line for both the bra and the ket. All possible states must be included in the diagrams to be correct, but we assume here that a resonant state contributes overwhelmingly to the wavefuntion. A total of 48 diagrams represent the different time ordered combinations that are possible for both the bra and ket parts of the wavefunction if three different laser frequencies are used.

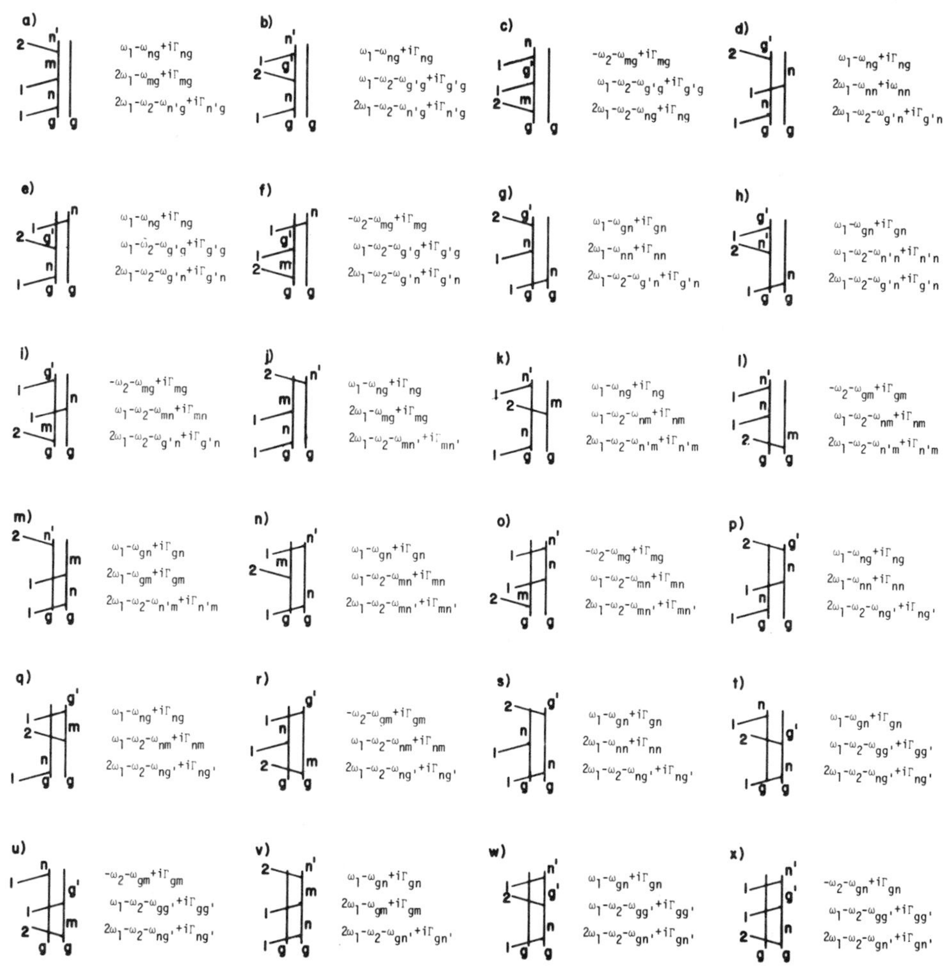

Fig. 30 Diagrams showing the time evolution of the bra (left-hand vertical line) and the ket (right-hand vertical line) for all possible interactions with two photons at ω_1 (*1*) and one at ω_2 (*2*) to form an output beam at $2\omega_1 - \omega_2$. Upward slanted lines represent absorption, while downward slanted lines represent emission. The letters represent the state that evolves after each interaction with a photon. Time evolves in the upward direction. The resonances that correspond to each diagram are given to the right.

These diagrams permit one to obtain the important terms in the denominator of $\chi^{(3)}$ that represent the resonant enhancements using only a few simple rules [300–306]. A resonant denominator term is produced after each interaction. Since there are three interactions that lead to the production of the final wave, there are three resonant denominators. Choosing one specific interaction, the particular term associated with that interaction consists of the sum of the frequencies of all the absorptions that have occurred up to that point minus the frequencies of all the emissions minus the frequency difference between the state on the left-hand side of the diagram (i.e., the bra) and the right-hand side (i.e., the ket) after the interaction being considered plus an imaginary term for the dephasing rate of the states after the interaction. Figure 30 illustrates the application of these guidelines. The concise notation ω_{mn} is used to represent the frequency difference between states m and n, $(\omega_m - \omega_n)$.

Few of the diagrams have three terms in the denominator that can become resonant simultaneously. There are many examples of antiresonant denominators where the real part cannot go to zero because of the relative sizes of the frequencies involved. For example, in c the real part of the first term can never become zero because both ω_2 and ω_{mg} must be positive numbers. One might expect that the first term in diagram h could become resonant $(\omega_1 - \omega_{gn} + i\Gamma_{gn})$ except that one must remember that ω_{gn} is negative since it represents the frequency difference between the ground state and a higher state, $(\omega_g - \omega_n)$. If we assume $\omega_1 > \omega_2$, there are four diagrams that contain a Raman resonance [i.e., one where $(\omega_1 - \omega_2)$ can match a vibrational resonance $\omega_{g'g}$ so $(\omega_1 - \omega_2 - \omega\omega_{g'g} + i\Gamma_{g'g}) \rightarrow i\Gamma_{g'g}$], $-b$, c, e, and f. These diagrams represent the CARS signals. If $\omega_1 < \omega_2$, these four diagrams become antiresonant, but diagrams t, u, w, and x then contain a Raman resonance. These diagrams represent the CSRS signal. The remaining diagrams represent two photon absorption processes and processes where the difference between the two frequencies is resonantly enhanced by the frequency difference between two excited states.

To represent $\chi^{(3)}$, all the terms represented by the diagrams must be summed to obtain a massive expression that has been given by Bloembergen et al. [302]. An important simplification of this expression occurs when it is realized that for most situations, $\Gamma_{ij} = \Gamma_{ig} + \Gamma_{gj}$ [302, 306]. The Γ_{ij} represent dephasing rates (as well as other rates that contribute to the damping) associated with intermediate states i and j and therefore determine the homogeneous linewidths of the transitions. The equality $\Gamma_{ij} = \Gamma_{ig} + \Gamma_{gj}$ simply states that the linewidth of the transition between states i and j is determined by the linewidths of the individual states i and j relative to the ground state $(\Gamma_{gj} = \Gamma_{jg})$. In this approximation, many of the terms in the equation for $\chi^{(3)}$ combine to form a simpler equation with only 12 terms [302, 307].

$$\chi^{(3)}_{\sigma\alpha\beta\gamma}(-\omega_3, \omega_1, \omega_1, -\omega_2)$$

$$= \frac{6N}{\hbar^3}\left(\frac{n^2+2}{3}\right)^4 \sum_{g',n,n',m}$$

$$
\times \left[\frac{2P^{\alpha}_{g'n}P^{\gamma}_{ng}}{(\omega_{ng} + \omega_2 - i\Gamma_{ng})(\omega_{g'g} - (\omega_1 - \omega_2) - i\Gamma_{g'g})} \right.
$$

$$
\times \left(\frac{P^{\sigma}_{gn'}P^{\beta}_{n'g'}}{(\omega_{n'g} - \omega_3 - i\Gamma_{n'g})} + \frac{P^{\beta}_{gn'}P^{\sigma}_{n'g'}}{(\omega_{n'g} + \omega_1 + i\Gamma_{n'g})} \right)
$$

$$
+ \frac{2P^{\gamma}_{g'n}P^{\alpha}_{ng}}{(\omega_{ng} - \omega_1 - i\Gamma_{ng})(\omega_{g'g} - (\omega_1 - \omega_2) - i\Gamma_{g'g})}
$$

$$
\times \left(\frac{P^{\sigma}_{gn'}P^{\beta}_{n'g'}}{\omega_{n'g} - \omega_3 - i\Gamma_{n'g}} + \frac{P^{\beta}_{gn'}P^{\sigma}_{n'g'}}{\omega_{n'g} + \omega_1 + i\Gamma_{n'g})} \right)
$$

$$
+ \frac{(P^{\alpha}_{mn}P^{\beta}_{ng} + P^{\beta}_{mn}P^{\alpha}_{ng})}{(\omega_{ng} - \omega_1 - i\Gamma_{ng})(\omega_{mg} - 2\omega_1 - i\Gamma_{mg})}
$$

$$
\times \left(\frac{P^{\sigma}_{gn}P^{\gamma}_{nm}}{\omega_{n'g} - \omega_3 - i\Gamma_{n'g}} + \frac{P^{\gamma}_{gn}P^{\sigma}_{nm}}{\omega_{n'g} - \omega_2 + i\Gamma_{n'g}} \right)
$$

$$
+ \frac{2P^{\alpha}_{g'n}P^{\gamma}_{ng}}{(\omega_{ng} - \omega_2 + i\Gamma_{ng})(\omega_{g'g} + (\omega_1 - \omega_2) + i\Gamma_{gg})}
$$

$$
\times \left(\frac{P^{\sigma}_{gn}P^{\beta}_{ng'}}{\omega_{n'g} + \omega_3 + i\Gamma_{n'g}} + \frac{P^{\beta}_{gn}P^{\sigma}_{ng'}}{\omega_{n'g} - \omega_1 - i\Gamma_{n'g}} \right)
$$

$$
+ \frac{2P^{\gamma}_{g'n}P^{\alpha}_{ng}}{(\omega_{ng} + \omega_1 + i\Gamma_{ng})(\omega_{g'g} + (\omega_1 - \omega_2) + i\Gamma_{g'g})}
$$

$$
\times \left(\frac{P^{\sigma}_{gn'}P^{\beta}_{n'g'}}{\omega_{n'g} + \omega_3 + i\Gamma_{n'g}} + \frac{P^{\beta}_{gn'}P^{\sigma}_{n'g'}}{\omega_{n'g} - \omega_1 - i\Gamma_{n'g}} \right)
$$

$$
+ \frac{(P^{\alpha}_{mn}P^{\beta}_{ng} + P^{\beta}_{mn}P^{\alpha}_{ng})}{(\omega_{ng} + \omega_1 + i\Gamma_{ng})(\omega_{mg} + 2\omega_1 + i\Gamma_{mg})}
$$

$$
\left. \times \left(\frac{P^{\sigma}_{gn'}P^{\gamma}_{n'm}}{\omega_{n'g} + \omega_3 + i\Gamma_{n'g}} + \frac{P^{\gamma}_{gn'}P^{\sigma}_{n'm}}{\omega_{n'g} + \omega_2 - i\Gamma_{n'g}} \right) \right] \tag{54}
$$

In this expression, the $\sigma\alpha\beta\gamma$ are Cartesian coordinates representing the polarizations of the four interacting fields at ω_3, ω_1, ω_1, and ω_2, N is the number density of the sample (we are assuming everything is in the ground state), and n is the index of refraction that is necessary to perform the local field corrections in the problem. The P_{ij}^{x} represent the dipole moment operator for the transition between states i and j for the x coordinate. A sum is performed over all possible states because for many situations, a particular state may not have a dominant contribution. The first four terms correspond to CARS resonances assuming $\omega_1 > \omega_2$. (When $\omega_2 > \omega_1$, these terms are antiresonant CSRS contributions.) The first and third terms come directly from diagrams c and b in Fig. 30, respectively, while the second and fourth terms are obtained by summing diagrams f, h, i together and d, e, g together, respectively. The fifth and sixth terms represent two-photon absorption. The fifth term comes directly from diagram a while the sixth term results from summing together diagrams j, k, l. The seventh through tenth terms represent CSRS resonances assuming $\omega_2 > \omega_1$. (When $\omega_1 > \omega_2$, these terms are antiresonant CARS contributions.) The seventh and ninth terms come directly from diagrams x and w, respectively, while the eighth and tenth terms are obtained by summing diagrams q, r, u together and p, s, t together, respectively. Finally the eleventh and twelfth terms represent antiresonant two-photon absorption. The eleventh term comes directly from diagram v, while the twelfth term results from summing together diagrams m, n, o.

The resonances involved in (54) can be neatly diagrammed by the energy level schemes shown in Fig. 31. These energy level diagrams are very deceptive because one would like to envisage them as a series of consecutive transfers. This desire is unreasonable as inspection of the levels for a CSRS process quickly indicates. The energy level schemes are just convenient methods to visualize the resonances involved.

The approximation that $\Gamma_{ij} = \Gamma_{ig} + \Gamma_{gj}$ does not always hold if collisional processes become very important. Under these conditions, 24 terms in the

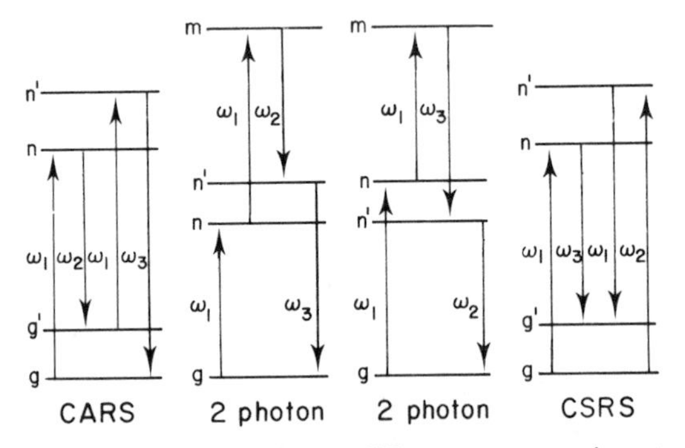

Fig. 31 Schematic diagrams that represent the different resonances important in CARS, two photon, and CSRS spectroscopy. The direction of the arrows should not be interpreted as absorption or emission. They instead reflect the states that are in resonance.

original expression for $\chi^{(3)}$ cannot be combined to make the 12 terms in (48) and new resonances are expected to appear. Diagrams *h, i, k, l, n, o, q,* and *r* in Fig. 30 all predict a resonance when $(\omega_1 - \omega_2)$ matches the difference between two excited state frequencies, even if neither state is involved in a previous interaction! These pressure induced extra resonances in four wave mixing have been called PIER 4 signals and they have been observed in sodium vapor diluted in a much higher pressure of helium [308].

CARS

Coherent anti-Stokes Raman scattering was first observed in the work of Maker and Terhune [284]. Its applications have expanded considerably in recent years [287–291, 300–324]. Anti-Stokes Raman scattering was obtained in a high-intensity, coherent, directional beam that was separated in space from the excitation beams and was easily detected [312]. The problem of fluorescence could be almost eliminated by several methods [288]. Since the anti-Stokes beam was coherent and directional, it could be isolated from any incoherent fluorescence by spatial filtering. In addition, the wavelength of the anti-Stokes beam is on the low side of the excitation wavelengths, where it is difficult to have fluorescence. The technique is also capable of easily providing high-resolution Raman spectra by using narrow-bandwidth lasers [325].

The analytical characteristics of CARS can be most easily understood by reference to (50), which gives the anti-Stokes intensity at resonance for a system where vibrations are the only contribution to $\chi^{(3)}$. It should first be noted that the intensity I_A depends quadratically on concentration. This dependence can be understood intuitively because the number of vibrational excitations created by the frequency difference of the two exciting laser beams $(\omega_L - \omega_S)$ is proportional to N, and the efficiency of scattering of a photon at ω_L off the created vibration is also proportional to N. As we see later, the dependence becomes linear at low values of N as the nonresonant part of $\chi^{(3)}$ becomes important. Similarly, the intensity depends quadratically on the high-energy laser power I_L and linearly on I_S. These dependences result because one photon at ω_L and one photon at ω_S are required to create the vibrational excitation, which then scatters with a second photon at ω_L to create the photon at ω_A. As we see earlier, caution must be exercised in using this intuitive argument too widely, however, because the generation of the coherent scatter is a parametric process in which the vibration is created and destroyed nearly simultaneously within the coherence time of the vibrational excitation [288]. The strong intensity dependences dictate that one should use intense laser sources for excitation and should focus them, although one is eventually limited by saturation of the vibrationally excited state [288] or breakdown of the material.

The important analytical variables in (50) are the path length in the material, the phase matching, and the laser frequencies and intensities. We see earlier that although $\omega_A = 2\omega_L - \omega_S$, the wave vector k_A is not necessarily equal to $2k_L - k_S$ because of the dispersion in the medium. One can write the equality expressing phase matching as $\omega_A/n_A = 2\omega_L/n_L - \omega_S/n_S$, where n_A, n_L, and n_S are the indices of refraction at the three frequencies. However, if beams are crossed at an appropriate angle (see Fig. 32), the equality can be established. From Fig. 32, one can use the law of cosines to obtain the following relationship for the phase matching angle θ between the beams at ω_L and ω_S [324]

$$\cos\theta = \frac{4n_L^2\omega_L^2 + n_S^2\omega_S^2 - n_A^2\omega_A^2}{4n_L n_S \omega_L \omega_S} \tag{55}$$

The size of θ limits the amount of spatial overlap that occurs between the beams. Although limited overlap limits the increase in the CARS signal, it simultaneously provides spatial resolution that can be quite valuable for practical applications of CARS. This angle cannot be a constant if the material under study exhibits dispersion and must be changed if a wavelength scan is performed. When the three beams are not phase matched, the intensity of the anti-Stokes beam oscillates as a function of path length in the material. This phenomenon is shown in Fig. 33 for several degrees of phase matching [288]. For a fixed path length, it is clear that the efficiency increases markedly as the optimum conditions are reached.

One of the most influential parameters is the dependence of I_A on $|\chi^{(3)}|^2$ [see (37)], which causes a quadratic concentration dependence. It also limits the sensitivity of the technique, causes an asymmetrical lineshape, and creates interactions between neighboring vibrational transitions [287–291]. These effects arise because both the real and imaginary parts of the third order susceptibility contribute to $|\chi^{(3)}|^2$. If the real and imaginary parts of

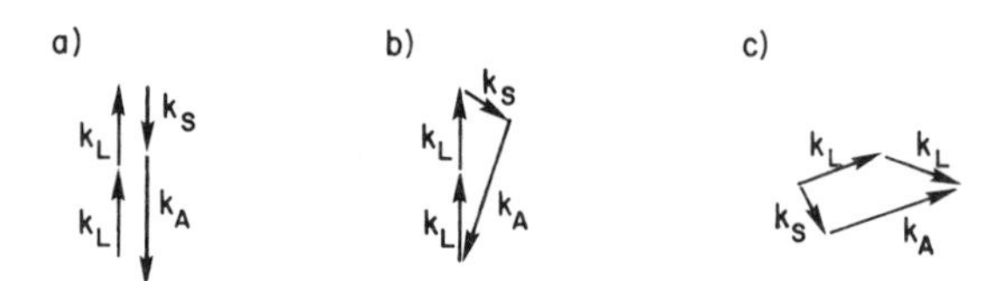

Fig. 32 Wave vectors for the three photons that interact in CARS (k_L, k_L, and k_S) to produce a fourth photon (k_A). (*a*) If the wave vectors are all collinear, the sum of the first three does not match the fourth. (*b*) For a proper angle between the beams at ω_L and ω_S, phase matching can be achieved where $2\vec{k}_L + \vec{k}_S = \vec{k}_A$. (*c*) It is also possible to split the beam at ω_L and cross three beams at different angles such that $\vec{k}_L + \vec{k}_L + \vec{k}_S = \vec{k}_A$. This configuration is called BOXCARS [302].

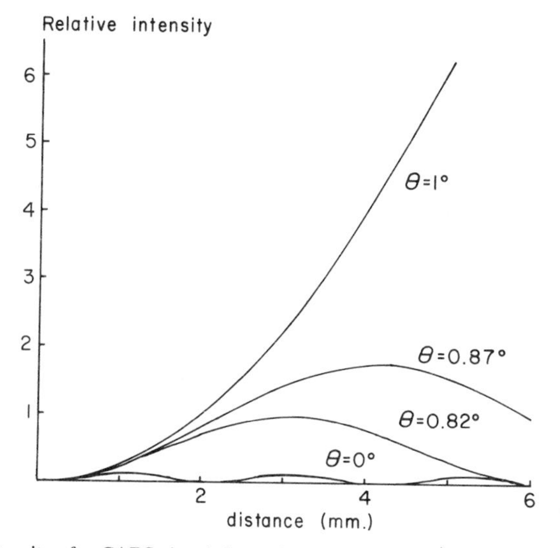

Fig. 33 The intensity of a CARS signal depends on the path length over which the four waves can interact and maintain proper phasing. If phase matching is not achieved, the CARS intensity oscillates with a period that depends on the degree of phase matching. If there is perfect phase matching, the intensity continues to grow with path length. This graph shows the increase in CARS intensity as a function of path length in a typical medium for four different crossing angles. The phase-matching angle for this illustration is $\theta = 1°$.

$\chi_5^{(3)}$ in (49) are written explicitly as χ_5^R and χ_5^I and the two photon contribution is ignored, then

$$
\begin{aligned}
|\chi^{(3)}|^2 &= (\chi_5^R + \chi_{NR} + i\chi_5^I)(\chi_5^R + \chi_{NR} - i\chi_S^I) \\
&= (\chi_5^R + \chi_{NR})^2 + (\chi_5^I)^2 \\
&= (\chi_5^R)^2 + (\chi_5^I)^2 + (\chi_{NR})^2 + 2\chi_5^R\chi_{NR}
\end{aligned} \tag{56}
$$

The frequency dependence of each term is sketched in Fig. 34 using (47) for $\chi_5^{(3)}$ and a frequency independent value for χ_{NR}. We have chosen values of $\Gamma = 2.4\ \text{cm}^{-1}$, $\omega_V = 991.5\ \text{cm}^{-1}$, and $\chi_5^I = 63.6 \times 10^{-14}\ \text{cm}^3/\text{erg}$, corresponding to toluene as a diluent for benzene [293, 326]. This selection is not rigorously correct since the measurement corresponded to the frequencies for a Raman gain experiment, but it is still suitable as an illustration. If χ_{NR} is small compared with the other contributions, the lineshape is symmetrical and a quadratic concentration dependence results. However, when χ_{NR} is comparable to or larger than the other contribution, as occurs at low concentrations, then the *frequency dependent* term in (56) is the cross term $2\chi_5^R\chi_{NR}$ [288]. This term can become negative and causes the changes in lineshape

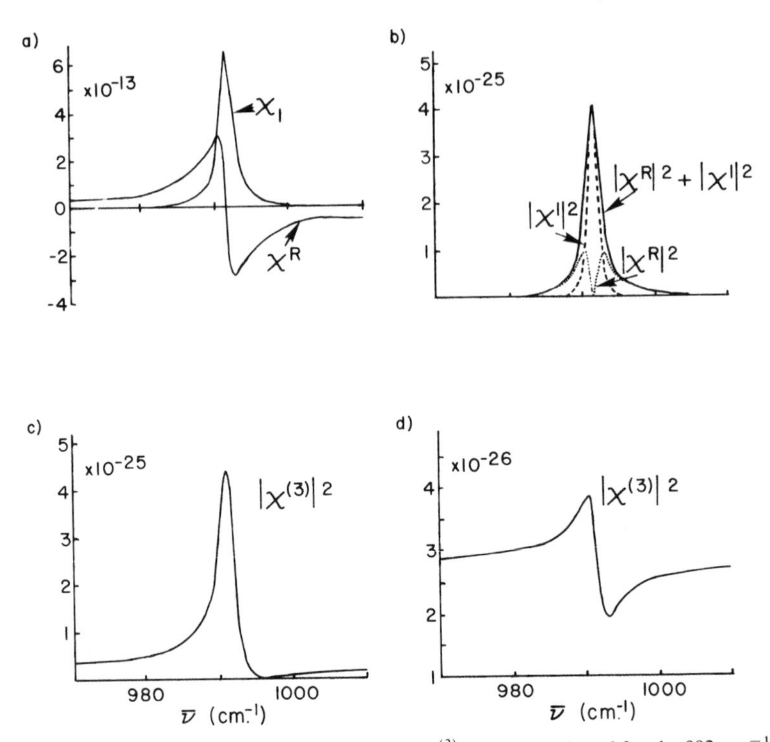

Fig. 34 (*a*) The real and imaginary components of $\chi^{(3)}$ have been plotted for the 992 cm^{-1} line of benzene using the values obtained by Owyoung and (20) for the definition of $\chi^{(3)}$. All ordinate axes are in cm^3/erg or (cm^3/erg)^2. (*b*) The $[(\chi^{\text{Real}})^2 + (\chi^{\text{Imag.}})^2]$ has been plotted, as well as the individual terms of that expression. (*c*) Equation (56) plotted for high benzene concentrations assuming a value for χ_{NR} that corresponds to toluene. At high concentrations, the cross-terms are unimportant and the lineshape is approximately equal to the expression in b. (*d*) Equation (56) plotted for low benzene concentrations in toluene. The cross-terms now predominate and the lineshape reflects the real part of $\chi^{(3)}$.

that appear in Fig. 34*d*. Note also that the concentration dependence becomes linear in this regime. If the concentration of analyte approaches zero, I_A does not go to zero but reaches a constant value determined by $|\chi_{NR}|^2$. The background is caused by all the nonresonant parts of the susceptibility and includes two photon and antiresonant contributions. It arises fundamentally because CARS is a parametric process. The process begins and ends with the vibrational and/or electronic state of the material unchanged. No energy dissipation occurs because no resonances are excited, although the polarization of the material is still driven at the difference frequency $(\omega_L - \omega_S)$. The lowest detectable concentration of an analyte therefore depends on how well one can observe changes in this background signal or how well it can be eliminated. Amplitude fluctuations of the excitation sources can be particularly severe because of the high-order intensity dependence of $I_L^2 I_S$.

In addition to the interaction between the vibrational and the nonresonant contributions to the susceptibility, there are interactions between nearby vibrational resonances because of the cross terms, and the lineshapes are changed. A more detailed discussion of the changes in lineshape appears elsewhere [288, 290].

An example of a commonly used system for performing CARS experiments is shown in Fig. 35 [288]. A high-power N_2 laser is commonly used to excite two dye lasers. This combination of lasers is relatively simple, reliable, and inexpensive, at least compared with other options. A N_2 laser with 4-mJ energies in 10-nsec pulses can excite 200-μJ energies from the two dye lasers, producing about a 20-kW peak power. One of the dye lasers is commonly fixed in wavelength while the second is scanned over the region of interest. The detector monitoring the output resonance must also be scanned simultaneously if its bandwidth is narrower than the region scanned. If ω_L is scanned, the detector must scan twice as rapidly to continue looking at $2\omega_L - \omega_S$. If ω_S is scanned, the detector must scan at the same speed. The beams are made parallel to each other and are separated by a distance that provides the correct phase matching. They are then both focused with a lens into the sam-

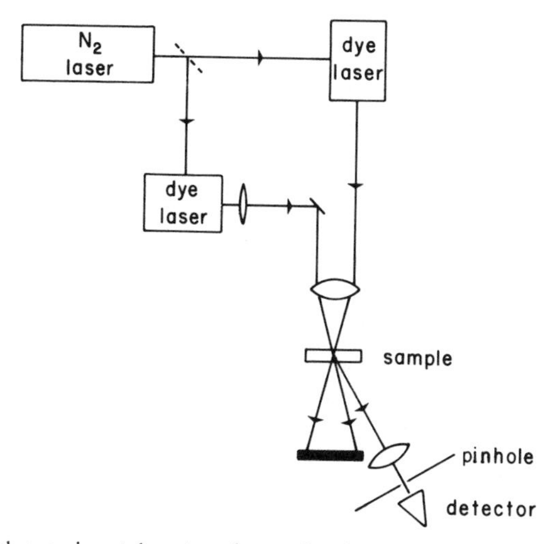

Fig. 35 A typical experimental system for performing CARS experiments. Two nitrogen pumped dye lasers are focused at an angle θ in the sample. A weak focusing lens is placed in the path of one beam to match the lasers' divergences. The anti-Stokes beam is spatially filtered and detected.

ple. The focal length of the lens is an important variable because it determines the spot size (or beam waist w_0), the path length over which the interaction occurs (the confocal parameter z_c), and the physical separation of the beams needed to achieve the correct crossing angle. The radius of the beam waist and the confocal parameter are given by (5). Typically a lens with a focal length of 15 to 25 cm is used. The angle of crossing required for phase matching is chosen according to the dispersion of the medium. For condensed phases it is typically 1 to 3 deg while for gases it is 0 deg. The beams emerge evenly spaced from the sample—ω_A, ω_L, ω_S, and ω_C.

We can make a simple estimate of the conversion efficiencies using (37) if we assume the interaction takes place in a cylindrical volume of radius w_0 and length z_c. Substituting (55) into (37), we find the following expression for the conversion efficiency [288]:

$$\frac{I_A}{I_S} = \frac{576\,\pi^4\omega_A^2\,|\chi^{(3)}|^2\,P_L^2}{c^4 n_A n_L^2 n_S \lambda^2} \tag{57}$$

where P_L is the peak power of ω_L and λ is the "average" wavelength determining the focal volume [see (5)]. If we assume $\chi^{(3)} \sim 2 \times 10^{-16}$ cm^3/erg for a gas at STP, we find a conversion efficiency of $1.5 \times 10^{-30}\,P_L^2$ or 6×10^{-8} for the 20-kW (or $\lambda \times 10^{11}$ erg/sec) lasers. If the 200-μJ beam at ω_S had 6×10^{14} photons, there would be 4×10^7 photons at ω_A. If the gas concentration dropped by 100 to 0.01 atm (the point where $\chi_{NR}^{(3)}$ begins to limit detection), the output would decrease to 4×10^3 photons at ω_A. These photons would all be contained within the CARS beam profile and thus could be manipulated easily. Nevertheless, it would be desirable to have higher signal levels. For gases, it is common to use higher power lasers, such as a doubled Nd:YAG or excimer excitation lasers, where powers of 200-kW are typical [327]. The beam quality of the Nd:YAG laser is sufficiently good that it can replace one of the two dye lasers.

The situation is different in the condensed phase, where the much higher density causes larger values for $\chi^{(3)}$. If we assume $\chi^{(3)} \sim 7 \times 10^{-13}$ cm^3/erg, then the conversion efficiency is typically $2 \times 10^{-23}\,P_L^2$. For 20-kW beams, the predicted conversion efficiency would be unrealistic because of saturation effects that are not discussed here. Thus the energies provided by the N_2 laser system are quite adequate for performing CARS in the condensed phase. If the beams are attenuated to 1 kW, the conversion efficiency becomes 0.2%. If the beam at ω_S had 3×10^{13} photons in a pulse, the output pulse would contain 6×10^{10} photons. At a dilution of 100, there are still 6×10^6 photons in the beam.

One must discriminate efficiently between the signal photons and the much stronger beams at ω_L and ω_S. Since the beam at ω_A is coherent and directional, it can be isolated from the other beams by a simple aperture assuming the divergence of the lasers is smaller than the phase matching angle and the sample does not scatter the beams. Filters, prisms, and/or monochromators can then be used to provide any additional discrimination required. These procedures are also very effective in eliminating fluorescence that is generated in the sample since it is incoherent. Begley et al. estimated a 10^9 rejection ratio for fluorescence [312]. A photomultiplier or a photodiode can be used as the detector. For more than 10^7 photons in the beam, photodiodes are adequate. This situation often occurs for condensed phase samples. Photomultipliers are used otherwise.

The alignment of the beams at ω_L and ω_S is crucial to the CARS experiment because the beams must be spatially overlapped in the very small focal spot. A pinhole can be used to ensure that both beams are aligned to pass through. It is particularly convenient to pass a portion of the excitation beam through an auxiliary pinhole that can be used as a continuous monitor of the spatial overlap. A second method utilizes a variation on the Foucault knife-edge test, in which two orthogonal knife edges define the focal spot [298]. The beams have also been aligned using glass capillaries containing the sample [316]. The strong focusing action of the small capillary makes the system sensitive to misalignments and provides a convenient sample container as well. In each method of alignment, it is important for the spatial profile of the laser to remain constant. Otherwise the variation causes fluctuation in the beam overlaps that produces larger fluctuations in the CARS output signal.

A second problem arises in the focusing because the two beams do not have the same divergences and therefore cannot be focused to the same axial position. To correct this problem a weakly focusing lens or set of lenses is introduced into the more divergent beam (usually the longer wavelength) to match divergences [298]. This procedure is important for obtaining good intensities.

Not only must the beams be well overlapped in space, they must also be in time. For short pulses, it becomes important to make provisions to ensure temporal overlap of the beams. In addition it is important for the temporal profile of the pulses to remain stable. Random fluctuations in the pulse envelope become magnified in the CARS output and can cause excessive noise. Nd:YAG pumped dye laser systems can generate large amounts of noise because of mode beating.

The noise in a CARS signal is an important problem because the output power depends quadratically on the intensity I_L and linearly on I_S. Variations in the pulse-to-pulse energy, spatial profile, temporal profile, or wavelength can cause fluctuations in I_A as large as 50 to 1000% if one does not take appropriate care. Even after these variations are reduced, it is desirable to establish a reference channel to measure and correct the signal for fluctuation. A slab of glass can serve as a convenient reference because it has a wide and relatively flat region where a CARS signal is produced by $\chi_{NR}^{(3)}$. A portion of the beams can be split off and used for the reference leg or the ω_L and ω_S beams can be refocused into the reference material after they emerge from the sample [327]. The CARS signal from the sample is separated from the other two beams before they are refocused. An elegant method developed by Oudar et al. uses a polarization technique to obtain two beams from the CARS signal—one of which is proportional to $\chi_{NR}^{(3)}$ [328]. This method uses the $\chi_{NR}^{(3)}$ of the sample itself as the reference channel. Further details of this technique are given in our discussion on methods of suppressing the contribution from the nonresonant susceptibility.

CARS SPECTRA AND APPLICATIONS

The spectra that result when ω_S is scanned relative to a fixed ω_L resemble conventional Raman spectra except for differences in relative intensity [CARS scales as $(d\alpha/dX)^4$ instead of $(d\alpha/dX)^2$], selection rules, and interferences between neighboring resonances. In addition, the relative intensity of lines depends on concentration for the same reason that the concentration dependence is quadratic at high concentrations but linear at low concentrations [288]. The instrumental resolution is usually better and can be much better than a conventional Raman spectrum because of the narrow bandwidths characteristic of the lasers used. The same qualitative structural information is contained in a CARS spectrum as in a conventional Raman experiment.

The limitation on effectively using the high Raman signal levels of CARS for performing trace analysis is that a background signal is also present. The background signal caused by the nonresonant susceptibility of the solvent scales with intensity in the same manner as the resonant component of interest. There are always small changes in the background at the CARS detection limit. The nonresonant background susceptibility is typically 10 to 100 times smaller than the peak value for the resonant susceptibility. The background intensity is thus about 100 to 10^4 times smaller than that of the resonant intensity. The ratio $|\chi_{NR}|/|\chi_R|$ of benzene, for example, is 0.038 [328]. One would typically expect the resonant signal to equal the nonresonant signal at a 4 mole % benzene concentration. Smaller concentrations can be measured since it is the changes in the background that determine the detection limit. Begley et al. were able to observe 0.3 wt % of benzene in toluene, but they did not quote any detection limit [303]. There also have been CARS experiments performed in aqueous solution, where one can typically detect 0.1 mole % of an analyte or about $0.05M$ concentrations [305, 307].

CARS has proven to be especially valuable for gas phase studies involving flames, plasmas, and molecular beams as well as static gas systems because of the high resolution, background rejection, and spatial sampling capabilities [288, 291, 300, 320–324, 327, 329–333]. Again, the minimum detectable limit of an analyte is determined by the nonresonant susceptibility of the background gas. (When there is no background gas, one is able to perform extremely low concentration measurements.) Hydrogen, for example, has been determined at a 10 ppm concentration in nitrogen [320]. Other CARS experiments have mapped the spatial distribution of H_2 in a natural gas flame [320]. In addition, the temperature profile in a flame was determined by measuring the relative intensity of the rotational lines of different gases, including H_2, O_2, N_2, and CO [288, 323].

The phase matching condition in gas samples can be met by using colinear beams. If a species distribution and/or temperature profiling is to be performed, the exciting beams can be simply moved to different portions of the system. The profiling can then provide two dimensional information about the system since the beams sample molecules all along their path. Eckbreth showed that phase matching can also be achieved by using a noncolinear alignment of the beams that has been called BOXCARS because of the box-like arrangement of the wave vectors [311]. This geometry is sketched in Fig. 32*c*. Only the limited region of overlap contributes to the CARS signal, so a third dimension is obtained in the CARS profiling. The signal levels decrease because there is a more limited region of interaction, but these decreases are made up by the detection sensitivity available with photomultipliers.

Multiplex CARS has the capability for rapidly obtaining a complete Raman spectrum [327, 334]. When the laser at ω_S is substituted by a pulsed source with a broad continuum of wavelengths, the anti-Stokes scatter contains all wavelength components simultaneously and can be spectrally analyzed with a monochromator using an imaging detector. A complete CARS spectrum can be obtained on a single laser firing with a time resolution determined by the pulsed source. Many applications can be envisaged for such a technique where it is necessary to follow the temporal evolution of a system.

Multiple resonances can also be examined by angularly resolved coherent Raman spectroscopy (ARCS) [330]. In this technique, a laser at ω_L is oppositely directed to a broad-band laser beam containing a spread of frequencies ω_S. A third beam at ω_L is crossed with these two beams and a fourth beam is generated at ω_A that propagates away from the interaction region on the opposite side from the third beam. Its angle relative to the first two beams depends on the phase matching condition. Thus different values for ω_A leave the interaction region at different angles and are therefore angularly resolved. A spectrum can be obtained by scanning the detector across the different angular positions.

A fundamental parameter of interest for CARS studies are the T_1 and T_2

vibrational relaxation times. The lifetime of a vibrational level is determined by T_1, while the lifetime of the vibrational coherence or phase is determined by T_2. T_1 can be measured by incoherent anti-Stokes Raman scattering and T_2 can be measured by coherent anti-Stokes Raman scattering [293]. For the CARS application, T_2 is the relevant parameter because the anti-Stokes scattering off the vibrational excitation must be coherent with the vibrational excitation. A vibrational excitation is created by irradiating with short laser pulses at ω_L and ω_S such that $(\omega_L - \omega_S) = \omega_V$. The rate of the vibrational phase relaxation is measured by generating a coherent anti-Stokes Raman signal with a third short laser pulse of ω_L that is delayed for adjustable times and inclined at a slight angle relative to the first pulse [310, 320]. The anti-Stokes signal from the third pulse can be distinguished from that created by the previous two pulses by its different angle. If the pulses are very short compared with T_2 (picosecond pulses are usually required for liquid studies), the anti-Stokes intensity decays exponentially with the delay time. A similar experiment can be used for measuring T_1 if the nondirectional, incoherent anti-Stokes Raman scatter generated by the second pulse is measured instead as a function of the delay time. Such methods have been used to measure T_1 and T_2 in a number of different materials [293, 310, 317].

RESONANCE CARS

Large enhancements of the CARS signals occur by using laser frequencies that are near to or match electronic resonances in the molecules of interest [298, 314, 316, 335–340]. The resonant enhancements obtained by working near electronic transitions have been known and used for many years in conventional Raman spectroscopy [282, 341]. Chabay et al. [298] and Hudson et al. [314] demonstrated significant increases in CARS signals for diphenyloctatetraene in benzene in the pre-resonance region near an electronic absorption. They estimated a lower detection limit of 5×10^{-5} *M*. Carreira et al. [316] studied the resonantly enhanced CARS signals of β-carotene in the resonance region and quoted a lower limit of 5×10^{-7} *M*. This represents an improvement of about 5 orders of magnitude over nonresonantly enhanced CARS limits. It was found, however, that the laser powers had to be sharply attenuated in the resonance regime in order to avoid saturating the system [316]. Both the intensities and shapes of the resonances change when the system enters saturation. The lower laser powers cause a large decrease in the CARS signal and require photomultiplier detection, but the absolute signal level in CARS is not a problem.

The lineshapes that one obtains in resonance CARS can change quite drastically because of the interference effects between the vibrational and electronic resonances [335–341]. Equation (54) contains the information that describes the lineshapes of resonance CARS. If one assumes that only one electronic state of ω_{eg} contributes to the resonant enhancement, the lineshape can be described by

$$\left| \frac{1}{(\omega_{eg} - \omega_1 - i\Gamma_{eg})\,[\omega_{g'g} - (\omega_1 - \omega_2) - i\Gamma_{g'g}]\,(\omega_{eg} - \omega_3 - i\Gamma_{eg})} \right|^2 \quad (58)$$

if the nonresonant part of the susceptibility can be neglected. If it cannot, it must also be included in the description as seen in (56). The lineshape can change from positive Lorentzians through dispersive shapes to negative Lorentzian as the frequency ω_1 is changed from the high-energy side of ω_e to the low-energy side [335–341].

The population of other electronic states is an important aspect of resonance CARS. Not only do the intensities, lineshapes, and linewidths change, additional resonances appear because of resonance CARS originating from the populated excited states [342–344]. Resonant enhancement occurs because of the participation from the higher electronic states, which become more densely spaced above the first excited state. The excited-state CARS spectra can therefore provide information about the excited-state CARS vibrational modes as well as the ground state.

Several other problems arise in resonance CARS. The high intensities within the focal region can cause heating and photodegradation. Additionally, because the CARS output is to the higher energy side of the excitation beams, it falls within the absorption profile of the sample and becomes attenuated [314]. This problem limits the upper end of the dynamic range of resonance CARS.

SITE SELECTIVE NONLINEAR SPECTROSCOPIES

One of the very exciting aspects of nonlinear spectroscopies, such as CARS, CSRS, and two photon processes, is the realization that site-selective, Doppler-free, and high-resolution spectroscopy can be performed [305, 306, 345]. The present laser methods that permit working within the inhomogeneous line profile all rely on sample emission or fluorescence. There are a majority of analytically important situations where the samples do not emit or their emission is subjected to quenching influences. It appears now that there is an entire family of methods using nonlinear spectroscopy that permit working within the inhomogeneous line profile as do the site-selective, Doppler-free, and high-resolution laser spectroscopies now known.

The high-resolution measurements that work within the inhomogeneous line are made by exciting the subset of the ensemble of spectroscopically active species that are resonant with an exciting laser. The excitation labels the specific subset. If only that subset can emit, the fluorescence spectrum can be simplified or narrowed greatly. The fluorescence spectrum contains the information about the lower lying energy levels of the labeled subset. (There are ways to read out that information other than fluorescence and these form the basis of other high-resolution methods.) Roles can be reversed by monitoring the fluorescence of a subset of the ensemble while scanning a laser over the excitation region. The resulting excitation spectrum contains the information about the upper levels of the subset selected by the fluorescence monitored. In either case, one wavelength is used to label the subset while the second wavelength reads the information about the labeled subset.

Let us consider the traditional CARS diagram sketched in Fig. 36. The dotted lines represent virtual levels in an arbitrary material. To perform site selective CARS spectroscopy, it is only necessary to realize that ω_a, ω_b, or ω_c can label a subset of the entire ensemble that contributes to the inhomogeneous envelope. Scanning a second laser frequency can provide the information about the subset selected by the first laser frequency. The third laser frequency can be used to provide additional labeling specificity and information, or it can be nonresonant. For example, if ω_a is tuned to a specific wavelength within the inhomogeneous envelope of level n and ω_b is scanned, the specific subset selected by ω_a can resonantly enhance $\chi^{(3)}$ twice when ω_b reaches resonance with level n'. Of course the other parts of the ensemble also contribute a resonant enhancement when ω_b matches the corresponding level n' in those sections of the ensemble, but this situation results only in a single resonant enhancement. One would see the envelope for the inhomogeneous width with a sharp line at the position of the subset selected by ω_a. Additional enhancement of that sharp line could be provided by tuning ω_c to level g' for the subset desired. There are many variations of this simple idea that could provide information about all the levels. The situation for CARS is described, but the same ideas hold for all the other possible nonlinear spectroscopies, such as two photon, CSRS, Raman gain, and Raman loss spectroscopies. One can scan ω_b and ω_c together to maintain resonance with state n' for a particular subset. This procedure would pick out level g' for the subset and provide narrowed information about the lower

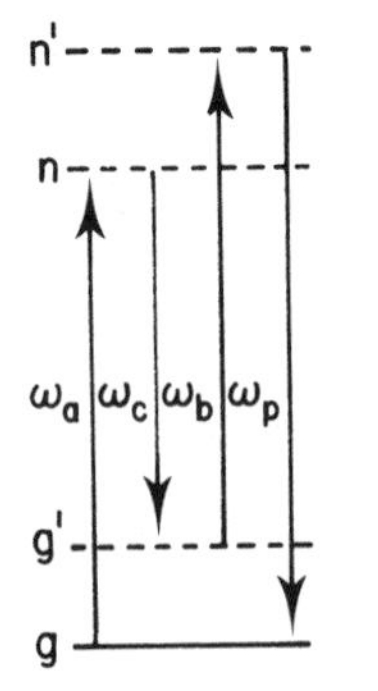

Fig. 36 The CARS energy level diagram for virtual levels. The letters refer to possible states that could be present.

levels. There are in fact a large number of possible options for the scanning strategy.

An underlying assumption of these ideas is that the selection or labeling of the subset is not lost before the information about that subset can be obtained. This assumption is identical to that required for traditional high-resolution laser spectroscopy.

The ideas presented for site selective and high resolution nonlinear spectroscopies are somewhat simplified and naive. Consider just the third term in (54)

$$\frac{P^{\sigma}_{gn'} P^{\gamma}_{n'g'} P^{\alpha}_{g'n} P^{\beta}_{ng}}{(\omega_{ng} - \omega_1 - i\Gamma_{ng})(\omega_{g'g} - (\omega_1 - \omega_2) - i\Gamma_{g'g})(\omega_{n'g} - \omega_3 - i\Gamma_{n'g})} \tag{59}$$

This term is resonant when $\omega_1 = \omega_{ng}$ and $(\omega_1 - \omega_2) = \omega_{g'g}$ (we ignore the third resonance for the time being). If we had an ensemble of condensed-phase molecules or ions in different conformations and environments or an ensemble of gas-phase molecules or atoms moving with different velocities, each would have a different energy ω_{ng} or $\omega_{g'g}$ that together form the inhomogeneous envelope. To describe this situation let us say there are a central frequency and a shifted frequency that depend on either the distortion or the strain from the mean and the strength of the interaction between that distortion and the energy levels if the sample is in a condensed phase or the velocity of the molecules or atoms if the sample is in a gas phase. Then the energy of an arbitrary level can be written as $(\omega^0_{ng} - T_{ng}s)$, where ω^0_{ng} is the mean frequency of the transition. For condensed phase situations, s represents the distortion or strain from the mean, while T_{ng} represents the strength of the interaction between that distortion and state n. For gas phase situations, s represents the velocity v, while T_{ng} represents ω^0_{ng}/c. Although this description is simplified for the condensed phase situation, it nevertheless contains the essential features.

For an individual subset of the ensemble then, our term becomes

$$\frac{P^{\sigma}_{gn'} P^{\gamma}_{n'g'} P^{\alpha}_{g'n} P^{\beta}_{ng}}{\begin{array}{l}(\omega^0_{ng} - T_{ng}s - \omega_1 - i\Gamma_{ng})(\omega^0_{g'g} - T_{g'g}s \\ \quad - (\omega_1 - \omega_2) - i\Gamma_{g'g})(\omega^0_{n'g} - T_{n'g}s - \omega_3 - i\Gamma_{n'g})\end{array}} \tag{60}$$

If we tune ω_1 to resonance with a subset of molecules such that $\omega_1 = \omega^0_{ng} - T_{ng}s$, the term becomes large depending on the size of ω_{ng}. It becomes larger when ω_2 is tuned so that $\omega_1 - \omega_2 = \omega^0_{g'g} - T_{g'g}s$ if s is the same s as the first resonance (i.e., the same labeled subset selected by ω_1). Other parts of the ensemble with different values of s do not have both resonances go to zero simultaneously and the contributions from these parts are smaller. The crucial question is how much smaller? The second resonance does not necessarily produce a dominant contribution from the selected subset [305]. To see this, we must perform the ensemble average of our term over all possible values of s. We assume a Gaussian distribution for s given by $(1/\sqrt{\pi}\sigma)e^{-s^2/\sigma^2}$, where σ represents either the distribution of strains or the mean velocity. Then we must calculate

$$\int_{-\infty}^{\infty} \frac{e^{-s^2/\sigma^2}\,ds}{(\omega_{ng}^0 - T_{ng}s - \omega_1 - i\Gamma_{ng})(\omega_{g'g}^0 - T_{g'g}s - (\omega_1 - \omega_2) - i\Gamma_{g'g})(\omega_{n'g}^0 - T_{n'g}s - \omega_3 - i\Gamma_{n'g})} \tag{61}$$

This expression is evaluated in several references [304, 305]. It can be broken into two complex error functions. If one sets ω_1 at resonance with $\omega_{ng} - T_{ng}s$ and plots $\chi^{(3)}$ as a function of $(\omega_1 - \omega_2)$, one finds that when $T_{g'g}$ and T_{ng} have the same sign the function has a width given predominantly by inhomogeneous broadening (this result assumes the inhomogeneous width is much larger than the homogeneous width, i.e., $\sigma \gg \Gamma$). There is no narrowing or site selection. The subset selected by ω_1 does not provide a dominant contribution to $\chi^{(3)}$ relative to the nonresonant subsets. For gas-phase samples, $T_{g'g}$ and T_{ng} must have the same sign because all levels shift in the same direction for a given velocity. For condensed-phase samples though, $T_{g'g}$ and T_{ng} can have different signs. For a given distortion s, level g' could increase in energy while level n could decrease in energy. The ensemble average would then produce a very different effect because in this case the selected subset does contribute a dominant effect. There is a sharp line with the homogeneous width at the position in the inhomogeneous profile selected by ω_1. The sharp line rides on the inhomogeneously broadened line from the remaining molecules in the ensemble. The intensity difference between the sharp line and broad line scales roughly as the ratio of the inhomogeneous to homogeneous widths. This dramatic difference in behavior can perhaps be understood best as arising from the relative phases of the contributions from the different subsets adding constructively or destructively [305]. It can be tracked to the relative signs of the term being integrated—Ts—and the imaginary term $i\Gamma$ for each of the factors in the denominator that can be resonant. If the integrand has poles only in the same half of the complex plane, one does not get a dominant contribution from the selected subset and there is no narrowing. The relative signs for Ts and $i\Gamma$ must differ among the factors in the denominator that are resonant [305, 306]. If Ts and $i\Gamma$ have the same sign in one resonant denominator, they must have a different sign for another resonant denominator. This situation corresponds to having poles in different halves of the complex plane and one does get a dominant contribution from the selected subset and there is narrowing.

With this knowledge, we can look at the other terms in (54) and determine the conditions under which narrowing is possible. Consider term 8, which causes CSRS resonances. This term becomes resonant when $\omega_1 = \omega_{n'g}$, $\omega_2 = \omega_{ng}$, or $(\omega_2 - \omega_1) = \omega_{g'g}$. Note that the sign of $i\Gamma_{n'g}$ is different from that of $\omega_{n'g}$ in the last factor of this term, whereas $i\Gamma_{ng}$ and $i\Gamma_{g'g}$ have the same signs as ω_{ng} and $\omega_{g'g}$ in the first two factors, respectively. For CSRS then, one obtains narrowing if $T_{n'g}$ and T_{ng} or $T_{g'g}$ have the same signs. Thus Doppler-free spectroscopy is possible for gas phase samples if one uses CSRS instead of CARS. One can therefore perform high-resolution nonlinear spectroscopy regardless of the relative signs of the T parameters, but one must select the type of spectroscopy judiciously.

The two photon processes given by the fifth and sixth terms behave in a similar manner. The fifth term contains three factors that all have different signs between ω_{jg} and $i\Gamma_{jg}$ and thus the T_{jg} parameters must differ in sign to obtain narrowing. The sixth term, however, contains a factor where $\omega_{n'g}$ and $i\Gamma_{n'g}$ have the same signs, but the other factors have different signs. Thus the $T_{n'g}$ parameter can have the same sign as the T parameters of the other two factors and still result in narrowing.

These ideas were demonstrated in a beautiful experiment by Decola et al. [346], who studied four wave mixing in benzoic acid crystals doped with 5×10^{-7} mole/mole of pentacene. The measurements were performed at 1.6°K. These workers observed narrowing in the CSRS spectra down to the homogeneous linewidth of an excited-state Raman transition. They also observed narrowing of the electronic resonances in both CARS and CSRS by a factor of almost 100. The CARS spectrum is shown in Fig. 37 along with absorption and fluorescence spectra that serve to identify the transitions. The resonances involved for each transition are also sketched in this figure. The last resonance comes from the benzoic acid crystal. One laser is tuned to the $0 \rightarrow 0$ electronic transition (ω_1), while the lower energy laser (ω_2) is scanned over the resonance region ($\omega_1 - \omega_2$). Resonances appear whenever ($\omega_1 - \omega_2$) matches a ground-state vibrational energy or an excited-state vibrational energy. They appear as three sets of pairs because a given vibrational mode has different energies in the ground and excited states. The transitions in which the vibrational mode is unchanged are favored for pentacene because pentacene has only small shifts in configuration space for the upper electronic state. The Franck-Condon factors that are part of the electric dipole matrix elements in $\chi^{(3)}$ are small if there are changes in the vibrational mode in the transition.

NONRESONANT BACKGROUND SUPPRESSION

The nonresonant background contribution from the solvent is a severe problem that prevents the application of CARS methods at low concentrations. Several methods have been developed to suppress the background.

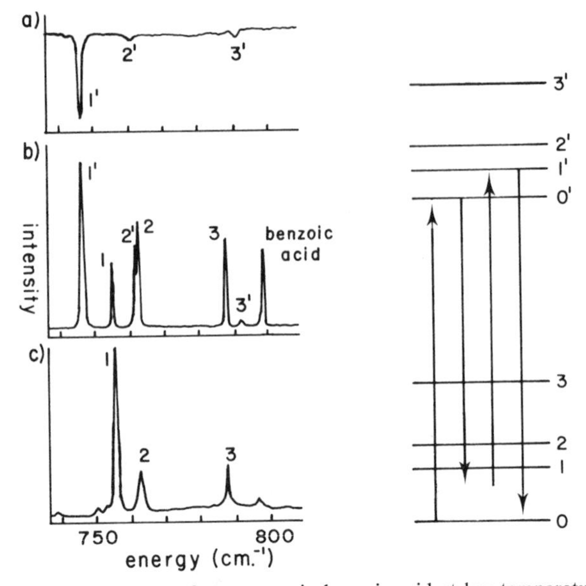

Fig. 37 (*a*) Absorption spectrum of pentacene in benzoic acid at low temperature. The *x*-axis has been adjusted to coincide with the scan in *b*. The numbered lines correspond to absorptions from state 0 to states 1′, 2′, and 3′ in the right-hand diagram. (*b*) CARS spectrum of pentacene in benzoic acid as the lowest energy laser of the two dye lasers is scanned. The numbers refer to which vibrational state is in resonance. The diagram shows the resonances for line *1*′. The highest energy laser remains fixed at the 0→0′ transition. The last line in the spectrum is a benzoic acid resonance line. (*c*) Fluorescence spectrum of pentacene in benzoic acid. The *x*-axis has been adjusted to match that in *b*. The lines are labeled according to the particular fluorescence transition from state 0′ to state 1, 2, or 3.

One of the most successful of these uses the differences in polarization characteristics of $\chi_{NR}^{(3)}$ and $\chi_{R}^{(3)}$ to eliminate the background [306, 328, 347, 348]. The development of the formalism for $\chi^{(3)}$ in this chapter neglects the tensorial nature of $\chi^{(3)}$. In fact, $\chi^{(3)}$ depends on the polarization of all four waves that are involved in the interaction. The information in the $\chi^{(3)}$ tensor is extensive and provides details about molecular symmetry that can be important for analytical purposes. We only need to realize that $\chi_{NR}^{(3)}$ and $\chi_{R}^{(3)}$ arise from different mechanisms and therefore have different tensorial characteristics. Polarization methods can therefore be used to separate the contributions.

Oudar et al. demonstrated this rejection capability in dilute benzene samples [328]. Two lasers are used with their polarizations oriented at an angle to each other. The polarization of the CARS signal generated by the nonresonant part is rotated relative to the two lasers by an amount that depends on the relative size of the tensorial components of $\chi_{NR}^{(3)}$ associated with the parallel and the perpendicular components of the exciting polarizations. By aligning a polarization analyzer perpendicular to the nonresonant CARS polarization, the nonresonant background can be suppressed. Reduction of the background intensity by 10^4 is possible with this method. Part of the simplicity of the method arises because $\chi_{NR}^{(3)}$ is usually real, so the output CARS signal remains linearly polarized. This is not the situation though if two photon processes contribute to $\chi_{NR}^{(3)}$.

The CARS signal generated by $\chi_R^{(3)}$, on the other hand, has tensorial characteristics different from those of $\chi_{NR}^{(3)}$ and therefore the polarization is rotated by a different angle. In particular, $\chi_R^{(3)}$ has both real and imaginary components so the polarization is not only rotated, but it also becomes elliptical. A component of the signal passes through the polarizer that is blocking the nonresonant contribution. There is a loss of signal because the part that is blocked by the polarizer is lost, but the contrast ratio of the signal to the background can be improved significantly. Benzene concentrations of 200 ppm in CCl_4 were detected by this method.

If a Glan–Thomson prism is used to analyze the CARS beam, the transmitted beam depends on just $\chi_R^{(3)}$, while the reflected beam depends on both $\chi_R^{(3)}$ and $\chi_{NR}^{(3)}$. If $\chi_{NR}^{(3)} \gg \chi_R^{(3)}$, the reflected beam intensity can be used as reference to correct the measured intensity of the transmitted beam for fluctuations of laser intensity and spatial and temporal overlap [328]. This method is particularly convenient because the sample under study serves as its own standard.

A clever way of suppressing the nonresonant background contribution from the solvent has been described by Lynch et al. [315] for measuring the scatter from cyclohexane in benzene. They used three laser frequencies at ω_0, ω_1, and ω_2 and generated a CARS beam at $\omega_0 + \omega_1 - \omega_2$ (if $\omega_1 = \omega_0$, this experiment reverts to a normal CARS experiment). They tuned ω_0 so that $\omega_0 - \omega_2$ corresponded to a negative value for the real part of the $\chi^{(3)}$ for benzene (see Fig. 34*a*). The magnitude of the negataive value could be adjusted by tuning $\omega_0 - \omega_2$ until it canceled the real, positive nonresonant part of $\chi^{(3)}$. If $\omega_0 - \omega_2$ was then fixed and $\omega_1 - \omega_2$ was varied, the CARS spectrum of cyclohexane could be obtained with the background contribution reduced by an order of magnitude because of the cancellation of the nonresonant susceptibility. This method would not be practical in its present form for most analytical applications because of the number of adjustments, but it remains a clever idea that could contribute to a solution.

The last method of suppressing the nonresonant background is to use time resolution techniques to distinguish the resonant and nonresonant parts [349, 350]. The nonresonant contribution dies away much more quickly than the resonant contribution, which relaxes with a characteristic time T_2. For condensed phase samples, the dephasing time T_2 is between a few tenths to several picoseconds, but for gas phase samples it is several hundred picoseconds depending on the temperature and pressure of the gas. Picosecond pulse width lasers are used to provide pulses at ω_L and ω_S, which together generate vibrational excitation in the medium. A third pulse, which has been split from the laser at ω_L and delayed a time τ_D, is then brought through the sample at a different angle. If τ_D is within the dephasing time T_2 of the sample, a CARS signal is generated because of the interaction of the third pulse with the vibrational population created by the first two pulses. The CARS signal generated by the first two pulses can be discriminated against because it leaves the sample at a different angle. Kanga and Sceats achieved a reduction of 100 in the nonresonant background by using a delay of 20 psec corresponding to the T_2 of the CS_2 sample [350]. The reduction was limited by trailing pulses that occurred after the major pulse. This suppression method would be most applicable for gas-phase samples where the dephasing times are much larger and the requirements on the lasers are not as stringent.

Raman Gain Spectroscopy or Stimulated Raman Scattering

Earlier we saw that if the frequency difference $(\omega_L - \omega_S)$ between the two waves propagating in a medium matched a vibrational frequency ω_V, that vibration could be driven to create an excited vibrational population. The creation of one vibrational quantum at ω_V causes the destruction of a photon at ω_L and the creation of an additional photon at ω_S. Raman gain spectroscopy exploits the increase of I_S when a vibrational resonance is matched. The important experimental problem that must be solved is how does one distinguish between the photons at ω_S from the excitation source and that created by scattering from photons at ω_L. An elegant solution to this problem developed by Owyoung is to modulate the excitation at ω_L and detect the induced modulation at ω_S [351]. In this way, one measures only the contribution from the Raman gain. The gain at the center of a vibrational resonance is given by the exponent in (53) or in a more general form by the exponent in (41). The first thing to notice about these expressions is that the Raman gain depends only on the imaginary component of $\chi^{(3)}$. Since the nonresonant part of $\chi^{(3)}$ is real, it can make no contribution to the gain. Consequently the background that limits the sensitivity of CARS is absent in a Raman gain measurement. Only the imaginary component is important for stimulated Raman because a real vibrational excitation must be created for scattering to occur from ω_L to ω_S, and therefore the energy loss described by the imaginary component controls the process [292]. Nonresonant processes do not create a real vibrational excitation.

Several additional characteristics are evident from (41) and (53). The magnitude of the Raman gain depends linearly on both excitation intensities, I_L and I_S. This linear relationship is maintained until other nonlinearities such as thermal blooming, become important. There is also a linear concentration dependence. No phase matching is required for the technique so that one does not have to worry about crossing angles. Additionally, because the scattering at ω_S is coherent and in the same direction as the excitation source at ω_S, one may use spatial filtering to reject the incoherent fluorescence that might be generated from the sample. Finally, the technique has a resolution that is determined by the bandwidth of the excitation lasers and thus is inherently a high-resolution technique.

Historically, stimulated Raman scattering or Raman gain spectroscopy has not been considered for analytical applications [293, 352–354]. Earlier experiments used a single laser source that generated incoherent Raman scattering. Eventually at high enough laser powers, enough photons were generated at ω_S by spontaneous Raman scattering to drive vibrations and develop stimulated Raman scattering. The expected Raman gain under these conditions is typically 10^{-2} $cm^{-1}/(MW)$ (cm^2) so that to generate 10^{10} Raman photons with one laser over a 1-cm path length would require a laser intensity of 2.3×10^9 W/cm^2 [285]. In practice, stimulated Raman scattering was observed in liquids at much lower power values because of self-focusing of the laser [284, 285]. Not all vibrational modes participate in stimulated Raman scattering either, because once the strongest or highest gain mode crosses the threshold, it depopulates the excitation beam, preventing any lower gain modes from reaching threshold [293]. The generation of stimulated Raman scattering can be very efficient and has been successfully implemented as a coherent, tunable infrared source [285, 353]. The high laser powers and the participation of only the highest gain modes has prevented the single laser approach from making an analytical contribution.

ANALYTICAL MEASUREMENTS USING RAMAN GAIN SPECTROSCOPY

The key to performing a sensitive measurement by Raman gain spectroscopy is to have a powerful laser at ω_L and a low-noise laser at ω_S. The laser system used by Owyoung is sketched in Fig. 38 [355]. A 100-mW c.w. dye laser (tunable from about 580 to 630 nm) was used for the beam at ω_L and a single mode, 10-mW He–Ne laser (632.8 nm) was used for the beam at ω_S. The dye laser output was electro-optically modulated at 25 kHz. It is very important to have excellent colinearity of the two excitation beams as they pass through the sample. Telescopes were therefore included in each beam to ensure that the beam divergences were matched at the dichroic mirror that combined the two beams. The weak transmission through the dichroic mirror was used to align the two beams properly and a pinhole was used to test for good collinearity. The stronger transmission through the dichroic mirror was tightly focused into the sample cell. The beam at 632.8 nm was isolated from the dye laser beam by a diffraction grating and was detected with a

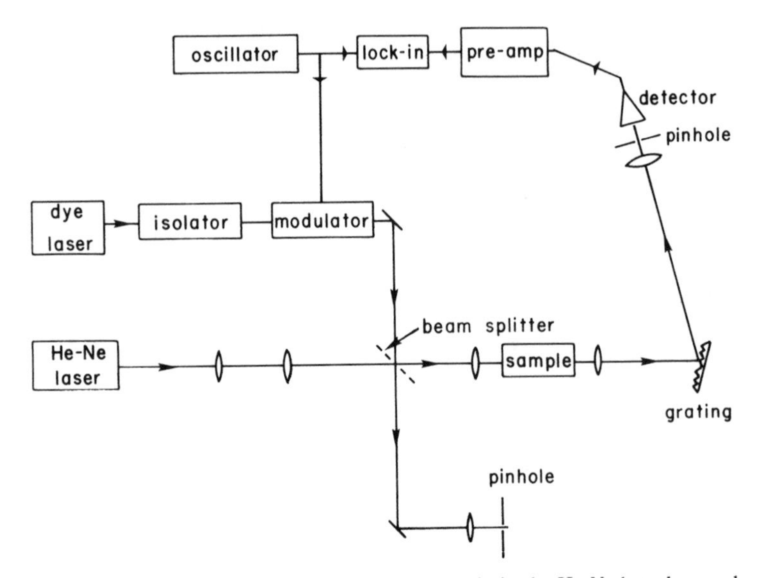

Fig. 38 Experimental system for measuring the Raman gain in the He-Ne laser beam when the frequency difference between the dye laser and He-Ne laser matches a vibrational resonance.

photodiode and a lock-in detector tuned to the modulation frequency. Because the detection limit is determined by the ability to detect small amplitude modulations at 632.8 nm, it is very important to prevent any scattering of the He–Ne laser from passing through the modulator, being reflected back through the modulator by the mirrors of the dye laser or other components, and being detected by the photodiode. To eliminate this problem, an optical isolator is placed in the dye laser optical path. Either a dispersing prism in conjunction with a pinhole spatial filter or a combination of a quarter wave Fresnal rhomb, Glan polarizer, and a pinhole spatial filter have been used [292, 355].

Raman spectra are recorded by scanning the dye laser while monitoring the He–Ne laser intensity at the modulation frequency. The spectra are very similar to those obtained by conventional Raman spectroscopy because of the direct proportionality between the Raman gain and Im $\chi^{(3)}$ as long as the gain is small. Spectra have been obtained for CCl_4, benzene, deuterated benzene, benzene in toluene, and 10% chlorobenzene in benzene [292, 351, 355]. The Raman gain for the chlorobenzene was 1.3×10^{-5} for the 992 cm^{-1} mode and a 40-mW dye laser power. The resolution was determined by the dye laser and was typically 0.5 cm^{-1}. A spectrum of 0.2 *M* benzene in CCl_4 was recorded with a *S/N* of about 25 at a time constant of 3 sec. The signal strengths observed were within a factor of 2 of those predicted theoretically.

The theoretical expression for the Raman gain expressed by (41) should be modified to include the focusing of the lasers into the sample. In this expression I_L represents the laser intensity in the medium. If we want to express I_L in terms of the laser power P_L before focusing and the focusing characteristics of the optics, an expression can be derived for the total Raman gain G, assuming a Gaussian beam profile, low gain, and focusing within the sample medium [292, 351, 355]

$$G = \frac{96\pi^4 P_L}{\lambda_L \lambda_S c n_S} \operatorname{Im} \chi_8^{(3)} \tag{62}$$

This expression differs by 4 from that of Owyoung because of the difference in the definitions of susceptibility [292].

The background noise that limits the lowest detectable concentrations arises primarily from three sources—fluctuations in the Raman gain itself, flucutations in the beam intensity at ω_S, and shot noise caused by the random nature of photons at ω_S (it is assumed that I_S is sufficiently high that electronic noise can be eliminated). Thus (63) expresses the signal-to-noise ratio expected for the technique [355]

$$\frac{S}{N} = \frac{G}{[(\Delta_L G)^2 + \Delta_S^2 + (2h\omega_S B/\eta P_S)]^{1/2}} \tag{63}$$

where Δ_L, Δ_S = fractional fluctuation of the power at ω_L and ω_S, respectively, within the detection bandwidth

G = Raman gain factor

B = detection bandwidth

η = detector quantum efficiency

P_S = power of laser at ω_S

The gains of interest are sufficiently small that the first term in the denominator can be ignored. The last term representing the shot noise limit determines the best achievable sensitivity, although in practice the laser fluctuations, ω_S, usually limit the sensitivity. The lowest gain that could be measured with $S/N = 1$, a detector with a 100% quantum efficiency, a 1-sec time constant, and a ω_S laser power of 100 mW incident on the detector would be about 2×10^{-9}. Owyoung demonstrated that the system sketched in Fig. 38 was capable of measuring a gain of 5.5×10^{-8} [355]. Using (62), one should be able to detect a concentration corresponding to $\operatorname{Im} \chi_8^{(3)} = 2 \times 10^{-15}$ cm^3/erg if the dye laser power is 50 mW. For the 992-cm^{-1} benzene peak, this corresponds to 0.035 M. Levine and Bethea used two synchronously pumped mode-locked dye lasers that had been well-stabilized for the ω_L and ω_S excitation [356, 357]. They demonstrated the capability of measuring gains of 2×10^{-9}, which represents shot noise limited performance. The higher peak power of the synchronously pumped dye lasers also provides a higher gain than the c.w. dye lasers of Owyoung. Levine and Bethea obtained a signal-to-noise ratio of 2×10^5 for neat benzene samples, which implies a detection limit of 5×10^{-5} M. This system has been used to measure monolayer coverages of metal surfaces [358, 359]. Thus the detection limits for Raman gain spectroscopy appear to exceed those of CARS methods.

A serious problem in Raman gain measurements arises from the disturbance of the optical characteristics of the sample by a single beam so the propagation of the second beam is disturbed as well. Typically the beam at ω_L heats the sample medium and causes thermal blooming, which changes the propagation of the beam at ω_S. A modulation of the ω_L can therefore be transferred to the beam at ω_S even if there are no Raman modes at $\omega_L - \omega_S$. Levine and Bethea reduced this problem markedly by frequency modulating the dye laser output [356]. The heating effects have only a small dependence on the laser frequency, but the Raman signal has a sharp dependence. By phase sensitive detection of the output, the Raman signal can be extracted from the thermal component.

The Raman gain expression in (62) shows the gain scales as P_L. By using the higher powers of a pulsed laser instead of a c.w. laser, it is possible to get much higher values of the Raman gain. Morris and co-workers used a pulsed nitrogen-pumped dye laser as the beam at ω_S and measured the Raman loss in the beam of an argon laser operating at 488 nm [360–362]. AC coupling was used in the detection electronics so only the spike imposed on the argon laser beam by the interaction with the dye laser in the sample was measured. A boxcar integrator sampled the output at that instant in time. The higher peak powers of pulsed dye lasers make the Raman gains or losses considerably larger. At the same time, however, the number of photons measured over the very short time the output is sampled are much lower so the shot noise contributions increase markedly. This system measured NO_3^- concentrations of 0.005 M and expectations were that the detection limit could be reduced by another factor of 10 [361]. The detection limits are determined by the thermal blooming caused by the dye laser and the scattering of the argon laser by small particles migrating through the beam. In this method, the thermal blooming can be temporally discriminated against because its signal has a much longer time response. By gating the boxcar on only during the laser pulse, the thermal blooming signal is minimized, although it cannot be eliminated.

Nestor used two pulsed nitrogen pumped dye lasers as the two excitation souces with a peak power of 0.34 and 1.7 kW at ω_L and ω_S, respectively [363].

A spectrum of the 992-cm^{-1} band of benzene was obtained by monitoring the loss in the beam at ω_L as ω_S was scanned. The pulsed dye lasers are much less stable than c.w. lasers and consequently it is more difficult to reach the shot noise limit. However, the gains are much larger than those of the c.w. laser. In addition, the nitrogen pumped dye lasers are easier to use, have a wider tuning range, and permit time resolutions of events <10 nsec. The wide tuning range is particularly attractive for taking advantage of resonance enhancement.

A completely different approach to detecting the Raman loss or gain signal uses photoacoustic methods to detect the energy deposited in the sample medium by the stimulated Raman scattering [364, 365]. Since the acoustical energy is generated by Raman resonances that thermally relax in the solvent bath, only the imaginary part of $\chi^{(3)}$ is measured, just as in the conventional Raman gain or loss measurement. The difference is that the photoacoustic Raman spectroscopy (PARS) [364] or optoacoustic Raman gain spectroscopy (OARS) [365] detects the vibrational energy directly and not indirectly through the effects on the ω_L or ω_S beams.

IMPROVEMENTS OR MODIFICATIONS OF RAMAN GAIN SPECTROSCOPY

There are a number of possibilities for improving the method. The resonant enhancements in $\chi^{(3)}$ that have been exploited for resonance Raman and CARS when the excitation wavelengths approach electronic transitions also increase the Raman gain. Multiple passes can be used to increase the sensitivity. Owyoung, for example, used a simple reflecting configuration to make five multiple passes [355]. One of the most effective ways to obtain multiple passes though is to place the sample inside the cavity of the lasers. Werncke et al. [366, 367] demonstrated the increased sensitivity obtained for both resonant enhancement and the use of an intracavity arrangement using a pulsed laser system to perform both Raman gain and inverse Raman spectroscopy. The system they used is sketched in Fig. 39. The ruby laser output

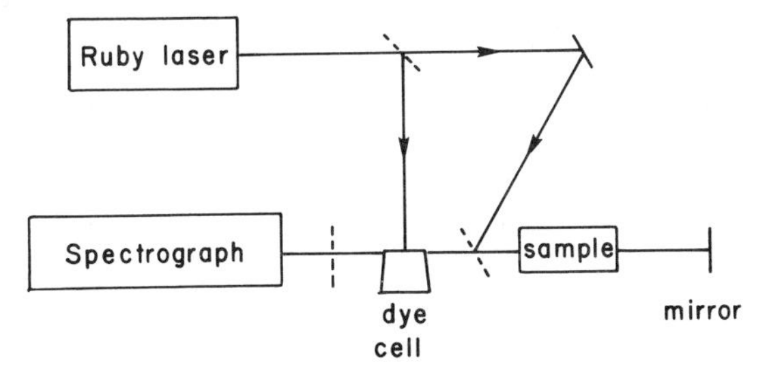

Fig. 39 Experimental system used by Werncke et al. to measure the Raman gain of a sample at the wavelengths of the dye emission. The dotted surfaces are beam splitters or dichroic mirrors. The output is recorded photographically on a spectrograph.

is beam split; part excites a broad-band dye laser and part is brought into the cavity. The ruby laser and broad-band dye emission are present simultaneously in the sample and Raman resonances cause gain in the broad-band dye emission at the wavelengths of the resonances. The broad-band emission with the superimposed Raman gain lines is recorded photographically with a monochromator. This system was used to record the Raman gain spectrum of 3,3′-diethyl thiadicarbocyanine iodide (DTDC) at a 4×10^{-4} M concentration in ethanol on a single laser shot. The ruby laser wavelength falls within the absorption spectrum of DTDC and causes a resonant enhancement of the gain. It is particularly striking that the Raman spectrum was obtained in 30 nsec.

Several clever methods have been used to null the probe beam at ω_S and transmit only the portion that results from vibrational resonances. One arrangement is sketched in Fig. 40. It is a modified Jamin interferometer, which splits the probe beam (ω_S) into two equal intensities and sends them on different paths through the sample [292]. They are recombined by the second plate, where they can interfere with each other. A second beam at ω_L is injected onto one of the two paths with the probe beam, where it also passes through the sample. The interferometer is adjusted to cause destructive interference of the probe beam, while the beam at ω_L is rejected with a filter. When the difference ($\omega_L - \omega_S$) matches a vibrational resonance, the intensity of one path of the probe light is increased because of the Raman gain. There are also refractive index changes that in turn cause additional phase changes. The condition for destructive interference is no longer met and a signal appears. In practice, the interferometer is not completely nulled so that a small portion of the probe beam is transmitted in the absence of a resonance. This procedure permits one to work in a linear region of gain.

In a second, related method a polarized probe beam at ω_S passes through a

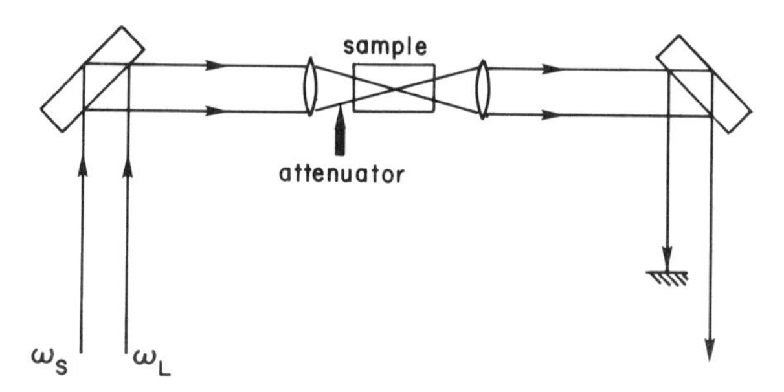

Fig. 40 A Jamin interferometer can be used to null the output of a laser at ω_S by properly adjusting the attenuator. Changes in the number of photons or the phase difference of one of the two beams at ω_S then causes an output signal.

sample and then the beam is blocked with a crossed polarization analyzer [291]. The probe polarization can be either linearly or elliptically polarized. A second polarized beam at ω_L is passed through the sample with a polarization direction that has components both aligned with and orthogonal to the probe beam polarization. If $(\omega_L - \omega_S)$ matches a Raman resonance, a Raman gain results with a component along the polarization direction of ω_L as well as a phase shift induced by a change in the index of refraction. As a result, a portion of the probe beam can pass through the analyzer and be detected. Again it is best to transmit a small additional portion of the probe beam through the analyzer in order to obtain a linear relationship with $\chi^{(3)}$.

These techniques have been used successfully to obtain Raman spectra of benzene and indeed give noise reduction to the shot noise limit [292]. The noise was suppressed by a factor of about 15 over spectra obtained by direct Raman gain spectroscopy and allowed a benzene concentration of about 0.006 *M* to be obtained. This detection limit could be improved by increasing the laser powers and integration times. These methods have the disadvantage, however, that it is difficult to incorporate multiple passes of the beam through the sample.

These two techniques have been used to make accurate measurements of both the real and imaginary components of the Raman contribution to $\chi^{(3)}$ as well as the value of the nonresonant susceptibility [326]. The Raman gain, as we mentioned earlier, depends only on the imaginary component of $\chi^{(3)}$. The real part of $\chi^{(3)}$ does not affect the gain, but it can affect the phase of a beam through induced changes in refractive index. Since both the nonlinear interferometer technique and the heterodyned polarization interferometry technique are sensitive to phase difference, the real part of $\chi^{(3)}$ can be accessed as well. The calibration required to get accurate values was obtained by comparison with accurately known induced refractive index changes for CS_2 [326]. Once one has accurately known values for the Raman part of the susceptibility, accurate values can be obtained for the differential peak cross section for Raman scattering using the relationship

$$\frac{d\sigma}{d\Omega} = \frac{3\hbar\omega_S^4\Gamma}{Nc^4}\,\mathrm{Im}\,\chi_8^{(3)} \tag{64}$$

(This equation is sometimes expressed in terms of $d^2\sigma/d\Omega d(\Delta\bar{\nu})$, which is the same as $(4c/\Gamma)\,(d\sigma/d\Omega)$ for a Lorentzian lineshape [368]. This method promises to be one of the most accurate ways of obtaining Raman cross-section values.

Inverse Raman Scattering or Raman Loss Spectroscopy

A Raman loss measurement is closely correlated with a Raman gain measurement. Since the intensity loss at ω_L accompanies the gain at ω_S and the production of a stimulated vibrational excitation, one could measure either effect. The loss is governed by (42) or (52), equations that are very similar to those for Raman gain. Therefore, all the remarks made in the preceding section apply equally well to Raman loss spectroscopy. In particular, the loss depends directly on the laser intensity at ω_S and the imaginary components of $\chi^{(3)}$.

Historically, there is a significant difference between the inverse Raman effect and simulated Raman scattering [299, 369, 370]. Stimulated Raman scattering was observed by focusing a very high power laser into a medium and examining the output frequencies. Spontaneous Raman scattering would create photons at the Stokes frequency. If the laser was intense enough, the intensity at the Stokes frequency would reach a threshold where additional photons could be scattered into the Stokes beam. Thus an incoming beam at ω_L created its own beam at ω_S, which in turn caused Raman gain. This process was very nonlinear and only the intense Raman line reached threshold. This is not the case for the inverse Raman effect, where a different approach was first taken. A very high power pulse laser was typically used as an excitation souce at ω_S and a pulsed broad-band continuum source was used as as source of all possible frequencies at ω_L. The continuum was monitored with a monochromator with photographic detection and the Raman loss signal appeared as an absorption line on the continuum. This procedure permitted Raman spectra to be obtained quickly with a time resolution determined by the pulse width of the laser. The absorption lines appeared on the anti-Stokes side of the excitation so that sample fluorescence was strongly discriminated against. It also gave a complete Raman spectrum as opposed to a stimulated Raman scattering experiment, where only the highest gain Raman line contributes. This approach was shown to be capable of measuring about 0.1 M benzene concentrations [299]. It is limited by the ability to observe an absorption on a photographic plate, which has been quoted to be about a 5% change in background intensity [367]. One can increase the amount of absorption by increasing the excitation intensity [see (52)], but this only helps up to the point where stimulated Raman scattering becomes important. When the stimulated Raman scattering becomes comparable to the excitation intensity, anti-Stokes Raman scattering is produced and eliminates any loss signal appearing at that portion of the continuum [299]. Thus the excitation intensity is an important variable in successfully performing the inverse Raman experiment with a single laser.

An important advance was the use of two lasers for performing the Raman loss measurement as is described earlier. Klein et al. performed an experiment of this kind in 1974 using a ruby laser for ω_L and a ruby laser pumped dye laser for ω_S [366]. The difference in the ruby laser intensity was measured before and after the sample and was displayed on an oscilloscope to determine the Raman loss. The experiment was otherwise identical with that recently performed by Nestor [363] and described in the preceding section.

Werncke et al. demonstrated the large enhancements that can be obtained if intracavity absorption techniques [371] or resonance enhancement [367] is applied to inverse Raman scattering. The intracavity enhancement was demonstrated by placing a benzene solution within the cavity of a dye laser pumped by the ruby second harmonic and operated in the broad-band mode with no dispersive elements. The ruby laser fundamental was also brought into the optical axis of the dye laser to furnish the strong excitation at ω_S. The lower detection limit for benzene was 10^{-3} *M*, 2 orders of magnitude lower than that obtained without intracavity techniques [371]. When the same experimental configuration was used for demonstrating resonance enhancement in an intracavity arrangement, it was found that a solution of 5×10^{-6} *M* DTDC could be detected [367].

Haushalter and Morris have studied resonant enhancement of AC-coupled inverse Raman spectra [362]. When one or both lasers are tuned into the absorption bands of a sample, thermal blooming becomes a more severe problem, while ground-state depopulation effects cause a strong decrease in the Raman signal. Nevertheless, resonances can be identified. The shapes of the Raman resonances are a function of the position in the band profile that is being excited in the same way as resonance CARS lineshapes change. Positive and negative Lorentzians and dispersive lineshapes are all possible depending on the relative positions of the laser frequency to the electronic transition.

Rikes

Most of the coherent Raman methods described rely on extracting a Raman signal from a larger frequency independent signal. A series of techniques have been discovered whose goal is to eliminate all signals except the Raman signal of interest but to retain the high signal levels of coherent Raman spectroscopies [372–375]. These methods are an outgrowth of the development of Raman induced Kerr effect spectroscopy (RIKES) introduced by Heiman et al. [372].

A diagram of a RIKE apparatus is shown in Fig. 41. The polarized output of an argon laser at ω_2 is passed through a sample cell and is then blocked by a crossed polarizer. If the sample is irradiated simultaneously by a second laser at ω_1 tuned so that the frequency difference $|\omega_1 - \omega_2|$ matches a vibrational frequency, an additional polarization is induced through $\chi_8^{(3)}$ and $\chi_{10}^{(3)}$ [see (29)] at ω_1 and ω_2. In our treatment up to this point, we ignore the tensor character of $\chi^{(3)}$, but RIKES in fact depends on the tensor character. Let us

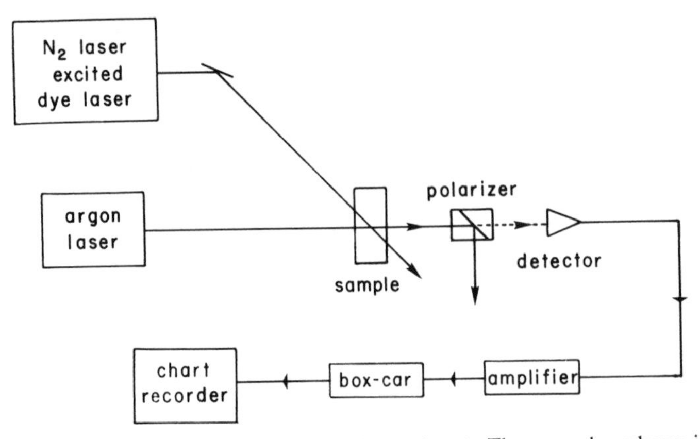

Fig. 41 The apparatus for measuring a RIKES experiment. The argon laser beam is linearly polarized and is blocked from reaching the detector by a crossed polarizer. A polarization rotation can be induced by a dye laser pulse that differs from the argon laser frequency by a vibrational frequency.

assume the argon probe laser at ω_1 is linearly polarized along an x-axis while the dye laser is circularly polarized. It is possible to induce a polarization along the y-axis because of terms of the form

$$P_y^{(3)}(z, t) = 6/8\chi_{yxyx}^{(3)} E_{1x} E_{1y}^* E_{2x} e^{i(k_2 z - \omega_2 t)} + \text{c.c.} \tag{65}$$

where E_{1x} is the amplitude of the electric vector at ω_1 in the x direction, and so on (other terms also contribute [373]). The important point is the presence of an induced polarization at ω_2 that is orthogonal to the original polarization and can therefore be transmitted through the crossed polarizer and be detected. Note that (65) is of the same form as the comparable equation for a Raman gain experiment. The two phenomena are in fact very closely connected, as is implied earlier in the section of Raman gain spectroscopy.

An important consideration in RIKES is the presence of a background that arises because of strain or residual birefringence in any of the optics of the probe beam. Fluctuations in that background limit the sensitivity of RIKES [374]. Additionally, the signal strength depends quadratically on $\chi^{(3)}$ and therefore on concentration. These problems can be reduced or solved by employing a heterodyning technique [374]. An additional signal is allowed to reach the detector by rotating the blocking polarizer slightly. This additional signal acts as a local oscillator that can interfere with the background and the Raman signal. Let E_{LO} and E_R represent the electric field amplitudes of the local oscillator and the Raman signal resulting from the interaction between the two excitation lasers. The resultant intensity is

$$I_T = \frac{n_2 c}{8\pi} |E_{LO} + E_R|^2$$

$$= \frac{n_2 c}{8\pi} (E_{LO}^2 + E_R^2 + E_{LO}E_R^* + E_{LO}^* E_R) \quad (66)$$

For low values of Raman signal and for $E_{LO} > E_R$, the cross terms in this expression exceed the term E_R^2. Since E_R is linearly proportional to $\chi^{(3)}$ and therefore to the concentration, the cross terms are also linearly proportional to concentration. The intensity from the cross term is synchronous with the Raman signal and can therefore be extracted from the total intensity by correlation with the pumping source at ω_1. This procedure is actually very similar to the interference we discuss in CARS experiments between the nonresonant background and a Raman line at low concentrations.

The actual relationship between the detected intensity and $\chi^{(3)}$ depends intimately on the polarizations of the two excitation beams at ω_1 and ω_2 and the polarization state of the local oscillator [372]. It is also possible to use three lasers since we are dealing with four wave mixing techniques. This modification increases the number of polarizations available to the experiments. There is clearly room for lots of creativity in selecting appropriate polarization conditions. We consider several of the combinations that have been examined. The original RIKES experiment utilized a circularly polarized pump laser and a linearly polarized probe laser [372]. The polarizers were arranged to prevent the probe laser from reaching the detector in the absence of a Raman resonance. The detected signal for this situation is proportional to the imaginary component of $\chi^{(3)}$ while the real part of $\chi^{(3)}$ does not contribute. Consequently, the nonresonant contribution to $\chi^{(3)}$ is not important. If the polarizer blocking the probe laser is slightly misaligned, a small amount of the probe laser can reach the detector, causing a linearly polarized local oscillator frequency that can interfere with the Raman signal. The intensity that arises from the cross terms between the local oscillator and the Raman signal [see (66)] depends on the real part of $\chi^{(3)}$. However, if the local oscillator is phase shifted so that it is 90° out of phase with the original probe beam, the intensity from the cross terms depends on the imaginary part of $\chi^{(3)}$. Obviously many other combinations are possible. The important point is that either the real or imaginary components of $\chi^{(3)}$ can be examined separately [313, 375].

RIKE-type experiments have been carried out in many different configurations. They have been generally limited in detection capability by the laser fluctuations either because of background light that leaks through the polarizers or the local oscillator component. It is estimated that benzene concentrations of 10^{-3} M should be detectable with improvements in laser stability, although only 10^{-2} M concentrations have been demonstrated [375].

Conclusions

The field of coherent Raman spectroscopy is very young, active, and exciting and new ideas are being developed rapidly. The signal levels are very large in almost all the techniques, but unfortunately the noise levels are also. When methods are found to reduce the noise levels, the coherent Raman techniques could have a major impact in analytical spectroscopy because they potentially have low detection limits and excellent selectivity for chemical analysis. At the present time and in the foreseeable future, conventional incoherent Raman techniques remain the method of choice for most samples requiring Raman data. Detection limits for common anions in aqueous solutions are typically about 10^{-4} M [376, 377] and can reach values of less than 100 ppb for resonant enhancement for some dyes [341, 378]. Resonance Raman measurements can be performed even on samples where the absorptions are in the UV with presently available methods [379]. Time resolution techniques have also been developed for rejection of fluorescence after pulsed excitation [380, 381]. There are situations though where conventional Raman spectroscopy is inadequate and coherent Raman techniques will become increasingly attractive because of their unique characteristics, including the ability to reject fluorescence interference, the high resolution determined by the laser bandwidth, the coherent and directional nature of the signal, and the strong signal levels. Applications that might require such capabilities include obtaining Raman spectra from fluorescent samples, performing high-resolution spectroscopy of low-pressure gases or flames, and measuring samples that have low concentrations of a Raman active material in the absence of other matrix materials. It also has been suggested that resonantly enhanced CARS would provide the necessary sensitivity for a liquid chromatography detector [319]. These applications, however, are reasonably restrictive or speculative and it is important to continue the development of the coherent Raman techniques to the point where they can become a significant new tool for the analysis of chemical samples.

Two Photon Spectroscopy

One of the processes that contributes to the CARS output frequency at $(2\omega_L - \omega_S)$ is a two-photon absorption that is represented in Fig. 31 and is described by (49) and (54). Two-photon spectroscopy has commonly been performed by measuring the incoherent population excited after two photons have been absorbed rather than by measuring the coherent output that is resonantly enhanced by two-photon resonances. The performance of the experiment then has many features that are similar to conventional single photon spectroscopy. The conventional absorption and fluorescence spectroscopy involves single photons whose energy is resonant with allowed transitions. For electric dipole transitions, the initial and final states must differ in parity to have appreciable transition moments. There is also a probability that two photons can be absorbed or emitted at half the energy of an allowed transition [382]. When two photons are involved in an electric-dipole-induced transition, the initial and final states must have the same parity to have an appreciable transition probability [383]. Two photon spectroscopy is also capable of accessing transitions where two electrons change state [384]. These transitions are completely forbidden in one-photon spectroscopy. Two photon spectroscopy therefore accesses different information from one photon spectroscopy and it is this featuare that makes it analytically interesting. The analogy has been made between one- and two-photon spectroscopy and infrared and Raman spectroscopy because both obtain complementary information about a molecular system [385].

The primary difficulty in performing two-photon spectroscopy arises from the very low transition probabilities. Since the transitions involve absorption or emission of two photons, the probability depends quadratically on the excitation intensity. If ΔP is the incident power absorbed and P_1 and P_2 are the incident powers at frequencies ω_1 and ω_2, one can write the expression

$$\Delta P = \frac{\delta l C P_1 P_2}{A} \tag{67}$$

where l is the path length, C is the concentration, A is the area of the focused laser, and δ is a constant characteristic of the absorption [387]. A typical value for δ is 10^{-50} cm^4 sec/(photon) (molecule) or 10^{-11} cm/(W) (M) for light at 333 nm. The resulting absorption strength is much smaller than that of an allowed single-photon transition. If (67) is rearranged to the form of Beer's law, the effective molar absorptivity is $\delta P/A$. For a 10^9 W/cm^2 intensity (values higher than this can cause sample destruction), the effective molar absorptivity is only 10^{-2} M^{-1} cm^{-1}, almost 6 orders of magnitude smaller than that of a typical single photon transition, but still about 6 orders of magnitude higher than that of a typical Raman transition [385]. It is clear that the detection limits for two-photon spectroscopy are poorer than those for one-photon spectroscopy.

The powers that appear in (67) are instantaneous powers. Most dye lasers have an output that changes in time either because the excitation source itself is changing in time or because mode-beating within the dye laser cavity causes changes that can occur on the picosecond time scale. Johnson and Lytle have studied how two-photon absorption depends on the dye laser characteristics [386]. A single mode dye laser, for example, has no mode-beating and the power is quite constant. The two-photon absorption follows (67). A multimode dye laser gives a larger two-photon signal than is expected from (67) because the experimentally measured power is usually an average power, not the instantaneous power required by (67). Short-term fluctuations in the dye laser output are more effective in producing two-photon absorption because of the nonlinear dependence on power.

After the sample is excited, the excited state must be detected. Any of the techniques described earlier can be used, including detection of the absorption directly or observing fluorescence, ionization, thermal, or acoustical signals. Fluorescence has been used almost exclusively in experimental studies. Although the small absorption strengths result in low fluorescence intensities, there are several aspects of two-photon transitions that make the spectroscopy easier [385, 387]. The excitation and fluorescence wavelengths are usually widely separated since the excited state has twice the energy of the excitation light. Wavelength discrimination becomes quite easy as a result. It is also easier to obtain lasers to excite the absorptions of interest. Many interesting absorptions occur in the ultraviolet, where convenient tunable lasers do not exist. These same absorptions appear in the visible region though for two-photon excitation, where there are many convenient lasers. In addition, since the wavelengths of the excitation light are usually much longer than the wavelength for allowed transitions of either the sample or its solvent, the attenuation of the excitation light in the sample can be neglected. The two-photon absorption itself is so weak that it can always be neglected. Attenuation at high concentrations is the limiting effect in conventional fluorescence spectroscopy, which restricts the linear dynamic range and is referred to as the inner filter effect. Wirth and Lytle [388] have pointed out that one can work in media that are optically dense at the one-photon transition wavelengths but transmit at the much longer wavelengths of the two-photon excitation. This of course assumes the sample emission is not quenched by the transition causing the optical density.

The two photon excitation spectrum of napthalene is shown in Fig. 42*b* along with the conventional single-photon excitation spectrum in Fig. 42*a* [385]. The most noteworthy fact is that the two spectra are completely different from each other because of the different parity requirements for each. The one-photon spectrum is dominated by the strong, parity allowed ${}^1A_g \rightarrow {}^1B_{2u}$ transition, but this same transition is very weak in the two photon spectrum. In addition, the vibronic states that couple to the electronic transition are different for the two spectroscopies, so the vibrational progressions differ. This example illustrates the complementary information obtained by the two-photon spectroscopy.

The value of δ that appears in (67) is not a simple constant. It represents a tensor that depends on the polarizations of the two photons involved in the

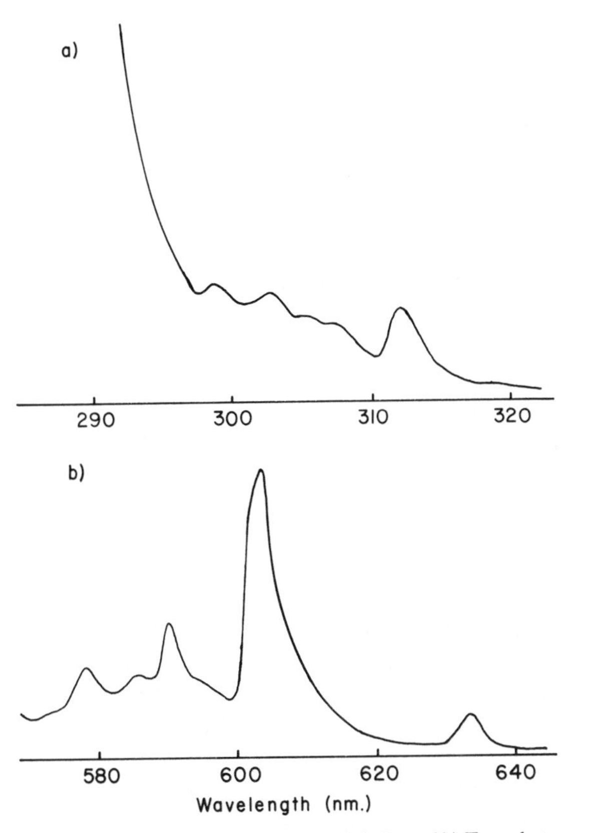

Fig. 42 (*a*) Single-photon excitation spectrum of naphthalene. (*b*) Two photon excitation spectrum of naphthalene. The wavelength axis has been drawn such that the total energy of the two-photon excitation matches that for the single photon excitation [385].

two-photon absorption [389–394]. The size of each element in that tensor depends on the symmetry of the molecular states, even for unoriented samples such as solutions. Thus two photons that are polarized in the same direction do not exert any influence on a totally symmetric molecule that will change its symmetry [397]. A transition to another totally symmetric state could therefore occur and the corresponding value of δ would be finite. Two photons polarized differently would act to change the symmetry so the component of the δ tensor with orthogonal polarizations would be very small for transitions between totally symmetric states. This simple example can be extended to the more general situation where an arbitrary molecule undergoes a transition between two states of definite symmetry. The size of each component of the δ tensor reflects the symmetries. If the system under study is a solution where all orientations of the molecules are possible relative to an externally defined polarization direction, the value for transition moment must be averaged. For a single laser that provides both photons for the two-photon absorption, it can be shown that

$$\delta = 2\delta_F \cos 2\theta + \delta_G (5 - \cos 2\theta) \tag{68}$$

where θ is the phase retardation associated with the exciting photons [389–391]. For $\theta = 0°$, the light is linearly polarized, whereas for $\theta = 90°$, the light is circularly polarized. Thus instead of having one value characteristic of an absorption, two values can be obtained from measuring the two-photon absorption for two different values of θ. The size of δ_G does not depend on the molecular symmetry, whereas δ_F does. The ratio δ_F/δ_G can therefore provide a number characteristic of molecular symmetry that is normalized and independent of most experimental conditions. These ideas have been used to identify the symmetry of molecular transitions and to study the changes that occur in molecular symmetry when derivatives are formed or when solvation occurs [394].

Sepaniak and Yeung [395] have studied the possibility of using two-photon excited fluorescence for liquid chromatography detection. They used a 4-W argon laser focused to a spot 10^{-4} cm^2 in area to excite two-photon fluorescence from the effluent of an LC column. The detection limits were comparable to those obtained with a UV absorbance detector for several oxadiazole dyes. The prime difficulty of course is the low transition probability for two photon absorption. These results indicate that substantially better results could be obtained by improving the peak intensity of the laser [395]. In the study of Sepaniak and Yeung the intensity was only 4×10^4 W/cm^2, far below the 10^9 W/cm^2 that can be obtained without particular difficulty.

Two-photon spectroscopy can be used in either atomic or molecular systems in either absorption or emission. The application to atomic fluorescence could be of interest because the atomic transitions are inherently stronger than the molecular transitions and many of the atomic transitions lie in the ultraviolet for direct one-photon excitation. A two-photon excitation would permit one to use visible lasers and would not be hampered by the absorption of excitation light by the flame environment, which becomes important in flame spectroscopy in the ultraviolet. The problem of wavelength discrimination can also be minimized by the large separation of the excitation and fluorescence wavelengths. Turk and co-workers have used a two-photon absorption to detect Li by optogalvanic spectroscopy at a detection limit of 0.5 ng/ml [246].

Another approach that can improve the small absorption probability of the two photon process is to have a real state intermediate between the initial and final states of the transition [385]. There is a resonant enhancement of the transition by the real state that increases the transition probability by many orders of magnitude. This situation is not one that can be generally applied to all systems because many do not have accessible states that would serve as intermediates.

References

1. P. W. Smith and T. W. Hänsch, *Phys. Rev. Lett.*, **26,** 740 (1971).
2. V. S. Letokhov, *Opt. Laser Technol.* 15 (February 1978).
3. R. Teets, J. Eckstein, and T. W. Hänsch, *Phys. Rev. Lett.*, **38,** 760 (1977).
4. K. C. Harvey, R. T. Hawkins, G. Meisel, and A. L. Schawlow, *Phys. Rev. Lett.*, **34,** 1973 (1975).
5. C. Wieman and T. W. Hänsch, *Phys. Rev. Lett.*, **36,** 1170 (1976).
6. R. Teets, R. Feinberg, T. W. Hänsch, and A. L. Schawlow, *Phys. Rev. Lett.*, **37,** 683 (1976).
7. M. Sargent, *Phys. Rev.*, **A14,** 524 (1976).
8. J. I. Steinfeld and P. L. Houston, *Laser and Coherence Spectroscopy*, Jeffrey I. Steinfeld, Ed., Plenum, New York, 1978.
9. B. R. Ware and W. H. Flygare, *Chem. Phys. Lett.*, **12,** 81 (1971).
10. W. H. Flygare, B. R. Ware, and S. L. Hartford, *Molecular Electro-Optics, Part I, Theory and Methods*, C. T. O'Konski, Ed., Dekker, New York, 1976.
11. B. A. Smith, B. R. Ware, and R. S. Weiner, *Proc. Natl. Acad. Sci. U.S.*, **73,** 2388 (1976).
12. Y. Yeh and H. Z. Cummins, *Appl. Phys. Lett.*, **4,** 176 (1964).
13. C. K. N. Patel, *Science*, **202,** 157 (1978).
14. J. F. Butler and J. O. Sample, *Cryogenics*, 661 (December 1977).
15. S. E. Harris, *Proc. IEEE*, **57,** 2096 (1969).
16. L. F. Mollenauer and D. H. Olson, *Appl. Phys. Lett.*, **24,** 386 (1974).
17. L. F. Mollenauer and D. H. Olson, *J. Appl. Phys.*, **46,** 3109 (1975).
18. C. K. N. Patel, *Fundamental and Applied Physics*, M. Feld, A. Javan, and N. Kurnit, Eds., Wiley, New York, 1973.
19. J. P. Goldsborough, *Laser Handbook*, Vol. 1, Part C, F. T. Arecchi and E. O. Schulz-Dubois, Eds., North-Holland, Amsterdam, 1972, p. 597.
20. R. A. Crane, *Appl. Opt.*, **17,** 2097 (1978).
21. R. S. Eng, K. W. Nill, and M. Whalen, *Appl. Opt.*, **16,** 3072 (1977).
22. E. D. Hinkley, *Opt. Quantum Electron.*, **8,** 155 (1976).
23. E. D. Hinkley, R. T. Ku, K. W. Nill, and J. F. Butler, *Appl. Opt.*, **15,** 1653 (1976).
24. S. H. Groves, K. W. Nill, and A. J. Strauss, *Appl. Phys. Lett.*, **25,** 331 (1974).
25. D. A. Leonard, *Nature*, **216,** 142 (1967).
26. K. W. Rothe, U. Brinkmann, and H. Walther, *Appl. Phys.*, **3,** 115 (1974).
27. P. T. Woods and B. W. Jolliffe, *Opt. Laser Technol.*, 25 (February 1978).

28. J. F. Butler, K. W. Nill, A. W. Mantz, and R. S. Eng, *New Applications of Lasers to Chemistry*, G. M. Hieftje, Ed., ACS Symposium Series, Washington, DC, 1978.
29. A. Rosencwaig, *Anal. Chem.*, **47,** 593A (1975).
30. K. P. Koch and W. Lahmann, *Appl. Phys. Lett.*, **32,** 289 (1978).
31. L. B. Kreuzer, *Anal. Chem.*, **46,** 237A (1974).
32. M. B. Robin, *J. Lumin.*, **12/13,** 131 (1976).
33. C. K. N. Patel and A. C. Tam, *Appl. Phys. Lett.*, **34,** 467 (1979).
34. C. K. N. Patel, A. C. Tam, and R. J. Kerl, *J. Chem. Phys.*, **71,** 1470 (1979).
35. S. Oda, T. Sawada, M. Nomura, and H. Kamada, *Anal. Chem.*, **51,** 686 (1979).
36. T. Sawada, S. Oda, H. Zhimizu, and H. Kamada, *Anal. Chem.*, **51,** 688 (1979).
37. S. Oda, T. Sawada, and H. Kamada, *Anal. Chem.*, **50,** 865 (1978).
38. W. Lahmann, H. J. Ludewig, and H. Welling, *Anal. Chem.*, **49,** 549 (1977).
39. E. Voigtman, A. Jurgensen, and J. Winefordner, *Anal. Chem.*, **53,** 1442 (1981).
40. J. F. McClelland and R. N. Kniseley, *Appl. Opt.*, **15,** 2658 (1976).
41. A. Rosencwaig, *J. Appl. Phys.*, **49,** 2905 (1978).
42. J. C. Roark, R. A. Palmer, and J. S. Hutchinson, *Chem. Phys. Lett.*, **60,** 112 (1978).
43. J. C. Murphy and L. C. Aaamodt, *J. Appl. Phys.*, **48,** 3502 (1977).
44. D. R. Siebert, G. A. West, and J. J. Barrett, *Appl. Opt.*, **19,** 53 (1980).
45. E. Nodov, *Appl. Opt.*, **17,** 1110 (1978).
46. K. Y. Wong, *J. Appl. Phys.*, **49,** 3033 (1978).
47. L. J. Thomas, M. J. Kelly, and N. M. Amer, *Appl. Phys. Lett.*, **32,** 736 (1978).
48. E. E. Marinero and H. Stuke, *Opt. Commun.*, **30,** 349 (1979).
49. J. M. Harris and N. J. Dovichi, *Anal. Chem.*, **52,** 695A (1980).
50. R. L. Swofford, *Lasers in Chemical Analysis*, G. M. Hieftje, J. C. Travis, and F. E. Lytle, Eds., Humana Press, New Jersey, 1981.
51. J. P. Gordon, R. C. C. Leite, R. S. Moore, S. P. S. Porto, and J. R. Whinnery, *J. Appl. Phys.*, **36,** 3 (1965).
52. R. C. C. Leite, R. S. Moore, and J. R. Whinnery, *Appl. Phys. Lett.*, **5,** 141 (1964).
53. C. Hu and J. R. Whinnery, *Appl. Opt.*, **12,** 72 (1973).
54. A. Hordvik, *Appl. Opt.*, **16,** 2827 (1977).
55. M. E. Long, R. L. Swofford, and A. C. Albrecht, *Science*, **191,** 183 (1976).
56. H. L. Fang and R. L. Swofford, *J. Appl. Phys.*, **50,** 6609 (1979).
57. N. J. Dovichi and M. M. Harris *Anal. Chem.*, **53,** 106 (1981).
58. N. J. Dovichi and J. M. Harris, *Anal. Chem.*, **51,** 728 (1979).
59. N. J. Dovichi and J. M. Harris, *Anal. Chem.*, **52,** 2338 (1980).
60. A. Yariv, *Quantum Electronics*, Wiley, New York, 1967.
61. G. E. Mieling and H. L. Pardue, *Anal. Chem.*, **50,** 1611 (1978).
62. H. W. Latz, H. F. Wyles, and R. B. Green, *Anal. Chem.*, **45,** 2405 (1973).
63. R. C. Spiker and J. S. Shirk, *Anal. Chem.*, **46,** 572 (1974).
64. N. C. Peterson, M. J. Kurylo, W. Braun, A. M. Bass, and R. A. Keller, *J. Opt. Soc. Am.*, **61,** 746 (1971).

65. R. J. Thrash, H. von Weyssenhoff, and J. S. Shirk, *J. Chem Phys.*, **55**, 4659 (1971).
66. R. A. Keller, E. F. Zalewski, and N. C. Peterson, *J. Opt. Soc. Am.*, **62**, 319 (1972).
67. G. H. Atkinson, A. H. Laufer, and M. J. Kurylo, *J. Chem. Phys.*, **59**, 350 (1973).
68. G. Horlick and E. G. Codding, *Anal. Chem.*, **46**, 133 (1974).
69. W. Brunner and H. Paul, *Opt. Commun.*, **12**, 252 (1974).
70. W. Brunner and H. Paul, *Opt. Quantum Electron.*, **10**, 139 (1978).
71. K. Tohma, *J. Appl. Phys.*, **47**, 1422 (1976).
72. W. T. Hill, R. A. Abreu, T. W. Hänsch, and A. L. Schawlow, *Opt. Commun.*, **32**, 96 (1980).
73. T. W. Hänsch, A. L. Schawlow, and P. E. Toschek, *IEEE J. Quantum Electron.*, **QE-8**, 802 (1972).
74. R. G. Bray, W. Henke, S. K. Liu, K. V. Reddy, and M. J. Berry, *Chem. Phys. Lett.*, **47**, 213 (1977).
75. J. S. Shirk, T. D. Harris, and J. W. Mitchell, *Anal. Chem.*, **52**, 1701 (1980).
76. T. D. Harris and J. W. Mitchell, *Anal. Chem.*, **52**, 1706 (1980).
77. J. P. Hohimer and P. J. Hargis, *Anal. Chem.*, **51**, 930 (1979).
78. J. W. Hosch and E. H. Piepmeier, *Appl. Spectrosc.*, **32**, 444 (1978).
79. E. H. Piepmeier, *Spectrochim. Acta*, **27B**, 431 (1972).
80. E. H. Piepmeier, *Spectrochim. Acta*, **27B**, 445 (1972).
81. N. Omenetto, L. P. Hart, P. Benetti, and J. D. Winefordner, *Spectrochim. Acta*, **28B**, 301 (1973).
82. N. Omenetto, P. Benetti, L. P. Hart, J. D. Winefordner, and C. Th. J. Alkemade, *Spectrochim. Acta*, **28B**, 289 (1973).
83. N. Omenetto, J. D. Winefordner, and C. Th. J. Alkemade, *Spectrochim. Acta*, **30B**, 335 (1975).
84. D. R. Olivares and G. M. Hieftje, *Spectrochim. Acta*, **33B**, 79 (1978).
85. B. L. Sharp and A. Goldwasser, *Spectrochim. Acta*, **31B**, 431 (1976).
86. H. L. Brod and E. S. Yeung, *Anal. Chem.*, **48**, 344 (1976).
87. W. M. Fairbanks, Jr., T. W. Hänsch, and A. L. Schawlow, *J. Opt. Soc. Am.*, **65**, 199 (1975).
88. V. I. Balykin, V. S. Letokhov, V. I. Mishin, and V. A. Semchishen, *JETP Lett.*, **26**, 357 (1977).
89. Arthur Schawlow, Personal communication.
90. K. Laqua, *Analytical Laser Spectroscopy*, N. Omenetto, Ed., Wiley, New York, 1979.
91. M. B. Denton and H. V. Malmstadt, *Appl. Phys. Lett.*, **18**, 485 (1971).
92. L. M. Fraser and J. D. Winefordner, *Anal. Chem.*, **43**, 1693 (1971).
93. L. M. Fraser and J. D. Winefordner, *Anal. Chem.*, **44**, 1444 (1972).
94. N. Omenetto, N. N. Hatch, L. M. Fraser, and J. D. Winefordner, *Anal. Chem.*, **45**, 195 (1973).
95. N. Omenetto, N. N. Hatch, L. M. Fraser, and J. D. Winefordner, *Spectrochim. Acta*, **28B**, 65 (1973).

96. S. J. Weeks, H. Haraguchi, and J. D. Winefordner, *Anal. Chem.*, **50,** 360 (1978).
97. N. Omenetto, L. M. Fraser, and J. D. Winefordner, *Appl. Spectrosc. Rev.*, **7,** 147 (1973).
98. S. J. Weeks, H. Haraguchi, and J. D. Winefordner, *Anal Chem.*, **50,** 360 (1978).
99. J. J. Horvath, J. D. Bradshaw, J. N. Bower, M. S. Epstein, and J. D. Winefordner, *Anal. Chem.*, **53,** 6 (1981).
100. S. J. Weeks and J. D. Winefordner, *Lasers in Chemical Analysis*, G. M. Hieftje, J. C. Travis, and F. E. Lytle, Eds., Humana Press, New Jersey, 1981.
101. J. W. Daily, *Appl. Opt.*, **17,** 1610 (1978).
102. B. Smith, J. D. Winefordner, and N. Omenetto, *J. Appl. Phys.*, **48,** 2676 (1977).
103. P. J. Th. Zeegers, W. P. Townsend, and J. D. Winefordner, *Spectrochim. Acta*, **24B,** 243 (1969).
104. J. W. Hosch and E. H. Piepmeier, *Appl. Spectrosc.*, **32,** 447 (1978).
105. M. B. Blackburn, J. M. Mermet, G. D. Boutilier, and J. D. Winefordner, *Appl. Opt.*, **18,** 1804 (1979).
106. R. B. Green, J. C. Travis, and R. A. Keller, *Anal. Chem.*, **48,** 1954 (1976).
107. D. A. Goff and E. S. Yeung, *Anal Chem.*, **50,** 625 (1978).
108. R. B. Green, R. A. Keller, G. G. Luther, P. K. Schenck, and J. C. Travis, *IEEE J. Quantum Electron.*, **QE-14,** 63 (1977).
109. D. J. Ehrlich, R. M. Osgood, G. C. Turk, and J. C. Travis, *Anal. Chem.*, **52,** 1354 (1980).
110. G. F. Kirkbright, *Analyst*, **96,** 609 (1971).
111. J. A. Gelbwachs, C. F. Klein, and J. E. Wessel, *IEEE J. Quantum Electron.*, **QE-14,** 121 (1978).
112. J. A. Gelbwachs, C. F. Klein, and J. E. Wessel, *Appl. Phys. Lett.*, **30,** 489 (1977).
113. J. P. Hohimer and P. J. Hargis, Jr., *Anal. Chem. Acta*, **97,** 43 (1978).
114. S. Neumann and M. Kriese, *Spectrochim. Acta*, **29B,** 127 (1974).
115. M. A. Bolshov, A. V. Zybin, L. A. Zybina, V. G. Koloshnikov, and I. A. Majorov, *Spectrochim. Acta*, **31B,** 493 (1976).
116. J. P. Hohimer and P. J. Hargis, Jr., *Appl. Phys. Lett.*, **30,** 344 (1977).
117. M. W. P. Cann, *Appl. Opt.*, **8,** 1645 (1969).
118. T. G. Matthews and F. E. Lytle, *Anal. Chem.*, **51,** 583 (1979).
119. R. J. Kelly, W. B. Dandliker, and D. E. Williamson, *Anal. Chem.*, **48,** 846 (1976).
120. D. M. Jameson, G. Weber, R. D. Spencer, and G. Mitchell, *Rev. Sci. Instrum.*, **49,** 510 (1978).
121. T. Imasaka and R. N. Zare, *Anal. Chem.*, **51,** 2082 (1979).
122. E. S. Yeung and M. J. Sepaniak, *Anal. Chem.*, **52,** 1465A (1980).
123. N. Ishibashi, T. Ogawa, T. Imasaka, and M. Kunitake, *Anal. Chem.*, **51,** 2096 (1979).
124. J. H. Richardson and S. M. George, *Anal Chem.*, **50,** 616 (1978).
125. F. E. Lytle and M. S. Kelsey, *Anal. Chem.*, **46,** 855 (1974).

126. T. Imasaka, H. Kadone, T. Ogawa, and N. Ishibashi, *Anal. Chem.*, **49,** 667 (1977).
127. J. H. Richardson, L. L. Steinmetz, S. B. Deutscher, W. A. Bookless, and W. L. Schmelzinger, *Anal. Biochem.*, **97,** 17 (1979).
128. R. E. Brown, K. D. Legg, M. W. Wolf, L. A. Singer, and J. H. Parks, *Anal. Chem.*, **46,** 1690 (1974).
129. D. D. Morgan, D. Warshawsky, and T. Atkinson, *Photochem. Photobiol.*, **25,** 31 (1977).
130. J. H. Richardson, *Anal. Biochem.*, **83,** 754 (1977).
131. M. A. Van Dilla, T. T. Trujillo, P. F. Mullaney, and J. R. Coulter, *Science*, **163,** 1213 (1969).
132. R. N. Zare and P. J. Dagdigian, *Science*, **185,** 739 (1974).
133. J. C. Wright, *Lasers in Chemical Analysis*, G. M. Heiftje, J. C. Travis, and F. E. Lytle, Ed., Humana Press, New Jersey, 1981.
134. R. A. O'Neil, L. Buja-Bijunas, and D. M. Rayner, *Appl. Opt.*, **19,** 863 (1980).
135. S. D. Lidofsky, T. Imasaka, and R. N. Zare, *Anal. Chem.*, **51,** 1602 (1979).
136. G. J. Diebold, N. Karny, and R. N. Zare, *Anal. Chem.*, **51,** 67 (1979).
137. N. Strojny and J. A. F. de Silva, *Anal. Chem.*, **52,** 1554 (1980).
138. N. Strojny and J. A. F. de Silva, *Lasers in Chemical Analysis*, G. M. Heiftje, J. C. Travis, and F. E. Lytle, Ed., Humana Press, New Jersey, 1981.
139. J. D. Winefordner, *New Applications of Lasers to Chemistry*, G. M. Hieftje, Ed., ACS Symposium Series, Washington, DC, 1978.
140. J. H. Richardson and M. E. Ando, *Anal. Chem.*, **49,** 955 (1977).
141. B. W. Smith, F. W. Plankey, N. Omenetto, L. P. Hart, and J. D. Winefordner, *Spectrochem. Acta*, **30A,** 1459 (1974).
142. M. F. Bryant, K. O'Keefe, and H. V. Malmstadt, *Anal. Chem.*, **47,** 2324 (1975).
143. T. F. Van Geel and J. D. Winefordner, *Anal. Chem.*, **48,** 335 (1976).
144. J. H. Richardson, B. W. Wallin, D. C. Johnson, and L. W. Hrubesh, *Anal. Chim. Acta*, **86,** 263 (1976).
145. D. M. Rayner and A. G. Szabo, *Appl. Opt.*, **17,** 1624 (1978).
146. V. I. Balykin, V. I. Mishin, and V. A. Semshishen, *Sov. J. Quantum Electron.*, **7,** 879 (1977).
147. A. B. Bradley and R. N. Zare, *J. Am. Chem. Soc.*, **98,** 620 (1976).
148. M. R. Berman and R. N. Zare, *Anal. Chem.*, **47,** 1200 (1975).
149. G. J. Diebold and R. N. Zare, *Science*, **196,** 1439 (1977).
150. M. J. Sepaniak and E. S. Yeung, *Anal. Chem.*, **49,** 1554 (1977).
151. M. J. Sepaniak and E. S. Yeung, *J. Chromatogr.*, **190,** 377 (1980).
152. E. S. Yeung, *Lasers in Chemical Analysis*, G. M. Hieftje, J. C. Travis, and F. E. Lytle, Ed., Humana Press, New Jersey, 1981.
153. C. M. O'Donnell and S. C. Suffin, *Anal. Chem.*, **51,** 33A (1979).
154. G. C. Blanchard and R. Gardner, *Clin. Chem.*, **24,** 808 (1978).
155. M. W. Burgett, S. J. Fairfield, and J. F. Monthony, *J. Immunol. Methods*, **16,** 211 (1977).
156. S. A. Levison, W. B. Dandliker, and D. Murayama, *Environ. Sci. Technol.*, **11,** 292 (1977).

157. E. F. Ullman, M. Schwarzberg, and K. E. Rubenstein, *J. Biol. Chem.*, **251,** 4172 (1976).
158. M. N. Kronick and W. A. Little, *J. Immunol. Methods*, **8,** 235 (1975).
159. T. Hirschfeld, *Appl. Opt.*, **15,** 2965 (1976).
160. T. Hirschfeld, *Appl. Opt.*, **15,** 3135 (1976).
161. B. E. Kohler, *Chemical and Biochemical Applications of Lasers*, C. B. Moore, Ed., Academic, New York, 1979.
162. F. J Gustafson and J. C. Wright, *Anal. Chem.*, **49,** 1680 (1977).
163. J. C. Wright, *Anal. Chem.*, **49,** 1690 (1977).
164. G. H. Dieke, *Spectra and Energy Levels of Rare Earth Ions in Crystals*, Interscience, New York, 1968.
165. D. R. Tallant and J. C. Wright, *J. Chem. Phys.*, **63,** 2075 (1975).
166. M. P. Miller, D. R. Tallant, F. J. Gustafson, and J. C. Wright, *Anal. Chem.*, **49,** 1474 (1977).
167. J. C. Wright and F. J. Gustafson, *Anal. Chem.*, **50,** A1147 (1978).
168. J. C. Wright, F. J. Gustafson, and L. C. Porter, *New Applications of Lasers to Chemistry*, G. M. Hieftje, Ed., ACS Symposium Series, Washington, DC, 1978.
169. F. J. Gustafson and J. C. Wright, *Anal. Chem.*, **51,** 1762 (1979).
170. M. V. Johnston and J. C. Wright, *Anal. Chem.*, **51,** 1774 (1979).
171. M. V. Johnston and J. C. Wright, *Anal. Chem.*, **53,** 1050 (1981).
172. M. V. Johnston and J. C. Wright, *Anal. Chem.*, **53,** 1054 (1981).
173. D. L. Perry, S. M. Klainer, H. R. Bowman, F. P. Milanovitch, T. Hirschfeld, and S. Miller, *Anal. Chem.*, **53,** 1048 (1981).
174. J. C. Wright, *Radiationless Processes in Molecules and Crystals*, Vol. 15, Springer-Verlag, Berlin, 1976.
175. L. A. Riseberg, *Phys. Rev.*, **28,** 789 (1972).
176. W. C. McColgin, A. P. Marchetti, and J. H. Eberly, *J. Am. Chem. Soc.*, **100,** 5622 (1978).
177. E. V. Shpol'skii, A. A. Il'ina, and L. A. Klimova, *Dokl. Akad. Nauk SSR*, **87,** 935 (1952).
178. E. V. Shpol'skii, *Sov. Phys. Usp.*, **3,** 372 (1960).
179. E. V. Shpol'skii, *Sov. Phys. Usp.*, **6,** 411 (1963).
180. R. J. Lukasiewicz and J. D. Winefordner, *Talanta*, **19,** 381 (1972).
181. M. Lamotta, A. M. Merle, J. Joussot-Dubien, and F. Dupry, *Chem. Phys. Lett.*, **35,** 410 (1975).
182. A. M. Merle, W. M. Pitts, and M. A. El-Sayed, *Chem. Phys. Lett.*, **54,** 211 (1978).
183. G. F. Kirkbright and C. G. DeLima, *Analyst*, **99,** 338 (1974).
184. R. Faroog and G. F. Kirkbright, *Analyst*, **101,** 566 (1976).
185. A. Olszowski, Z. Ruziewicz, and H. Chojnacki, *J. Mol. Struct.*, **28,** 5 (1975).
186. M. Lamotte and J. Joussot-Dubien, *J. Chem. Phys.*, **61,** 1892 (1974).
187. S. A. Kozlov and D. M. Grebenshehikov, *Opt. Spectrosc.*, **44,** 75 (1978).
188. Y. Yang, A. P. D'Silva, V. A. Fassel, and M. Iles, *Anal. Chem.*, **52,** 1350 (1980).

189. Y. Yang, A. P. D'Silva, and V. A. Fassel, *Anal. Chem.*, **53**, 894 (1981).
190. P. Tokousbalides, E. L. Wehry, and G. Mamantov, *J. Phys. Chem.*, **81**, 1769 (1977).
191. R. C. Stroupe, P. Tokovsbalides, R. B. Dickinson, E. L. Wehry, and G. Mamantov, *Anal. Chem.*, **49**, 701 (1977).
192. P. Tokovsbalides, E. R. Hinton, R. B. Dickinson, P. V. Bilotta, E. L. Wehry, and G. Mamantov, *Anal. Chem.*, **50**, 1189 (1978).
193. J. R. Maple, E. L. Wehry, and G. Mamantov, *Anal. Chem.*, **52**, 920 (1980).
194. J. R. Maple and E. L. Wehry, *Anal. Chem.*, **53**, 266 (1981).
195. E. L. Wehry and G. Mamantov, *Anal. Chem.*, **51**, 643A (1979).
196. R. B. Dickinson and E. L. Wehry, *Anal. Chem.*, **51**, 778 (1979).
197. A. Szabo, *Phys. Rev. Lett.*, **25**, 924 (1970).
198. J. H. Everly, W. C. McColgin, K. Kawaoka, and A. P. Marchetti, *Nature*, **251**, 215 (1974).
199. A. P. Marchetti, W. C. McColgin, and J. H. Eberly, *Phys. Rev. Lett.*, **35**, 387 (1975).
200. I. I. Abram, R. A. Auerbach, R. R. Birge, B. E. Kohler, and J. M. Stevenson, *J. Chem. Phys.*, **63**, 2473 (1975).
201. J. C. Brown, M. C. Edelson, and G. J. Small, *Anal. Chem.*, **50**, 1394 (1978).
202. J. C. Brown, J. A. Duncanson, and G. J. Small, *Anal. Chem.*, **52**, 1711 (1980).
203. J. C. Wright, *Appl. Spectrosc.*, **34**, 151 (1980).
204. W. Kaiser, J. P. Maier, A. Seilmeier, *J. Mol. Struct.*, **59**, 249 (1980).
205. A. Seilmeier, W. Kaiser, and A. Laubereau, *Opt. Commun.*, **26**, 441 (1978).
206. A. Seilmeier, W. Kaiser, A. Laubereau, and S. F. Fischer, *Chem. Phys. Lett.*, **58**, 225 (1978).
207. F. J. Knorr and J. M. Harris, *Anal. Chem.*, **53**, 272 (1981).
208. C. V. Shank and E. P. Ippen, *Dye Lasers*, F. P. Schäfer, Ed., Springer-Verlag, Berlin, 1977.
209. J. M. Harris, R. W. Chrisman, and F. E. Lytle, *Appl. Phys. Lett.*, **26**, 16 (1975).
210. C. K. Chan and S. O. Sari, *Appl. Phys. Lett.*, **25**, 403 (1974).
211. H. Mahr and M. D. Hirsch, *Opt. Commun.*, **13**, 96 (1975).
212. E. A. Bailey and G. K. Rollefson, *J. Chem. Phys.*, **21**, 1315 (1953).
213. F. E. Lytle, J. F. Eng, J. M. Harris, T. D. Harris, and R. E. Santini, *Anal. Chem.*, **47**, 571 (1975).
214. E. R. Menzel and Z. D. Popovic, *Rev. Sci. Instrum.*, **49**, 39 (1978).
215. G. M. Hieftje, G. R. Haugen, and J. M. Ramsey, *Appl. Phys. Lett.*, **30**, 463 (1977).
216. G. M. Hieftje and G. R. Haugen, *Anal. Chem.*, **53**, 755A (1981).
217. L. M. Bollinger and G. E. Thomas, *Rev. Sci. Instrum.*, **32**, 1044 (1961).
218. W. R. Wave, *Creation and Detection of the Excited State*, Vol. 1A, A. A. Lamola, Ed., Dekker, New York, 1971.
219. F. E. Lytle and M. S. Kelsey, *Anal. Chem.*, **46**, 855 (1974).
220. R. P. Van Duyne, D. L. Jeanmaire, and D. F. Shriver, *Anal. Chem.*, **46**, 213 (1974).

221. R. S. Meltzer and R. M. Wood, *Appl. Opt.*, **16,** 1432 (1977).
222. J. M. Harris, R. W. Chrisman, F. E. Lytle, and R. S. Tobias, *Anal. Chem.*, **48,** 1937 (1976).
223. J. M. Harris and F. E. Lytle, *Rev. Sci. Instrum.*, **48,** 1469 (1977).
224. U. P. Wild, A. R. Holzwarth, and H. P. Good, *Rev. Sci. Instrum.*, **48,** 1621 (1977).
225. V. J. Koester and R. M. Douben, *Rev. Sci. Instrum.*, **49,** 1186 (1978).
226. G. R. Haugen and F. E. Lytle, *Anal. Chem.*, **53,** 1554 (1981).
227. J. M. Ramsey, G. M. Heiftje, and G. R. Haugen, *Appl. Opt.*, **18,** 1913 (1979).
228. P. M. Rentzepis, *Science*, **202,** 174 (1978).
229. M. A. Dugray, *Progress in Optics*, Vol. XIV, E. Wolf, Ed., North-Holland, Amsterdam, 1976.
230. E. P. Ippen, C. V. Shank, and R. L. Woerner, *Chem. Phys. Lett.*, **46,** 20 (1977).
231. B. H. Ripin, U. Feldman, and G. A. Doschek, *Rev. Sci. Instrum.*, **48,** 935 (1927).
232. D. W. Phillion, D. J. Kuizenga, and A. E. Siegman, *Appl. Phys. Lett.*, **27,** 85 (1975).
233. D. Ricard, W. H. Lowdermilk, and J. Ducuing, *Chem. Phys. Lett.*, **16,** 617 (1972).
234. C. V. Shank, E. P. Ippen, and O. Tesschke, *Chem. Phys. Lett.*, **45,** 291 (1977).
235. W. T. Barnes and F. E. Lytle, *Appl. Phys. Lett.*, **34,** 509 (1979).
236. G. S. Hurst, M. H. Nayfeh, and J. P. Young, *Appl. Phys. Lett.*, **30,** 229 (1977).
237. G. S. Hurst, M. H. Nayfeh, and J. P. Young, *Phys. Rev.*, **15A,** 2283 (1977).
238. S. D. Kramer, C. E. Bemis, J. P. Young, and G. S. Hurst, *Opt. Lett.*, **3,** 16 (1978).
239. R. B. Green, R. A. Keller, G. G. Luther, P. K. Schenck, and J. C. Travis, *Appl. Phys. Lett.*, **29,** 727 (1976).
240. J. C. Travis and J. R. DeVoe, *Lasers in Chemical Analysis*, G. M. Heiftje, J. C. Travis, and F. E. Lytle, Eds., Humana Press, New Jersey, 1981.
241. K. C. Smyth, R. A. Keller, and F. F. Crim, *Chem. Phys. Lett.*, **55,** 473 (1978).
242. W. B. Bridges, *J. Opt. Soc. Am.*, **68,** 352 (1978).
243. E. F. Zalewski, R. A. Keller, and R. Engleman, *J. Chem. Phys.*, **70,** 1015 (1979).
244. J. C. Travis, P. K. Schenck, G. C. Turk, and W. G. Mallard, *Anal. Chem.*, **51,**1516 (1979).
245. E. Erez, S. Laui, and E. Miron, *IEEE J. Quantum Electron.*, **QE-15,** 1328 (1979).
246. G. C. Turk, J. C. Travis, J. R. DeVoe, and T. C. O'Haver, *Anal. Chem.*, **51,** 1890 (1979).
247. R. B. Green, R. A. Keller, P. K. Schenck, and J. C. Travis, *J. Am. Chem. Soc.*, **98,** 8517 (1976).
248. G. C. Turk, J. C. Travis, J. R. DeVoe, and T. C. O'Haver, *Anal. Chem.*, **50,** 817 (1978).
249. G. C. Turk, *Anal. Chem.*, **53,** 1187 (1981).

250. G. C. Turk, W. G. Mallard, P. K. Schenck, and K. C. Smyth, *Anal. Chem.*, **51,** 2408 (1979).
251. J. E. Lawler, A. I. Ferguson, J. E. M. Goldsmith, D. J. Jackson, and A. L. Schawlow, *Phys. Rev. Lett.*, **42,** 1046 (1979).
252. D. S. King, P. K. Schenck, K. C. Smyth, and J. C. Travis, *Appl. Opt.*, **16,** 2617 (1977).
253. P. M. Johnson, *Acc. Chem. Res.*, **13,** 20 (1980).
254. P. M. Johnson, *J. Chem. Phys.*, **64,** 4143 (1976).
255. I. N. Kryazev, Yu. A. Kudryautsev, N. P. Kuz'mina, and V. S. Letokhov, *Sov. Phys. JETP,* **49,** 650 (1979).
256. D. M. Lubman, R. Naaman, and R. N. Zare, *J. Chem. Phys.*, **72,** 3034 (1980).
257. L. Zandee and R. B. Bernstein, *J. Chem. Phys.*, **71,** 1359 (1979).
258. L. Zandee and R. B. Bernstein, *J. Chem. Phys.*, **70,** 2574 (1979).
259. J. H. Brophy and C. T. Rettner, *Chem. Phys. Lett.*, **67,** 351 (1979).
260. S. Leutwyler and U. Even, *Chem. Phys. Lett.*, **81,** 578 (1981).
261. D. Proch, D. M. Rider, and R. N. Zare, *Chem. Phys. Lett.*, **81,** 430 (1981).
262. G. J. Fisanick, T. S. Eichelberger, B. A. Heath, and M. B. Robin, *J. Chem. Phys.*, **72,** 5571 (1980).
263. J. O. Berg, D. H. Parker, and M. A. El-Sayed, *Chem. Phys. Lett.*, **56,** 411 (1978).
264. D. H. Parker, S. J. Sheng, and M. A. El-Sayed, *J. Chem. Phys.*, **65,** 5534 (1977).
265. D. H. Parker, J. O. Berg, and M. A. El-Sayed, *Chem. Phys. Lett.*, **56,** 197 (1978).
266. A. Herrmann, S. Leutwyler, E. Schumacher, and L. Wöste, *Chem. Phys. Lett.*, **52,** 418 (1977).
267. D. L. Feldman, R. K. Lengel, and R. N. Zare, *Chem. Phys. Lett.*, **52,** 413 (1977).
268. A. D. Williamson and R. N. Compton, *Chem. Phys. Lett.*, **62,** 295 (1979).
269. L. Zandee, R. B. Bernstein, and D. A. Lichtin, *J. Chem. Phys.*, **69,** 3427 (1978).
270. D. A. Lichtin, L. Zandee, and R. B. Bernstein, *Lasers in Chemical Analysis,* G. M. Heiftje, J. C. Travis, and F. E. Lytle, Eds., Humana Press, New Jersey, 1981.
271. D. H. Parker and P. Avouris, *J. Chem. Phys.*, **71,** 1241 (1979).
272. D. H. Parker, R. Paudolfi, P. R. Stannard, and M. A. El-Sayed, *Chem. Phys.*, **45,** 27 (1980).
273. K. Krogh-Jespersen, R. P. Rava, and L. Goodman, *Chem. Phys.*, **44,** 295 (1979).
274. J. Murakami, M. Ito, and K. Kaya, *Chem. Phys. Lett.*, **80,** 203 (1981).
275. C. M. Klimcak and J. E. Wessel, *Anal. Chem.*, **52,** 1233 (1980).
276. A. D. Williamson, R. N. Compton, and J. H. D. Eland, *J. Chem. Phys.*, **70,** 590 (1979).
277. M. B. Robin and N. A. Kuebler, *J. Chem. Phys.*, **69,** 806 (1978).
278. D. H. Parker and M. A. El-Sayed, *Chem. Phys.*, **42,** 379 (1979).
279. R. Frueholz, J. Wessel, and E. Wheatley, *Anal. Chem.*, **52,** 281 (1980).

280. T. Hirschfeld, *Anal. Chem.*, **52,** 297A (1980).
281. K. Siomos, G. Kourouklis, and L. G. Christophorov, *Chem. Phys. Lett.*, **80,** 504 (1981).
282. H. E. Van Wart and H. A. Scheraga, *Methods in Enzymology*, Vol. 49, Part G, Academic, New York, 1978, p. 67.
283. V. S. Letokhov, *Opt. Laser Technol.*, 129 (June 1978).
284. P. D. Maker and R. W. Terhune, *Phys. Rev.*, **137,** A801 (1965).
285. Y. R. Shen, *Rev. Mod. Phys.*, **48,** 1 (1976).
286. R. W. Hellwarth, *Prog. Quantum Electron.*, **5,** 1 (1977).
287. M. D. Levenson, *Phys. Today*, p. 44 (May 1977).
288. W. M. Tolles, J. W. Nibler, J. R. McDonald, and A. B. Harvey, *Appl. Spectrosc.*, **31,** 253 (1977).
289. A. B. Harvey, J. R. McDonald, and W. M. Tolles, *Progress in Analytical Chemistry*, Vol. 8, Plenum, New York, 1976, p. 211.
290. M. W. Tolles and R. D. Turner, *Appl. Spectrosc.*, **31,** 96 (1977).
291. A. B. Harvey, *Anal. Chem.*, **50,** 905A (1978).
292. A. Owyoung, *IEEE J. Quantum Electron.*, **QE-14,** 192 (1978).
293. M. Maier, *Appl. Phys.*, **11,** 209 (1976).
294. W. K. H. Panofsky and M. Phillips, *Classical Electricity and Magnetism*, 2nd ed., Addison-Wesley, Reading-London, 1962, pp. 403–424.
295. J. D. Jackson, *Classical Electrodynamics*, 5th ed., Wiley, New York, 1962, pp. 488–493.
296. J. A. Armstrong, N. Bloembergen, J. Ducuing, and P. S. Pershan, *Phys. Rev.*, **127,** 1918 (1962).
297. S. J. Cyvin, J. E. Rauch, and J. C. Decius, *J. Chem. Phys.*, **43,** 4083 (1965).
298. I. Chabay, G. K. Klauminzer, and B. S. Hudson, *Appl. Phys. Lett.*, **28,** 27 (1976).
299. A. Lau, W. Werncke, M. Pfeiffer, K. Lenz, and H. J. Weigmann, *Sov. J. Quantum Electron.*, **6,** 402 (1976).
300. S. Druet and J. P. Taran, *Chemical and Biochemical Applications of Lasers*, Academic, New York, 1979, p. 187.
301. S. Y. Yee, T. K. Gustafson, S. A. J. Druet, and J. P. E. Taran, *Opt. Commun.*, **23,** 1 (1977).
302. N. Bloembergen, H. Lotem, and R. T. Lynch, *Indian J. Pure Appl. Phys.*, **16,** 151 (1978).
303. S. A. J. Druet, B. Attal, T. K. Gustafson, and J. P. Taran, *Phys. Rev.*, **A18,** 1529 (1978).
304. T. K. Yee and T. K. Gustafson, *Phys. Rev.*, **A18,** 1597 (1978).
305. S. A. J. Druet, J. P. E. Taran, and Ch. J. Bordé, *J. Phys.*, **40,** 819 (1979).
306. J. L. Oudar and Y. R. Shen, *Phys. Rev.*, **A22,** 1141 (1980).
307. H. Lotem, R. T. Lynch, and N. Bloembergen, *Phys. Rev.*, **A14,** 1748 (1976).
308. Y. Prior, A. R. Bogdan, M. Dagenais, and N. Bloembergen, *Phys. Rev. Lett.*, **46,** 111 (1981).
309. T. Yajima, *J. Phys. Soc. Jap.*, **21,** 1583 (1966).
310. A. Laubereau, G. Wochner, and W. Kaiser, *Phys. Rev.*, **13A,** 2212 (1976).
311. A. C. Eckbreth, *Appl. Phys. Lett.*, **32,** 421 (1978).

312. R. F. Begley, A. B. Harvey, and R. L. Byer, *Appl. Phys. Lett.*, **25,** 387 (1974).
313. E. Wiener-Avnear, S. Chandra, and A. Compaan, *Appl. Phys. Lett.*, **32,** 286 (1978).
314. B. Hudson, W. Hetherington, S. Cramer, I. Chabay, and G. K. Klauminzer, *Proc. Natl. Acad. Sci. U.S.*, **73,** 3798 (1976).
315. R. T. Lynch, S. D. Kramer, H. Lotem, and N. Bloembergen, *Opt. Commun.*, **16,** 372 (1976).
316. L. A. Carreira, T. C. Maguire, and T. B. Malloy, *J. Chem. Phys.*, **66,** 2621 (1977).
317. C. H. Lee and D. Ricard, *Appl. Phys. Lett.*, **32,** 168 (1978).
318. G. V. Azizbekyan, N. N. Badalyan, N. I. Koroteev, K. A. Nersesyan, M. A. Khurshudyan, and Yu. S. Chilingaryan, *Sov. J. Quantum Electron.*, **7,** 1086 (1977).
319. L. B. Rogers, J. D. Stuart, L. P. Goss, T. B. Malloy, and L. A. Carreira, *Anal. Chem.*, **49,** 959 (1977).
320. P. Regnier and J. P. E. Taran, *Appl. Phys. Lett.*, **23,** 240 (1973).
321. P. Regnier and J. P. E. Taran, *Laser Raman Gas Diagnostics*, Plenum, New York, 1974, p. 87.
322. J. J. Barrett and R. F. Begley, *Appl. Phys. Lett.*, **27,** 129 (1975).
323. F. Moya, S. A. J. Druet, and J. P. E. Taran, *Opt. Commun.*, **13,** 169 (1975).
324. B. R. Hudson, *New Applications of Lasers to Chemistry*, by G. Hieftje, Ed., ACS Symposium Series, Washington, DC, 1978.
325. M. A. Henesian, L. Kulevskii, and R. L. Byer, *J. Chem. Phys.*, **65,** 5530 (1977).
326. A. Owyoung and P. S. Peercy, *J. Appl. Phys.*, **48,** 674 (1977).
327. A. B. Harvey and J. W. Nibler, *Appl. Spectrosc. Rev.*, **14,** 101 (1978).
328. J. L. Oudar, R. W. Smith, and Y. R. Shen, *Appl. Phys. Lett.*, **34,** 758 (1979).
329. L. A. Rahn, L. J. Zych, and P. L. Mattern, *Opt. Commun.*, **30,** 249 (1979).
330. G. Laufer, R. B. Miles, and D. Santavicca, *Opt. Commun.*, **31,** 242 (1979).
331. P. Huber-Wälchli, D. M. Guthals, and J. W. Nibler, *Chem. Phys. Lett.*, **67,** 233 (1979).
332. J. J. Valentini, D. S. Moore, and D. S. Bomse, *Chem. Phys. Lett.*, in press.
333. A. B. Harvey, *Anal. Chem.*, **50,** 905A (1978).
334. W. B. Roh, P. Schreiber, and J. P. E. Taran, *Appl. Phys. Lett.*, **29,** 174 (1976).
335. L. A. Carreira, L. P. Goss, and T. B. Malloy, *J. Chem. Phys.*, **69,** 855 (1978).
336. L. A. Carreira and L. P. Goss, *Advances in Laser Chemistry*, A. H. Zewail, Ed., Springer-Verlag, Berlin, (1978).
337. P. K. Dutta and T. G. Spiro, *J. Chem. Phys.*, **69,** 3119 (1978).
338. P. K. Dutta, R. Dallinger, and T. G. Spiro, *J. Chem. Phys.*, **73,** 3580 (1980).
339. R. Igarashi, Y. Adachi, and S. Maeda, *J. Chem. Phys.*, **72,** 4308 (1980).
340. C. M. Roland and W. A. Steele, *J. Chem. Phys.*, **73,** 5924 (1980).
341. R. T. Lynch, H. Lotem, and N. Bloembergen, *J. Chem. Phys.*, **66,** 4250 (1977).
342. R. König, A. Lau, and H. J. Weigmann, *Chem. Phys. Lett.*, **69,** 87 (1980).
343. A. Lau, R. König, and M. Pfeiffer, *Opt. Commun.*, **32,** 75 (1980).
344. J. P. Devlin and M. G. Tockley, *Chem. Phys. Lett.*, **56,** 608 (1978).
345. S. A. J. Druet, J. P. E. Taran, and Ch. J. Bordé, *J. Phys.*, **41,** 183 (1980).

346. P. L. DeCola, J. R. Andrews, R. M. Hochstrasser, and H. P. Trommsdorff, *J. Chem. Phys.*, **73,** 4695 (1980).
347. S. A. Akhmanov, A. F. Bunkin, S. G. Ivanov, and N. I. Koroteev, *Sov. Phys. JETP,* **47,** 667 (1978).
348. N. I. Koroteev, M. Endemann, and R. L. Byer, *Phys. Rev. Lett.*, **43,** 398 (1979).
349. C. H. Lee and D. Ricard, *Appl. Phys. Lett.*, **32,** 168 (1978).
350. F. M. Kanga and M. G. Sceats, *Opt. Lett.*, **5,** 126 (1980).
351. A. Owyoung and E. D. Jones, *Opt. Lett.*, **1,** 152 (1977).
352. C. D. Decker, *Appl. Phys. Lett.*, **33,** 323 (1978).
353. D. Cotter and D. C. Hanna, *IEEE J. Quantum Electron.*, **QE-14,** 184 (1978).
354. P. Lallemand, P. Simova, and G. Bret, *Phys. Rev. Lett.*, **17,** 1239 (1966).
355. A. Owyoung, *Opt. Commun.*, **22,** 323 (1977).
356. B. F. Levine and C. G. Bethea, *Appl. Phys. Lett.*, **36,** 245 (1980).
357. B. F. Levine and C. G. Bethea, *IEEE J. Quantum Electron.*, **QE-16,** 85 (1980).
358. J. P. Heritage, J. G. Bergman, A. Pinczuk, and J. M. Worlock, *Chem. Phys. Lett.*, **67,** 229 (1979).
359. J. P. Heritage and D. L. Allara, *Chem. Phys. Lett.*, **74,** 507 (1980).
360. M. D. Morris, D. J. Wallan, G. P. Ritz, and J. P. Haushalter, *Anal. Chem.*, **50,** 1796 (1978).
361. J. P. Haushalter, G. P. Ritz, D. J. Wallan, K. Dien, and M. D. Morris, *Appl. Spectrosc.*, **34,** 144 (1980).
362. J. P. Haushalter and M. D. Morris, *Anal. Chem.*, **53,** 21 (1981); J. P. Haushalter, C. E. Buffett, and M. D. Morris, *Anal. Chem.*, **52,** 1284 (1980).
363. J. R. Nestor, *J. Chem. Phys.*, **69,** 1778 (1978).
364. J. J. Barrett and M. J. Berry, *Appl. Phys. Lett.*, **34,** 144 (1979).
365. C. K. N. Patel and A. C. Tam, *Appl. Phys. Lett.*, **34,** 760 (1979).
366. V. J. Klein, W. Werncke, A. Lau, G. Hunsalz, and K. Lenz, *Exp. Tech. Phys.*, **22,** 565 (1974).
367. W. Werncke, A. Lau, M. Pfeiffer, H. J. Weigmann, G. Hunsalz, and K. Lenz, *Opt. Commun.*, **16,** 128 (1976).
368. Y. Kuto and H. Takuma, *J. Chem. Phys.*, **54,** 5398 (1971).
369. R. A. McLaren and B. P. Stoichaff, *Appl. Phys. Lett.*, **16,** 140 (1970).
370. E. S. Yeung, *J. Mol. Spectrosc.*, **53,** 379 (1974).
371. W. Werncke, J. Klein, A. Lau, K. Lenz, and G. Hunsalz, *Opt. Commun.*, **11,** 159 (1974).
372. D. Heiman, R. W. Hellwarth, M. D. Levenson, and G. Martin, *Phys. Rev. Lett.*, **36,** 189 (1976).
373. M. D. Levenson and J. J. Song, *J. Opt. Soc. Am.*, **66,** 641 (1976).
374. J. J. Song, G. L. Eesley, and M. D. Levenson, *Appl. Phys. Lett.*, **29,** 567 (1976).
375. G. L. Eesley, M. D. Levenson, and W. M. Tolles, *IEEE J. Quantum Electron.*, **QE-14,** 45 (1978).
376. A. G. Miller, *Anal. Chem.*, **49,** 2044 (1977).
377. K. W. Cunningham, M. C. Goldberg, and E. R. Weiner, *Anal. Chem.*, **49,** 70 (1977).

378. L. Van Hauerbeke and M. A. Herman, *Anal. Chem.*, **51,** 932 (1979).
379. R. P. Van Duyne and K. D. Parks, *Chem. Phys. Lett.*, **76,** 196 (1980).
380. R. P. Van Duyne, D. L. Jeanmaire, and D. F. Shriver, *Anal. Chem.*, **46,** 213 (1974).
381. G. R. Haugen and F. E. Lytle, *Anal. Chem.*, **53,** 1554 (1981).
382. M. Geoppart Mayer, *Ann. Phys.*, **9,** 273 (1931).
383. W. Kaiser and C. G. Garrett, *Phys. Rev. Lett.*, **9,** 455 (1961).
384. M. W. McClain and R. A. Harris, Excited States, Vol. 3, E. C. Lin, Ed., Academic, New York, 1977.
385. M. J. Wirth and F. E. Lytle, *New Applications of Lasers to Chemistry*, G. M. Hieftje, Ed., ACS Symposium Series 85, Washington, DC, 1978.
386. M. C. Johnson and F. E. Lytle, *J. Appl. Phys.*, **51,** 2445 (1980).
387. W. M. McClain, *Acc. Chem. Res.*, **7,** 129 (1974).
388. M. J. Wirth and F. E. Lytle, *Anal. Chem.*, **49,** 2054 (1977).
389. P. R. Monson and W. M. McClain, *J. Chem. Phys.*, **53,** 29 (1970).
390. W. M. McClain, *J. Chem. Phys.*, **55,** 2789 (1971).
391. M. J. Wirth, A. Koskelo, and M. J. Sanders, *Appl. Spectrosc.*, **35,** 14 (1981).
392. M. J. Wirth, A. C. Koskelo, C. E. Mohler, and B. L. Lentz, *Anal. Chem.*, **53,** 2045 (1981).
393. R. M. Hochstrasser, H. N. Sung, and J. E. Wessel, *Chem. Phys. Lett.*, **24,** 168 (1974).
394. R. M. Hochstrasser and H. N. Sung, *J. Chem. Phys.*, **66,** 3276 (1977).

Chapter **III**

ULTRAVIOLET AND VISIBLE LASER PHOTOCHEMISTRY

Ted R. Evans

Research Laboratories, Eastman Kodak Company
Rochester, New York

Flash photolysis, invented by Norrish and Porter more than 30 years ago [1, 2], has yielded a wealth of information concerning the spectral properties of transient species and the mechanisms of photochemically initiated reactions. Many improvements have been made on the original design, the most dramatic being the replacement of the discharge lamp with a laser for sample excitation. Lasers, which produce high-intensity, monochromatic pulses whose duration can be controlled from microseconds to several hundred femtoseconds, are ideal flash-photolytic excitation sources.

The apparatus for nanosecond flash photolysis has the same three basic elements as that for conventional flash photolysis [3, 4]: (1) an excitation source, (2) a sample chamber, and (3) a detection system.

1 APPARATUS

Nanosecond Pulsed Lasers

In the original flash photolysis experiments, the fundamental line, 694.3 nm, from a Q-switched ruby laser [5, 6] was used, and the need for an ultraviolet source was rapidly fulfilled by use of a frequency-doubling device to obtain output at 347.1 nm [7]. The frequency-doubled ruby laser is one of the most common laser sources for flash photolysis in the ultraviolet region, but a number of other lasers provide outputs in this region (Table 1). In the past few years rapid advances have been made in the development of ultraviolet gas lasers [8]. The rare gas exciplex lasers [8, 9] produce radiation at 193 (ArF), 248 (KrF), 308 (XeCl), and 351 nm (XeF). The energies available at these wavelengths are sufficient for single-shot transient detection. Raman scattering can be used to fill in some of the gaps in the ultraviolet, although with a significant loss in intensity [11] (Table 2). Other lasers that can be used in the ultraviolet are the nitrogen laser [12–14] at 337 nm and the tripled (353 nm) and quadrupled (266 nm) neodymium laser [15–18]. Radiation suitable for flash photolysis in the 400 to 900 nm region can be obtained from dye lasers [19–22] pumped either with a flashlamp or with any of the above-mentioned laser sources. These lasers fulfill most wavelength requirements for excitation of broad-band absorbing materials.

The pulse duration of a solid-state laser is generally controlled by a Pockels cell Q-switch [23] and is usually 20 to 30 nsec (Fig. 1*a*). Shorter pulses (2 to

Table1. Wavelengths Available from Common Pulsed Lasers

Laser	*Output (nm)*
ArF	193
KrCl	223
KrF	248
Neodymium-YAG, quadrupled	266
XeCl	308
N_2	337
Ruby, doubled	347
XeF	351
Neodymium-YAG, tripled	355
Neodymium-YAG, doubled	532
Ruby, fundamental	694
Neodymium-YAG, fundamental	1064

Table 2. Frequency Shifts for Raman Media

Raman Medium	ΔE (cm^{-1}) [11]
Liquid N_2	2326
CH_4	2916
D_2	2991
H_2	4155

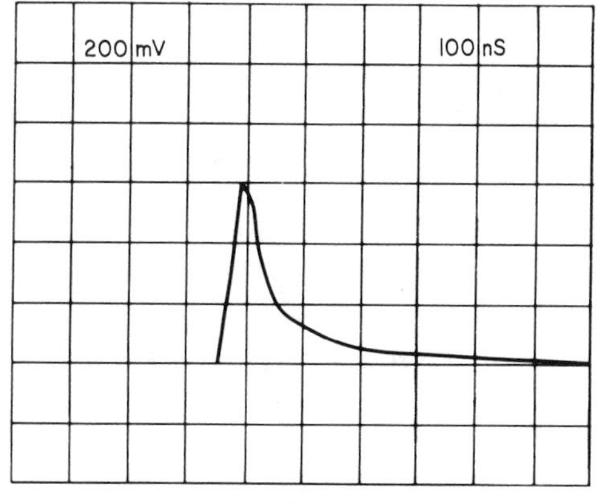

(*a*)

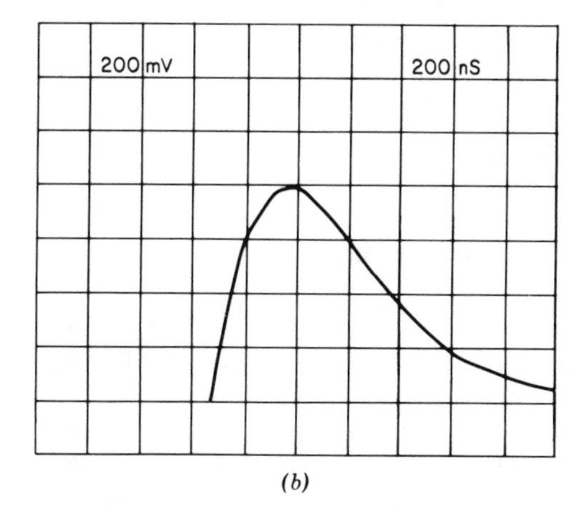

(*b*)

Fig. 1 (*a*) Pulse profile for Q-switched ruby laser. (*b*) Pulse profile for flash-pumped dye laser.

20 nsec) can be obtained by either Q-switching in the pulse-transmission mode (PMT) [15, 23] or by pulse slicing with an external Pockels cell [23].

Gas lasers are self-terminating with pulses of 10 nsec [24, 25] or less [26, 27] for nitrogen and 10 to 100 nsec for discharged pumped rare gas halide lasers [8].

The pulse duration of flashlamp-pumped dye lasers is dependent only on the duration of the exciting discharge lamp. Commercially available flash-pumped dye lasers have pulses of 100 nsec to 2 μsec (Fig. 1*b*) [28].

Sample Compartments

Since lasers are highly collimated (3 to 5 mradians divergence), the sample can be placed at a position remote from the laser. This allows flexibility in the experimental arrangement since the sample is not confined to one particular position. The geometry of the excitation source and the monitoring source can be either a collinear [8, 29] or right-angle [7, 8, 30–33] arrangement (Fig. 2). The major factor in choosing between the two is the optical density of the ground state at the laser wavelength.

Right-angle excitation produces a nonlinear concentration gradient of transients across a 1-cm cell. The degree of nonlinearity depends on the optical density of the absorbing species (Fig. 3). For single-pass arrangements a compromise must be made between the maximum yield of transients and a homogeneous distribution of excited states throughout the cell. If the initial optical density is too high, all the transients are produced at the face of the cell and few are observed in the monitoring path for right-angle observation. If the optical density is too low, few transients are produced, and the resultant measurements is difficult. The results shown in Fig. 3 were obtained by calculating the amount of light absorbed as a function of penetration through a 1-cm cell with the assumption that the Beer-Lambert law pertains.

Depending on the experiment and the experimental arrangement, optimum results will be obtained for different starting optical densities.

Transient absorption spectra are best obtained under conditions of maximum transient concentration in the path of the monitoring beam. For the right-angle geometry and viewing in the center 0.5 cm of the cell, the most advantageous ground-state optical density is ~1.0 (Fig. 3*b*). The transient concentrations in the center portion of the cell fall rapidly with higher optical densities, and thus the collinear arrangement must be used.

The requirements for kinetic measurements are different from those for spectral measurements. Although high transient concentrations are necessary, the uniformity of the transient population throughout the monitoring region is also important. The best ground-state optical density for right-angle kinetic monitoring is 0.2 to 0.4. The transient yield can be increased, at low optical densities, by using an off-axis reflector behind the sample cell and in-

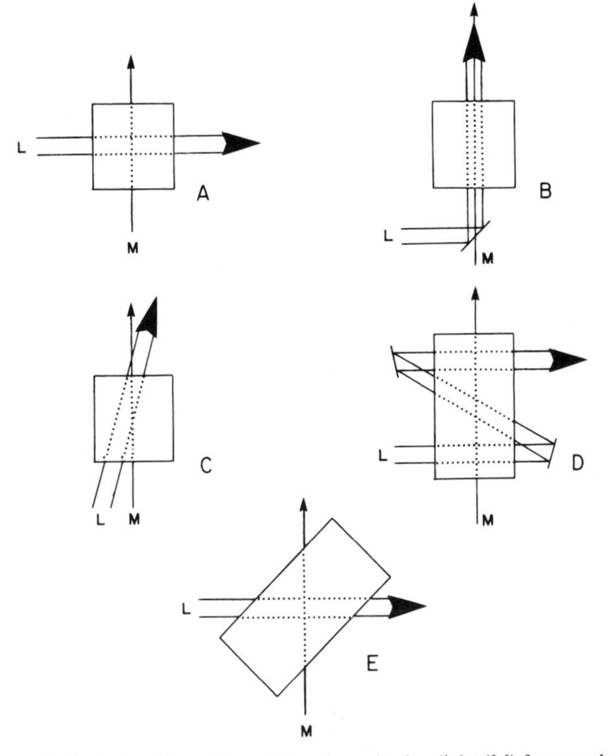

Fig. 2 Geometrical relationships of laser (L) and monitoring light (M) for sample excitation.

creasing the cell length (Figs. 2 and 4). Care must be taken that no light is reflected back into the laser, as this can cause severe damage. Samples with high optical density can be excited with a thin cell mounted at 45° to both the exciting source and the monitoring beam (Fig. 2*E*) or in one of the collinear arrangements (Figs. 2*B* and *C*). The collinear arrangement in Fig. 2*B* is accomplished by using a dichroic mirror to turn the laser beam. This has the advantage over the off-axis excitation (Fig. 2*C*) that the laser beam and the monitoring beam can be adjusted to overlap exactly. Long cells with samples with low optical density can be used in the collinear arrangement, yielding greater transient absorption than is obtained with the right-angle geometry. Bebelaar [15] has concluded that cross excitation is advantageous if the ground-state optical density is ~ 0.4 OD units, whereas samples with optical densities >0.4 are best examined in the collinear arrangement.

The relative dimensions of the excitation and monitoring beams need to be carefully controlled. If the height of the opening (Fig. 5) is larger than or

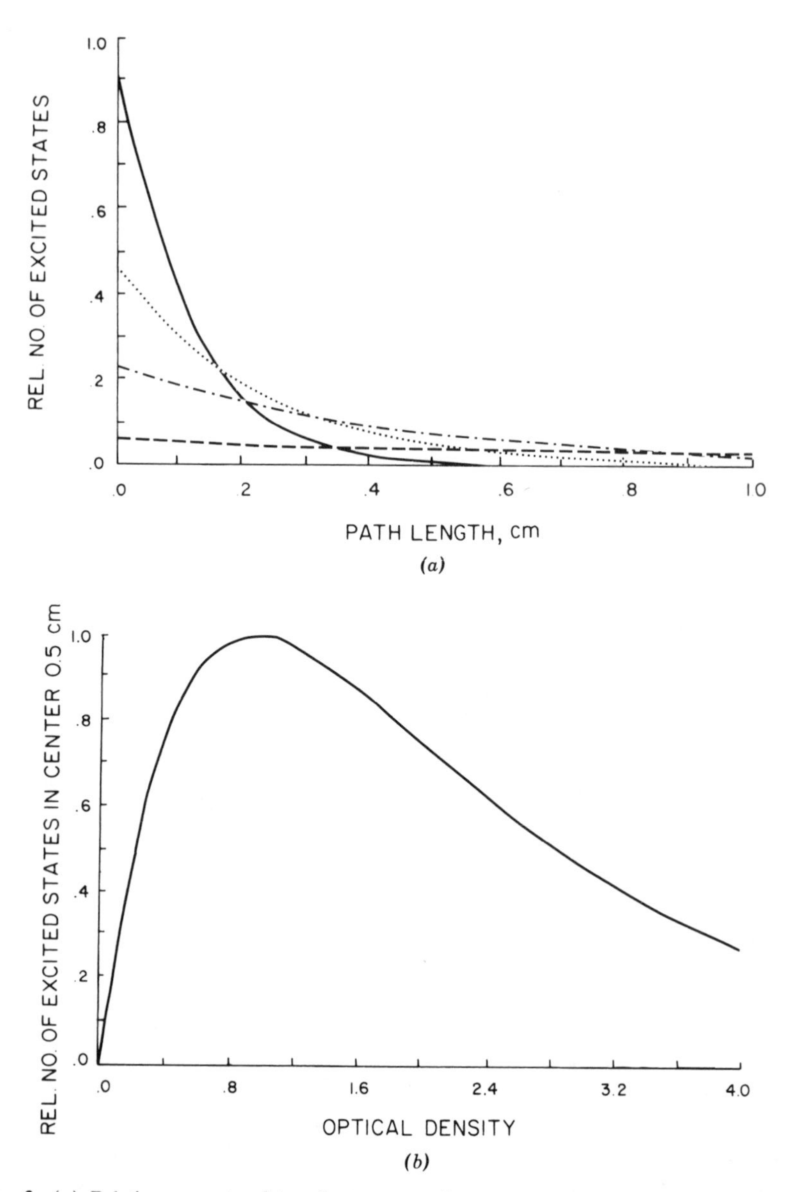

Fig. 3 (*a*) Relative amounts of transients versus distance across a 1-cm cell. (*b*) Relative amounts of transients observable in the center 0.5 cm of a 1-cm cell as a function of optical density.

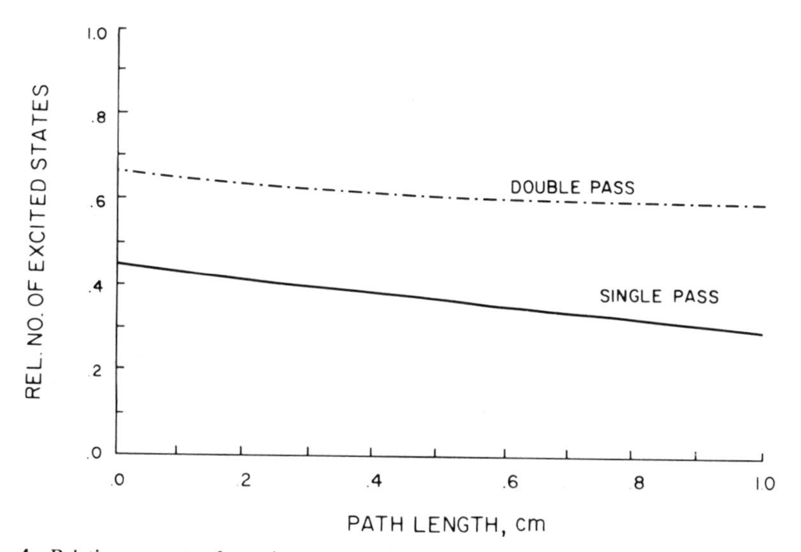

Fig. 4 Relative amounts of transients versus distance across a 1-cm cell for single- and double-pass arrangements. Optical density = 0.2; values have been multiplied by 10.

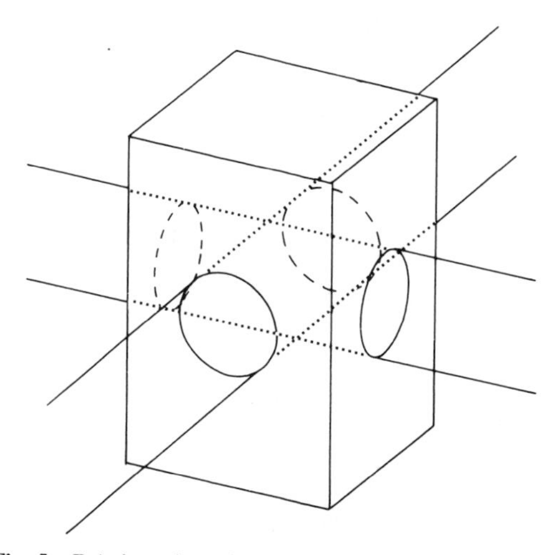

Fig. 5 Relative orientations of laser and monitoring beams.

misaligned with the excitation source, the optimum reading is reduced [13, 15]. Although the height of the monitoring source can be identical to that for the laser, it is usually less and the source is placed near the entrance of the laser beam [15, 34].

The intensity of a laser is rarely uniform across the beam and produces inhomogeneous populations of transients. Inhomogeneities can be reduced by using an off-axis mirror behind the sample or by first passing the laser through a frosted-glass plate [34].

Monitoring Light Sources and Detectors

Transient characterization requires two types of measurement, kinetic and spectral. Kinetic measurements are generally obtained with a high-pressure xenon lamp in combination with a monochromator and a photomultiplier tube. For samples with relatively small optical density changes, the signal-to-noise ratio observed with any given photomultiplier/light source combination is approximately

$$S = 0.43(I_0 \Phi \Delta t)^{1/2} \epsilon_t C_t d \qquad (1)$$

where I_0 is the monitoring light source intensity, Φ is the quantum efficiency of the photocathode, Δt is the resolution time, and ϵ_t, C_t, and d are the extinction coefficient, the concentration, and the effective path length for the transient absorption, respectively. Examination of (1) shows that the signal levels are more effectively increased by increasing either the concentration or the effective path length rather than the monitoring light intensity or the photocathode efficiency.

High light intensities are advantageous because of the increase in the signal-to-noise ratio and because scattering and fluorescence are then relatively less important. Since scattering and fluorescence radiate in all directions, the light intensity, within the plane of operation, decreases as a function of r^{-2}. Therefore the placement of the photomultiplier relatively far from the sample reduces the effect of scattering and fluorescence and still allows good signal-to-noise ratios if the monitoring light is well collimated. Scattering and fluorescence can also be reduced by inserting a baffle between the sample holder and the photomultiplier, thereby eliminating off-axis emissions.

High light intensities can be achieved either by using a long-lived flash-lamp or by the intensification of continuous high-pressure xenon arcs [8, 35–42]. These light sources are nonlinear over long time periods but are approximately flat for the region of interest, 10 to 1 μsec. Lasers can be used as the monitoring source for kinetic measurements. Since continuous dye lasers operate throughout the visible and near-infrared regions, there is no absolute limit to their use. In practice, the use of lasers for the monitoring beam has been limited to helium–neon at 628.8 nm [43] and pulsed diode lasers at 820 nm [44].

Transient absorption spectra can be, and usually are, measured point-by-point with the kinetic apparatus. Spectra can also be obtained with a spectroscopic arrangement composed of a pulsed monitoring light, which is synchronized with the exciting light, and a spectrograph. Ideally, the light source for spectroscopic investigations should have pulses whose durations and synchrony with the exciting light can be systematically varied and whose output is continuous throughout the spectrum.

Novak and Windsor [7] and Porter and Topp [45] have used a laser-initiated spark gap as a white-light source. The flash durations from the spark gap varied with the gas [7]. The durations were 30 nsec for an oxygen lamp, 200 nsec for argon, and >1 μsec for xenon. Porter and Topp later replaced the spark-initiated discharge with a solution of 1,1′,4,4′-tetraphenylbutadiene, which emits a broad band extending from 400 to 600 nm [46]. Others have extended this work to include various laser dyes and scintillators [4, 47, 48] (Table 3).

Lin and Stolen [49] have used a fiber waveguide to generate a continuum with dye-laser excitation. The radiation from the dye laser is wavelength broadened by nonlinear optical processes, yielding a continuum 100 to 200 nm wide. Since the spectral width of the continuum is dependent on the length of the fiber, broad-band output is concomitant with a delay, with respect to the exciting pulse. A 19.5-m silica fiber produced a 95-nsec delay.

Two lasers have often been used to obtain time-resolved transient absorp-

Table 3. Fluorescent Dyes for Monitoring Sources

Fluorescent Dye	*Solvent*	*Useful Wavelength Range (nm)*	λ_{max} *(nm)*
p-Terephenyl	Cyclohexane	315–390	340[a]
2,5-Diphenyloxazole	Ethanol	335–440	360[a]
1,1′,4,4′-Tetraphenylbutadiene	Cyclohexane	390–480	455[a,b]
2-Amino-7-nitrofluorene	Benzene	452–650	520[a]
	2-Propanol	500–1000	690[a]
4-Dimethylamino-4′-nitrostilbene	Isobutanol	530–1100	720[a]
Rhodamine 6G	Methanol	560–680	[c]
Rhodamine B	Ethanol	560–680	580[c]

[a]Ref. 47.
[b]Ref. 46.
[c]Ref. 48.

tion spectra [50–53]. The usual arrangement consists of a primary laser that excites the sample and also pumps a dye laser whose output is used as a probe (Fig. 6). The time resolution is accomplished by the difference in the distances that the probe and the primary pulse traverse. A 10-nsec delay between the probe and primary pulses can be achieved by moving the mirrors M_1 and M_2 enough to increase the probe path to 300 cm greater than that for excitation. One advantage in using a dye laser for the monitoring probe is that the intensity is sufficient for detection by a ballistic thermopile or a photodiode. Since the monitoring probe is so intense, the fluorescence from the sample can be ignored, and the absorption in the region of fluorescence can be obtained. The I_0 is obtained by blocking the excitation pulse with a shutter and measuring the energy transmitted through the unexcited sample cell.

Distortions of internal fluorescence have also been used to obtain transient spectra [45, 54, 55]. Masuhara and Mataga have shown that the fluorescence distortion is appreciable for rear or side viewing of the fluorescence from materials where (a) the ground-state optical density is relatively high (about 1.0), (b) the excitation intensity is high (about 10^{20} photons/pulse for a 20-nsec pulse width), and (c) there is overlap of the $S_i^* \rightarrow S_n^*$ absorptions with the fluorescence emission. The emission at low light intensities is compared with that observed at high excitation intensities to obtain the transient absorption spectra.

For kinetic measurements and point-by-point spectral determinations,

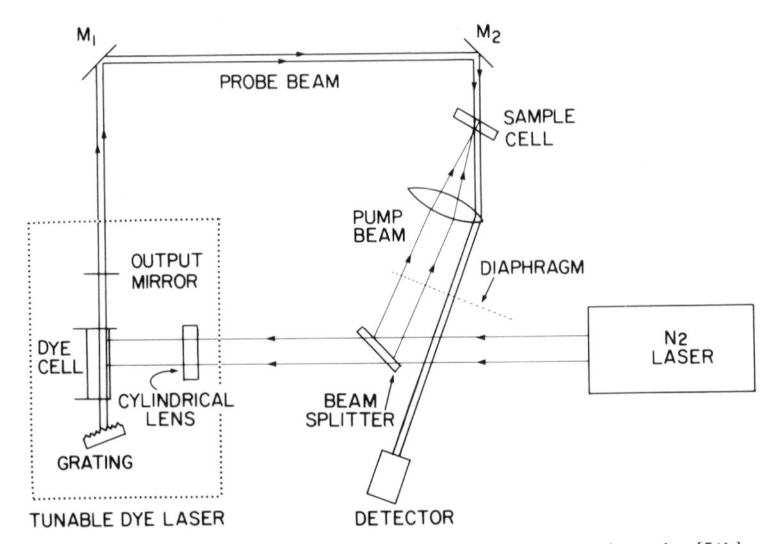

Fig. 6 Experimental arrangement for measuring excited-state absorption [51b].

photomultiplier tubes, wired for rapid response, are the most generally applicable devices [42, 56–58]. The output from the photomultiplier can be either recorded oscillographically and photographed, or interfaced to a transient recorder. In turn, the transient recorder can be interfaced to a microprocessor or a computer for subsequent analysis [59–61].

There has been increasing use of optical multichannel analyzers for spectroscopic applications [62–65] (Fig. 7). The optical multichannel analyzer consists of a vidicon tube or a diode array, either of which will register intensities as a function of distance across the face of the tube. When the incoming light is dispersed by a spectrograph, often a monochromator with a large exit slit, the resultant intensity versus wavelength output is usually stored in a multichannel analyzer. Since these tubes can be electronically gated, usually

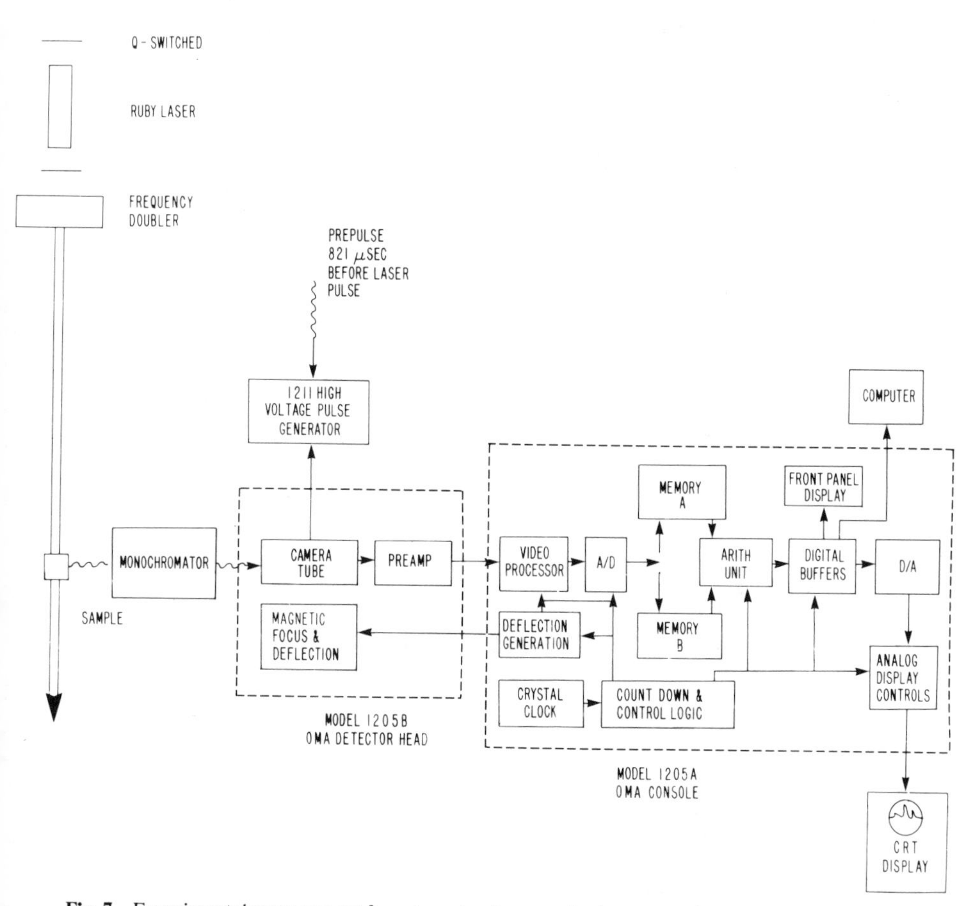

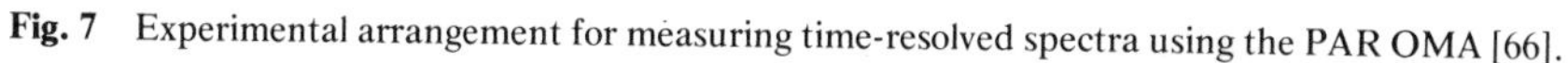

Fig. 7 Experimental arrangement for measuring time-resolved spectra using the PAR OMA [66].

triggered by one of the output signals from the laser, time-resolved spectra can be obtained for both emission and absorption. The PAR pulse generator can gate the OMA [66] with a variable delay and gate width. By using the capacitor charge pulse from the laser, 821 μsec before the laser pulse, to trigger the high-voltage pulse generator, the time-resolved emission spectra such as those shown in Fig. 8 can be obtained.

One severe limitation to optical spectroscopy is the lack of direct structural information obtained from the measurement of spectra. The absorption spectra must be related to structure by independent means. Spectral measurements in the infrared provide much more structural information, but the extinction coefficients in these regions are only a few hundred, and infrared detectors have slow response times and are rather insensitive. Bethune et al. [67] recently demonstrated the use of metal vapors to convert 400-cm^{-1} wide infrared absorption spectra into visible spectra that could

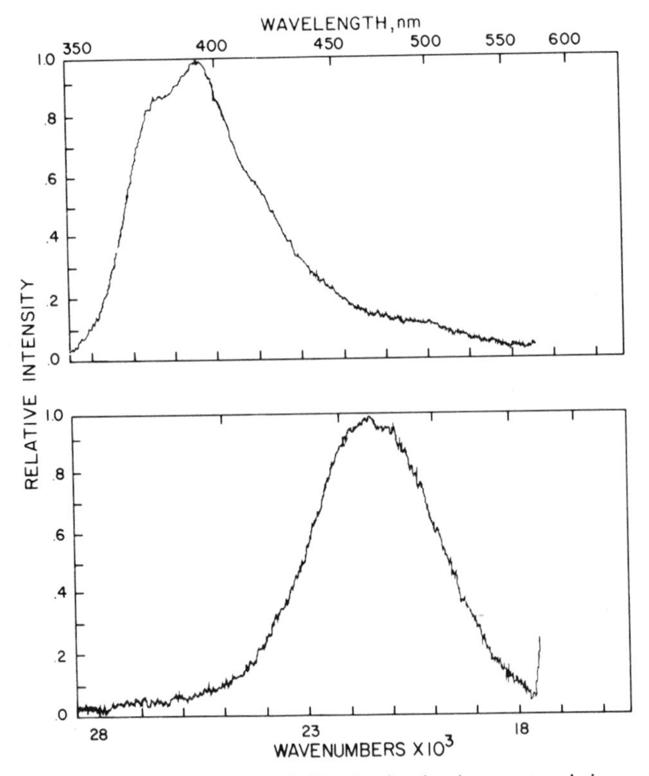

Fig. 8 Emission spectra from 1,8-naphthalimide showing (*top*) prompt emission and (*bottom*) delayed emission ensuing from triplet-triplet annihilation.

then be processed as visible light. The basis of these experiments is the nonlinear technique known as two-photon resonant optical four-wave mixing. This is shown pictorially in Fig. 9. The pump frequency, two-photon, and the infrared light are combined in a metal vapor cell and by means of the third-order nonlinear susceptibility produces the visible output.

Other recent complementary techniques that promise to provide structural information about transient species are Raman and stimulated Raman spectroscopy [69–73]. The sensitivity of Raman spectroscopy can be improved by resonance intensity enhancements. By use of resonance Raman spectroscopy, concentrations of $\sim 10^{-6}$ M can be detected. When two lasers are synchronized, one exciting the ground state and the other tuned to the red edge of the transient, time-resolved Raman spectra can be obtained (Fig. 10).

Fluorescence is often a serious interference with Raman techniques. The nonlinear Raman scattering technique called coherent anti-Stokes Raman scattering (CARS) [74] can also be applied for time-resolved measurements [75–77]. Since the scattering is measured to the short-wavelength region of the probe laser rather than to the long-wavelength region, fluorescence does not interfere.

The most direct method to obtain structural information about solid

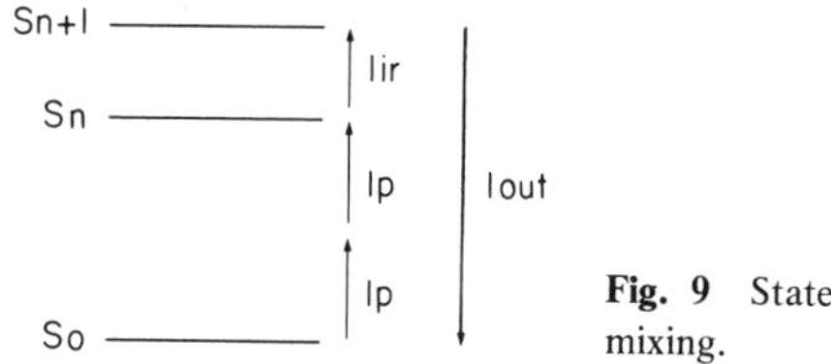

Fig. 9 State diagram for two-photon resonant four-wave mixing.

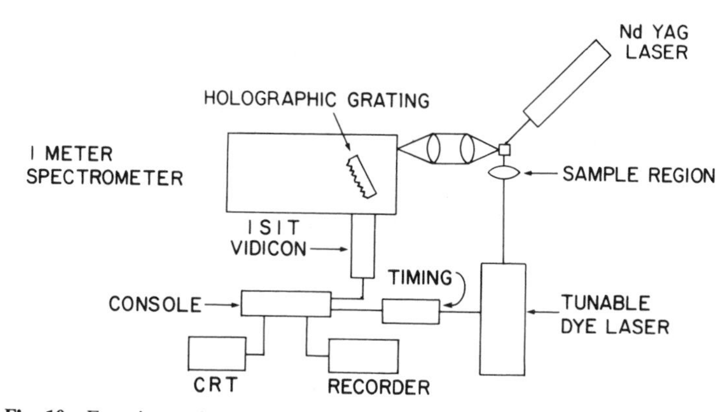

Fig. 10 Experimental arrangement for time-resolved Raman measurements [73].

samples is X-ray diffraction. It has been known for some time that the interaction of terawatt lasers with matter produced X-ray emission. This emission is now used as a diagnostic probe for molecular motions during irradiation [78].

In addition, transient photoconductive techniques can be used to obtain information concerning charge generation in photochemical reactions [79–82], and photoacoustic detection has been applied to transient species [83, 84].

Information concerning free radicals can be obtained from time-resolved electron spin resonance (ESR) and chemically induced nuclear polarization (CIDNP) experiments [85–88]. The current time resolution is in microseconds rather than nanoseconds.

Laser Intensity Measurements

For most measurements it is necessary to monitor the exciting laser pulse intensity as well as the signal of interest. We use a pair of photodiodes, before and after the sample, into which the laser beam is directed by means of quartz beam splitters (Fig. 11). The intensity of the pulse is reduced by neutral-density filters, and monochromicity is ensured by use of an interference filter. The output from these photodiodes is sent to amplifiers and then to sample-and-hold circuits that give a readout in millivolts [89–91]. This reading is then calibrated by means of a commercial ballistic [28, 92] thermopile for high-power operation (Fig. 12). At low-power operation we have found ferrioxalate to be a suitable actinometer [92–94]. We have also found

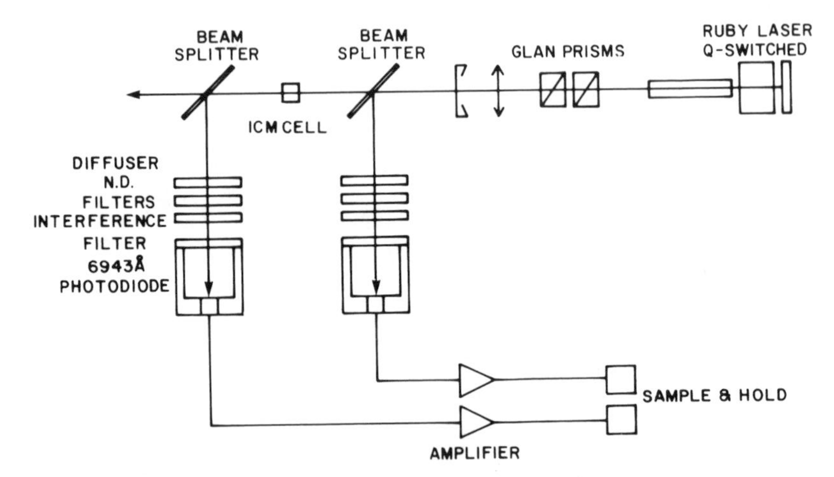

Fig. 11 Beam-splitter arrangement for laser intensity measurements.

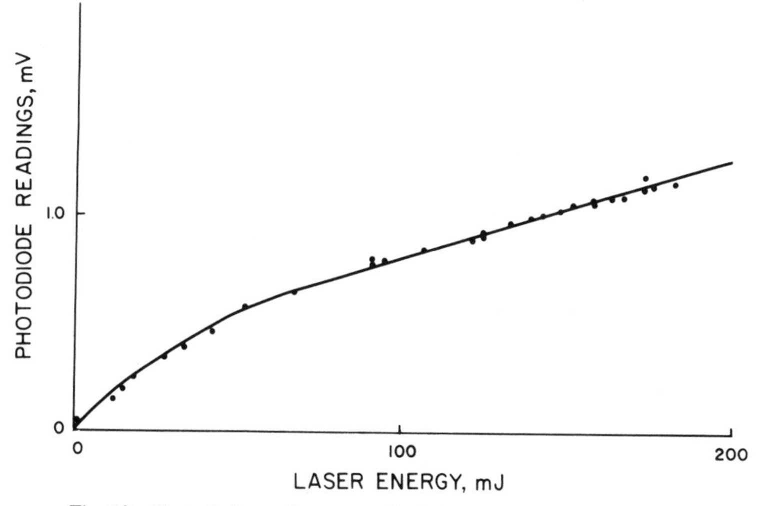

Fig. 12 Photodiode reading versus ballistic thermopile measurements.

the triplet–triplet absorptions of aromatic hydrocarbons to be suitable actinometers [95] (Fig. 13).

Naqvi [96] have devised a molecular actinometer based on the rapid decay of a dye. If the singlet state decays rapidly and no long-lived transients are formed, photostationary kinetics can be applied,

$$\frac{d[S^{*1}]}{dt} = IG_0[S_0] - \frac{[S^{*1}]}{\tau} = 0 \tag{2}$$

where I is the laser intensity, G_0 is the absorption cross-section for the ground state, and τ is the lifetime of the first excited singlet state $S^{1}*$. Since $[S^{*1}] + [S_0] = [S_g]$, the starting material concentration before irradiation, the equation can be rearranged to give

$$[S^1*] = \frac{G_0 I\tau[S_g]}{1 + G_0 I\tau} \tag{3}$$

With Δ denoting the difference in optical densities of the sample before and during the laser pulse, $\alpha = G_0 I$, $k = 1/\tau$, the following Michaelis–Menton type of equation can be obtained:

$$\Delta = \frac{\delta_0 \alpha}{k + \alpha} \tag{4}$$

This can be treated in the same manner as the Michaelis–Menton equation to yield α, which gives the laser intensity.

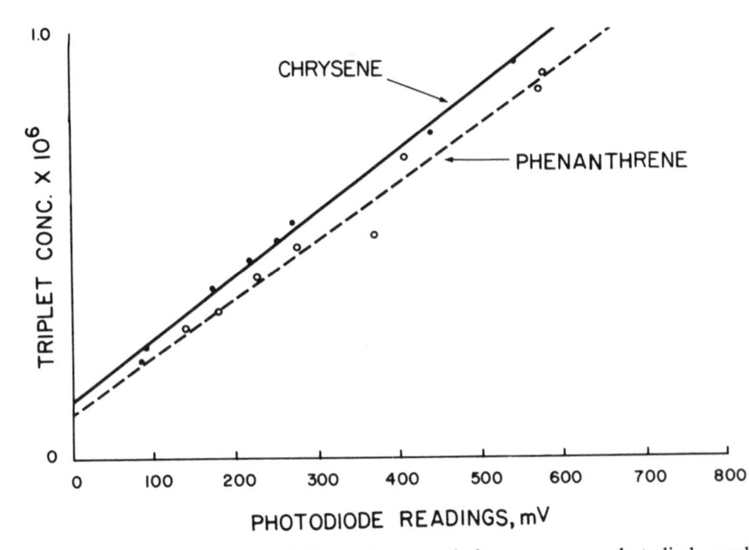

Fig. 13 Triplet-triplet absorption of phenanthrene and chrysene versus photodiode readings.

Sample Holder

Liquid samples that do not need to be degassed are usually contained in conventional Suprasil spectrometer cells with path lengths of 0.1 to 1 cm. We have found that normal quartz or Pyrex cells often emit when excited with high-intensity ultraviolet lasers.

For examination of photostable materials in the absence of air, a simple Suprasil cell sealed onto a degassing bulb is used. In general, materials of interest to photochemists are photoreactive and a fresh sample must be used for each flash. To obviate the necessity for a large number of single cells, we modified an existing design [97] for a multiple-sample cell (Fig. 14). By tipping the apparatus up, new material can be introduced into the optical cell, and by tipping in the other direction, the spent sample can be drained into the waste reservoir. By increasing the length of the tube connecting the apparatus to the optical cell, the multiple-cell holder can be used in a low-temperature Dewar flask.

2 CHEMISTRY INDUCED BY LASER EXCITATION

Using lasers as excitation sources offers tremendous advantages to photochemists, but there are also some pitfalls. With nanosecond excitation, and its expensive picosecond brother, instant photochemistry can be effected. Photostationary techniques always require indirect methods to identify the

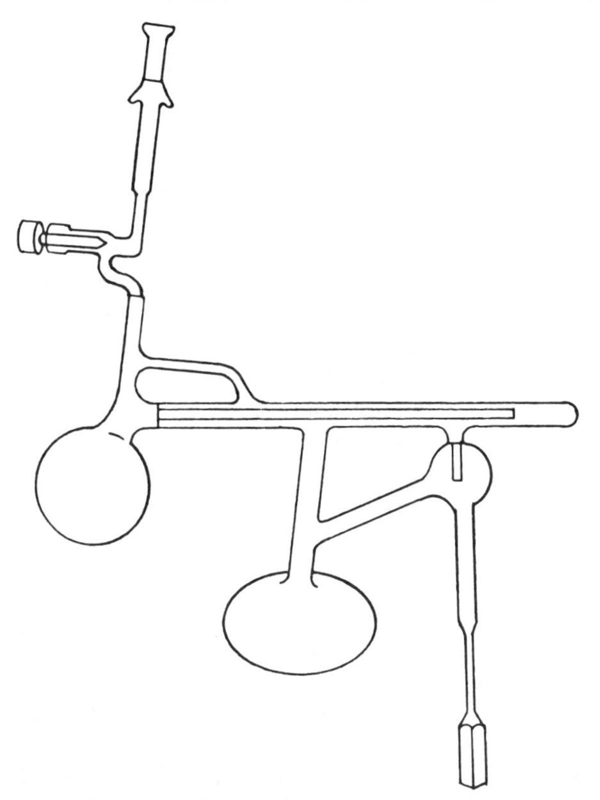

Fig. 14 Multiple-sample cell for degassed samples.

reactive states and the intermediates preceding the final stable products. Laser excitation often allows the direct observation of the reactive state, and its change with reaction conditions can be monitored.

Flash photolysis has great application in the study of monophotonic photochemical events, but the use of lasers certainly transcends this single use. Chemists now have a light source capable of inducing unique chemical events and, we hope, unique chemical products.

Kinetics

Analyzing kinetic data, such as that obtained from flash-photolysis experiments, is complicated by the fact that a variety of kinetic and experimental patterns may pertain. There are also several types of measurements that can be made and various experimental situations. One basic chemical reaction scheme is

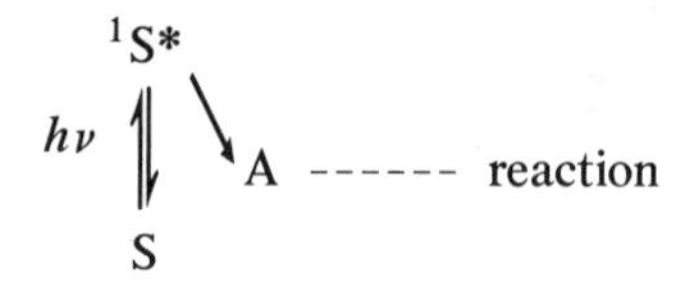

where the ground-state molecule is excited to the first excited singlet state, which then yields a longer-lived state A, which could be a triplet, a radical, or some other metastable species. Since the lifetime of A is generally orders of magnitude greater than the few nanoseconds required for $^1S^*$ to decay, only the kinetic behavior of A and its products are observable with normal nanosecond excitation. After $^1S^*$ has decayed, but before appreciable amounts of A have reacted, the concentration of A can be defined as $[A]_0$, the zero-time value. The ground-state concentration before excitation $[S]_g$ is depleted by the amount of A formed, leaving the ground-state concentration at zero time, $[S]_0$. If the only reaction of A is to re-form S, then the following mass-balance equations apply:

$$[S]_g - [A]_0 = [S]_0 \tag{5}$$

$$[A]_0 = [A]_t + [S]_t \tag{6}$$

where $[A]_t$ is the concentration of A at any time and $[S]_t + [S]_0$ is the concentration at ground state at any time.

Four main kinetic types are commonly encountered, both as transient decay and as product formation or ground-state re-formation: (1) first-order, (2) double-exponential, (3) second-order, and (4) mixed first- and second-order. The appropriate differential equations can be obtained by inspection, but obtaining the integrated equations can occasionally be difficult. There are two approaches to this problem. Most solutions can be obtained by normal integration of the differential [98, 99], but coupled first-order reactions can also be solved by matrix methods [98–102]. Once the integrated form for $[A]_t$ is obtained, the corresponding equation for $[S]_t$ can be obtained by applying the mass-balance equations (5) and (6). The exactly analogous equations for $[A]_t$ can be obtained for $[S]_t$ by using the substitutions $[S]_g - [S]_0 - [S]_t = [A]_t$ and $[S]_g - [S]_0 = [A]_0$.

There have been several treatments directed toward the separation of the first- and second-order decay rates from each other [103–104]. For first-order decay, A → S (k_1), the differential equation is

$$\frac{d[A]_t}{dt} = -k_1\,dt \tag{7}$$

giving the integrated form

$$\ln [A]_t = -k_1 t + C \tag{8}$$

Since $[A]_t = [A]_0$ at $t = 0$, $C = \ln [A]_0$ or $[A]_t = [A]_0 e^{-k_1 t}$. For second-order decay, $2A \rightarrow S$ (k_2),

$$\frac{d[A]_t}{dt} = -k_2[A]_t^2 \tag{9}$$

or

$$\frac{d[A]_t}{[A]_t^2} = -k_2\,dt \tag{10}$$

giving an integration

$$\frac{-1}{[A]_t} = -k_2 t + C \tag{11}$$

which with the usual conditions of $[A]_t = [A]_0$ at $t = 0$ gives

$$\frac{1}{[A]_t} = \frac{1}{[A]_0} + k_2 t \tag{12}$$

Mixed first- and second-order decays, $A \rightarrow S$ (k_1) and $2A \rightarrow 2S$ (k_2), lead to the differential

$$\frac{-d[A]_t}{dt} = k_1[A]_t + k_2[A]_t^2 \tag{13}$$

or

$$\frac{d[A]_t}{k_1[A]_t + k_2[A]_t^2} = -dt \tag{14}$$

which gives upon integration

$$\frac{1}{k_2}\,\frac{\ln [A]_t}{k_1 + k_2[A]_t} = -t + C \tag{15}$$

Solution for the integration constant and arrangement leads to

$$\frac{1}{[A]_t} = \left(\frac{1}{[A]_0} + \frac{k_2}{k_1}\right)e^{k_1 t} - \frac{k_2}{k_1} \tag{16}$$

or

$$[A]_t = \frac{k_1[A]_0}{(k_1 + k_2[A]_0)e^{k_1 t} - k_2[A]_0} \tag{17}$$

Finally, multiple exponential behavior encompasses a wide variety of chemical mechanisms. A trivial but common case is one in which two unrelated transients absorb in the same region. Any kinetic measurement will then measure the sum of the two individual decays. Nontrivial examples are coupled first-order reactions,

$$A_1 \rightleftharpoons A_2 \rightleftharpoons A_3 \rightleftharpoons \text{etc.}$$

which have solutions of the form

$$[A_i] = \Sigma a_i e^{-\lambda_i t} + C \tag{18}$$

The rate terms λ_i and the preexponentials can be complex functions of the rate constants and the mass-balance terms. These equations and their corresponding solutions have been the subject of a large body of literature [96–102].

Two coupled first-order reactions that are common in photochemistry are the consecutive reactions

$$A \xrightarrow{k_1} B \xrightarrow{k_2} C$$

and

$$\begin{array}{l} A \xrightarrow{k_1} B \xrightarrow{k_2} C \\ \big\downarrow k_3 \quad k-1 \\ D \end{array}$$

The former is a common sequential reaction type, and the latter can describe exciplex- or other complex-forming reactions.

The differential equations for A → B → C (k_1, k_2) are

$$\frac{d[A]_t}{dt} = -k_1[A]_t - k_2[B]_t \tag{19}$$

With the usual initial conditions of $[A]_t = [A]_0$, $[B]_t = 0$, and $[C]_t = 0$ at $t = 0$, the first differential has the normal solution

$$[A]_t = [A_0]e^{-k_1 t} \tag{20}$$

When this value for $[A]_t$ is substituted into the second differential, the following nonhomogeneous linear equation is obtained:

$$\frac{d[B]_t}{dt} + k_2[B]_t = k_1[A]_0 e^{-k_1 t} \tag{21}$$

This equation can be solved by standard methods:

$$[B]_t = e^{-h}(e^h k_1[A]_0 e^{-k_1 t}\, dt + C) \tag{22}$$

$$h = \int k_2[\mathrm{B}]_t \, dt \tag{23}$$

giving

$$[\mathrm{B}]_t = \frac{k_1[\mathrm{A}]_0}{k_2 - k_1}(e^{-k_1 t} - e^{-k_2 t}) \tag{24}$$

The kinetics associated with excimer and exciplex decay have been derived many times [105–106]. The integrated forms are complex but can be obtained by application of either standard integral techniques or matrix methods. The resultant equations are

$$[\mathrm{B}]_t = a_3(e^{-\lambda_1 t} - e^{-\lambda_2 t}) \tag{25}$$

and

$$[\mathrm{A}]_t = a_1 e^{-\lambda_1 t} + a_2 e^{-\lambda_2 t} \tag{26}$$

where

$$a_1 = \frac{(\lambda_1 - k_3 - k_2)[\mathrm{A}]_0}{\lambda_1 - \lambda_2} \tag{27}$$

$$a_2 = \frac{(k_3 + k_2 - \lambda_2)[\mathrm{A}]_0}{\lambda_1 - \lambda_2} \tag{28}$$

$$a_3 = \frac{k_2[\mathrm{A}]_0}{\lambda_2 - \lambda_1} \tag{29}$$

and

$$2\lambda_{1,2} = k_1 + k_2 + k_3 + k_4 \pm [(k_1 + k_4 - k_2 - k_3)^2 + 4k_1k_3]^{1/2} \tag{30}$$

If $k_3 = k_4 = 0$ the equations reduce to those obtained for the sequential reaction A → B → C.

These equations are illustrative of various kinetic schemes that are often necessary to describe photochemical reactions, but more complex schemes have been reported [107–109].

In general, the concentrations are not the directly measured quantities; the directly measured quantity is usually the optical density. By making the substitution $D = (\text{concentration})\epsilon l$, where ϵ is the extinction coefficient and l is the path length, into any of the above equations, the relationship between the optical density D and the kinetic parameters can be obtained.

Multiphotonic Effects

Although the efficiencies of monophotonic reactions are lower when lasers are used, multiphotonic reactions clearly require high intensities. Multiphotonic reactions initiated in the ultraviolet and visible regions can be categorized as (1) photochemistry of excited states, (2) bimolecular reactions, and (3) photochemistry from states produced by simultaneous two-photon absorption. The most widely observed multiphotonic reactions in the visible and ultraviolet region are those that proceed by way of excited-state absorption of light. The sequence

$$S_0 \xrightarrow{h\nu} S^* \xrightarrow{h\nu} X^{**} \rightarrow \text{products}$$

generally leads to electron-transfer reactions, since little else can compete with internal decay of upper excited states. Often, but not always, the electron is transferred to solvent, the end result being photoionization.

Endicott et al. [110] found that the quantum yield for photoreduction of aqueous solutions of $Co^{3+}(NH_3)_6$ increased by nearly 3 orders of magnitude over that obtained from a low-intensity source when a focused nitrogen laser was used as the excitation source. They proposed that the biphotonic reaction proceeds through an intermediate ligand-field excited state (Fig. 15). In this case the electron is transferred from the amine ligand to the central metal ion.

Naturally, high intensities do not always increase the quantum yields of the photoreaction. Excited-state absorption can lead to wastage if monophotonic chemistry is attempted with a laser. Donohue [111] has reported on lanthanide separations by use of the various lines of the exciplex laser. The experiment is based on the differences in absorption of aqueous solutions of the lanthanides in the ultraviolet and the fact that the various ionic states have different solubilities and can undergo photoredox reactions.

$$Eu^{3+}(H_2O) \xrightarrow{193\text{ nm}} Eu^{2+} + H^+ + OH$$

$$Eu^{3+} \xrightarrow{SO_4^{2-}} EuSO_4\downarrow$$

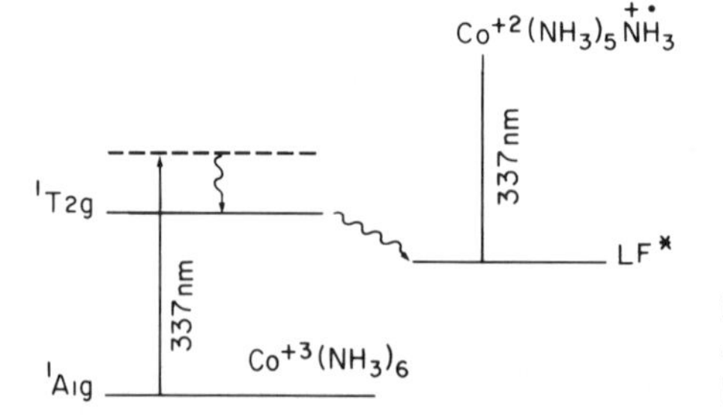

Fig. 15 Proposed reaction scheme for multiphoton excitation of $Co^{3+}(NH_3)_6$ [110].

The quantum yields obtained for laser excitation were lower by 2 or 3 than those obtained from low-intensity sources, owing to excited-state absorption.

In addition to the observation of the effects of intensity on product quantum yields, multiphoton absorption can be detected in several ways. Normal absorption spectroscopy can accurately measure the ground-state spectra because the light intensity is low enough that the ground-state population is not perturbed. At low excitation intensities the optical density of a sample is independent of intensity and the emission intensity is proportional to the excitation intensity. However, pulsed laser excitation provides enough photons for the ground state to be partly or completely bleached. The state produced after absorption can itself absorb during the laser pulse. The high intensities have three consequences. The absorption of the sample is not independent of light intensity, since the optical density is then

$$\mathrm{D} = \epsilon_0[\mathrm{S}_0]_t + \epsilon_1[\mathrm{S}^{*1}] + \epsilon_2[\mathrm{S}^{*3}]_0 \tag{31}$$

If the extinction coefficients of the excited states are less than that of the ground state, bleaching occurs. This is, of course, the condition required for passive dye mode-locking and Q-switching [112–119]. If $\epsilon_1 + \epsilon_2 > \epsilon_0$, enhanced absorption occurs. Explicit solutions for the optical density as a function of excitation power can be solved for the limited set of conditions that there are no long-lived states and the interconversion of excited states is very rapid. The latter assumption is often true, but, in general, triplet-state or product formation also occurs [118–120].

Hercher et al. have derived an equation for the transmission of the laser pulse through an optically thick sample [112].

$$\ln\left(\frac{T_0}{T}\right) = (\gamma - 1)\ln\left(\frac{\gamma + I_0/I_0}{\gamma + TI_0/I_s}\right) \tag{32}$$

In this equation γ is the ratio of ground-state extinction coefficient to that for the excited state(s) and I_s is the saturation flux. For a two-level system $I_s = 1/\epsilon\tau$. T_0 is the low-intensity transmission value.

Excited-state extinction coefficients can be estimated by using this technique, but reaction schemes requiring more intermediates need to be solved by computer iteration modeling [120].

Excited-state absorption (ESA) is of importance for laser dyes [112–119, 121, 122]. The probability of stimulated emission must be greater than that for absorption if the dye is to lase at any given wavelength. Excited singlet- and triplet-state absorptions are probably the greatest cause of the lack of lasing in materials with good fluorescence properties.

Another consequence, and monitor, of excited-state absorbances is the nonlinear response of fluorescence versus excitation energy, termed fluorescence quenching [118a, 123–129]. Since photons absorbed by excited states do not lead to fluorescence, the quantum yield for fluorescence decreases as the excitation energy increases, and the proportion of ESA versus ground-state absorption increases.

The quantum yields for monophotonic reactions are often adversely affected by high excitation fluxes. One example is the measurement of the intersystem-crossing efficiency by flash photolysis.

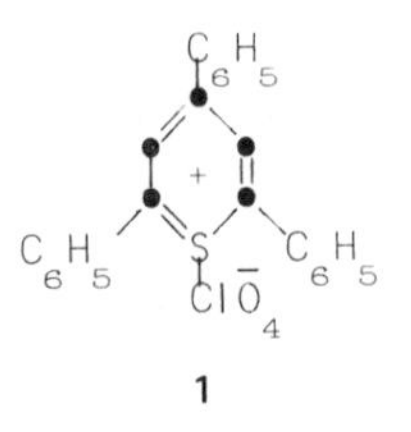

The dye 2,4,6-triphenylthiopyrylium perchlorate (**1**) has been used in the measurement of intersystem-crossing yields by flash photolysis [130].

The intersystem-crossing yields can be obtained in numerous ways. However, several of these techniques fail for pyrylium and thiopyrylium compounds because of a combination of factors. Since photolytic formation of excited states causes a corresponding decrease in the ground-state concentration, this decrease, in combination with a suitable actinometer, can be used to directly measure the quantum yields for formation of excited states. In the simplest case, with only singlet and triplet states produced, mass balance requires that

$$[S_0]_0 = [S_1]_t + [^1S^*]_t + [^3S^*]_t \tag{33}$$

where $[S_0]_0$ is the ground-state concentration before the excitation, and the other bracketed terms are the excited-state concentrations at some time t after excitation. For determination of the intersystem-crossing yield, ϕ_{TM}, the ground-state concentration must be measured after all excited singlets decay but before significant portions of the triplets decay. Since the rate constants for these two processes for **1** differ by 4 orders of magnitude; the rate of decay of the excited singlet state is 7.46×10^8 sec^{-1} and that of the triplet state is 8.1×10^4 sec^{-1}, this measurement is not difficult. The ground-state concentration is measured at 400 nsec, when all excited singlets but only ~3% of the triplets have returned to ground state. At 400 nsec,

$$[S_0]_0 - [S_0]_{400\ \text{nsec}} = [^3S^*]_{400\ \text{nsec}} \tag{34}$$

and

$$[^3S^*]_{400\ \text{nsec}} \simeq [^3S^*]_0 \tag{35}$$

where $[^3S^*]_0$ is the triplet concentration at time zero. The intersystem crossing efficiency is then given by

$$\phi_{TM} = \frac{[^3S^*]_0}{[^1S^*]_0} \tag{36}$$

where $[^1S^*]_0$ is the initial concentration of the excited singlet state. This concentration can be inferred by using an actinometer at concentrations where the optical density is identical with that of the unknown, but for which the intersystem-crossing yield has previously been measured.

Aromatic hydrocarbons are convenient actinometers, because many have known intersystem-crossing yields and extinction coefficients for triplet absorptions. We have used both phenanthrene, $\phi_{TM} = 0.8$, $\epsilon_T = 2.7 \times 10^4$ at 496 nm, and chrysene, $\phi_{IC} \simeq 0.70$, $\epsilon_T = 4.8 \times 10^4$ at 580 nm [95], as actinometers. Both give identical results, within experimental error. With the assumption that at identical optical densities of ground-state absorption for the actinometer and the unknown and at equal excitation intensities $[^1S^*]_0 = [^1A^*]_0$, then

$$\phi_{TM} = \frac{[^3S^*]_0}{[^3A^*]_0} \phi_{TM}^A \tag{37}$$

where $[^1A^*]_0$ and $[^3A^*]_0$ are the initial concentrations of the singlet and triplet states of the actinometer, and ϕ_{TM}^A is the intersystem-crossing yield of the actinometer.

In practice, it is difficult to do two experiments at exactly the same laser intensity. Therefore the triplet concentration of the actinometer is determined relative to the laser intensity, and then its triplet concentration is calculated from the empirical relationship for the pertinent intensities for the sample. The frequency-doubled ruby beam, at 347 nm, is first directed into a photodiode by means of a quartz beam splitter. The remainder of the pulse is used for sample excitation.

The concentration of phenanthrene triplets versus relative laser energy is shown in Fig. 16, and the resultant quantum yields for intersystem crossing versus intensity are shown in Fig. 17 for **1**.

The results in Fig. 17 illustrate two points: The apparent quantum yields are intensity dependent, and the measurements obtained from bleaching are wavelength dependent. The wavelength dependence indicates transient absorption in the region of ground-state absorption. In the case of no transient absorption, the ground-state concentration, immediately after excitation, is given by

$$[S]_0 = \frac{OD_\lambda}{\epsilon_{0,\lambda}} \tag{38}$$

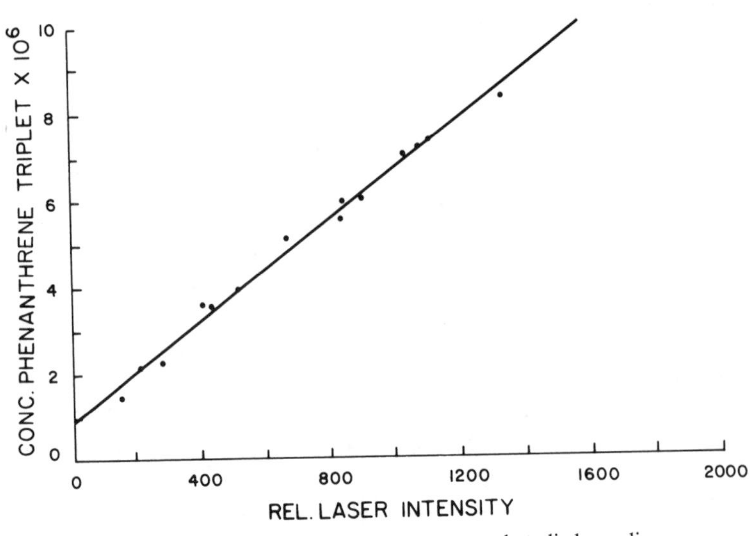

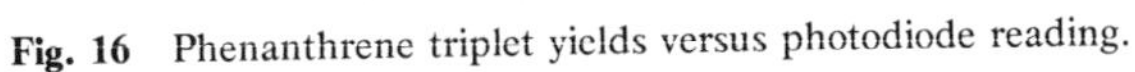
Fig. 16 Phenanthrene triplet yields versus photodiode reading.

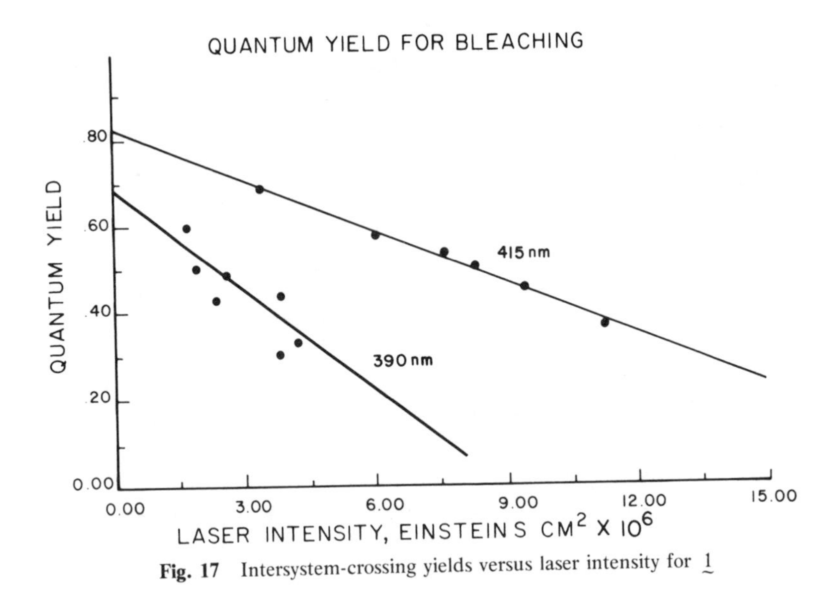

Fig. 17 Intersystem-crossing yields versus laser intensity for $\underset{\sim}{1}$

If transient absorption occurs, then

$$OD_\lambda = \epsilon_0[S]_0 + \epsilon^*[S^*] \tag{39}$$

and

$$\frac{OD_\lambda}{\epsilon_0} = [S]_0 + \frac{\epsilon^*}{\epsilon_0}[S^{*3}]_0 \tag{40}$$

The apparent intersystem-crossing yield is then given by the expression

$$\phi_{TM} = \frac{[^3S^*]_0\ 1 - \epsilon_T/\epsilon_0}{[^3A^*]}\phi^A_{TM} \tag{41}$$

and

$$\frac{\phi_{TM}}{\phi^A_{TM}} = 1 - \frac{\epsilon_T}{\epsilon_0} \tag{42}$$

The intensity effect results because the above equations assume that the ground depopulation is negligible for both sample and actinometer. If ground state is the only absorbing species of the excitation wavelength, for both the actinometer and the sample, it can be shown that [131]

$$\phi_{TM} = \phi^A_{TM}\frac{\epsilon_A \log(1 - n_s)}{\epsilon_s \log(1 - n_A)} \tag{43}$$

where n is the fraction of molecules converted to the triplet state. However, the important procedure is to measure monophotonic events at low intensities and then extrapolate to zero intensity.

The most common consecutive two-photon reaction is photoionization [120, 132–146]. In the early work the lowest excited singlet state was thought unimportant as an intermediate in the reaction. Both Ottolenghi [134] and Thomas et al. [133] found that additives that quenched the fluorescence did not affect the yield of radical cation produced. They therefore postulated either a simultaneous two-photon absorption [133] or an intermediate "semiionized" state [134]. The error in this analysis is that the two events are not linearly correlated with each other.

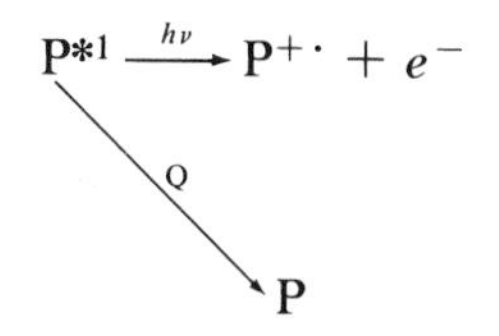

During the laser pulse the photon flux is very high, and the diffusional quenching by an external agent, unless in very high concentrations, simply cannot compete with photoionization. However, when the photon flux is low, photoionization is unimportant and quenching can compete with fluorescence.

That the sequential mechanism is correct has been shown in several ways. Detailed modeling of all the processes for the sequential mechanism has shown that this model is sufficient to explain the experimental observations [120]. If the lifetime of the first singlet state is dramatically reduced, the yield of photoionization is decreased. Hall and Kenny-Wallace have shown that a fifteenfold reduction of the pyrene lifetime reduces the photoionization yield by a factor of 2 [136].

The aromatic hydrocarbons pyrene and perylene (Fig. 18) both show nonlinear absorption at high laser intensities [147]. If either simultaneous two-photon or the "semi-ionized" state were involved, the nonlinear absorption would be unaffected by reducing the excited-state singlet lifetime.

The fluorescence of perylene is quenched by NaI, and the lifetime of the singlet state can be reduced from 6.5 nsec to 1 nsec and 90 psec by the addition of 0.1 and 1.0 *M* salt, respectively. The results in Table 4 show that the nonlinearity of the transmission is almost completely eliminated by providing a channel for excited state return to ground state. Finally, the double-pulse experiments with two 347-nm pulses have shown that pyrene ionizes from an intermediate state whose lifetime is >40 nsec [147].

That the sequential mechanism is operating for laser-initiated photoionization does not contradict earlier conventional flash experiments that suggested the major ionization route is from the triplet state [148–156]. Un-

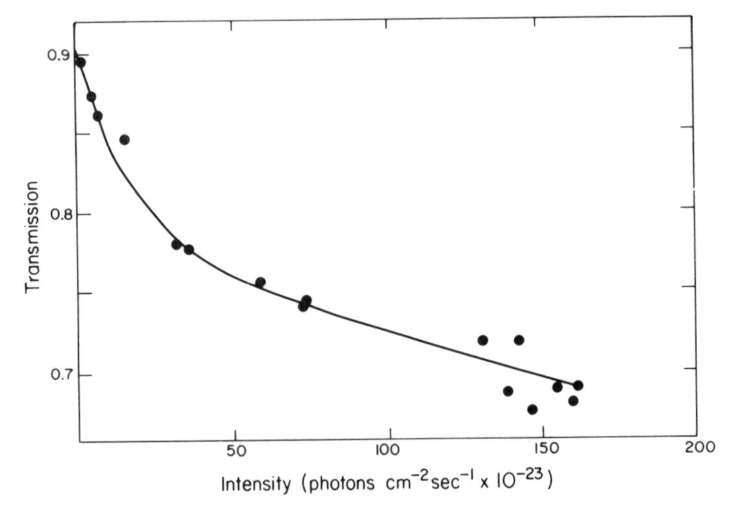

Fig. 18 Intensity-dependent transmission for perylene.

Table 4. Changes in Laser Transmission of Perylene Solutions with Added Sodium Iodide

Conditions	*Energy (mJ)*	*Percent Change in Transmittance,* $(1 - T/T_0) \times 100$
Aerated	186	51.0
0.1 *M* Sodium iodide	188	25.0
1.0 *M* Sodium iodide	187	3.6

fortunately, triplet-state participation is difficult to establish in the laser experiments [139, 144].

Clearly, chemical products can result from biphotonic photoionization. Repeated flashes of a methanolic solution of anthracene by a frequency-doubled ruby laser lead to measurable amounts of 9-methoxyanthracene by the mechanism shown in Fig. 19 [147].

Unfortunately in multiphotonic reactions the order of the photoreaction cannot be easily determined. At low light intensities the chemical yield for the reaction reflects the order of intensity, that is, $Y_R \alpha BI^n$, but the yield of photoreaction is very low and often undetectable. At very high intensities saturation occurs and the chemical yield is independent of intensity.

Thus $Y_R \alpha BI^n \longrightarrow bI^0$ as the intensity increases and may appear to be linear over a limited range of intermediate intensities. Lachish et al. [145] have published solutions for the three-level scheme (Fig. 20).

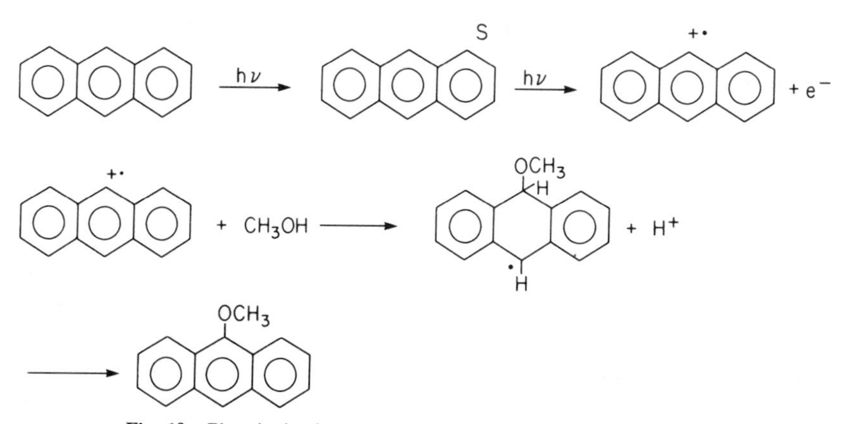

Fig. 19 Photoionization and reaction of anthracene with methanol.

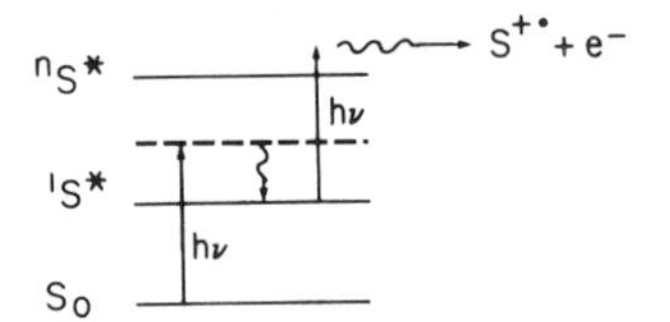

Fig. 20 Three-state scheme for photoionization.

$$[S^{+\cdot}] = [S_O]\left(1 + \frac{e^{-\lambda_2\tau} - e^{-\lambda_1\tau}}{\lambda_2 - \lambda_1}\right) \tag{44}$$

where

$$\begin{aligned} 2\lambda_{1,2} &= \phi_1\epsilon_1\bar{I} + \phi_2\bar{I} + k_1 \\ &\pm [(\phi_1\epsilon_1\bar{I} + \phi_2\epsilon_2\bar{I} + k_1)^2 - 4\phi_1\phi_2\epsilon_1\epsilon_2\bar{I}]^{1/2} \end{aligned} \tag{45}$$

and ϕ_1 is the probability of forming the first excited state, τ is the width of the square-wave pulse, and $\bar{I}$ is the photon flux. At low intensities $k_1 \gg \phi_1\epsilon_1\bar{I} + \phi_2\epsilon_2\bar{I}$, (45) reduces to (46):

$$[S^{+\cdot}] = [S_O]\,\frac{\phi_1\epsilon_1\phi_2\epsilon_2}{k_1}\bar{I}^2 \tag{46}$$

At very high intensities $\lambda_1\tau$, $\lambda_2\tau \gg 1$, $e^{-\lambda_1\tau}$, $e^{-\lambda_2\tau} \longrightarrow 0$ and the yield becomes independent of intensity. If the molecule is being pumped faster than it returns to ground state and if $\phi_1\epsilon_1$ and $\phi_2\epsilon_1$ are very different, there is a region where intensity behavior is linear [137, 141]. The pulse duration and pulse shape are important in determining the profile of the chemical yield versus integrated intensity curves [146]. Clearly, linear relationships of yield with intensity cannot be taken as proof that a monophotonic reaction is occuring [141].

The high intensity available from lasers can be used to enhance bimolecular reactions of metastable species over their unimolecular reactions. A general scheme for this type of reaction is

$$A \xrightarrow{h\nu} X$$

$$X \xrightarrow{k_1} B$$

$$2X \xrightarrow{k_2} P$$

Clearly, the ratio of products B and P depends on the transient concentration of X, which in turn depends on the light intensity. If $k_2[X] \gg k_1$, then P is the major product, whereas at low intensities $k_1 \gg k_2[X]$ and, in the above scheme, B is the major product.

One example of this type of reaction is illustrated by the fluorescence properties of naphthalic anhydride [147]. We have found that 1,8-naphthalic

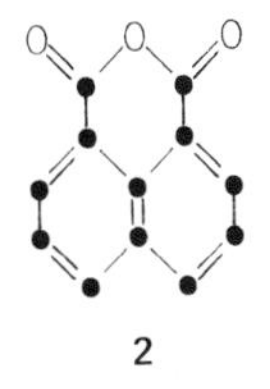

anhydride (**2**), exhibits unique properties when excited by a laser source. Under normal excitation (450-W xenon lamp), **2** shows normal fluorescence in acetonitrile (Fig. 21). However, when excited with a nitrogen laser (337 nm), a 10^{-4} M solution of **2** in degassed acetonitrile shows dramatically different fluorescence characteristics (Fig. 22). The long-wavelength and the delayed short-wavelength emissions disappear when either air or 10^{-3} M 1,3-cyclohexadiene is present. When **2** is excited with a frequency-doubled (347 nm) pulsed ruby laser, both prompt and delayed emission are observed. The delayed emission is identical with that observed by excitation with the nitrogen laser.

Attenuation of the nitrogen laser with glass slides shows that all of the long-wavelength and part of the short-wavelength emission is biphotonic (Fig. 23). In the short-wavelength region, the fluorescence is composed of both prompt and delayed fluorescence, and so a rather poor correlation with the square of the intensity is to be expected. For excitation with the ruby

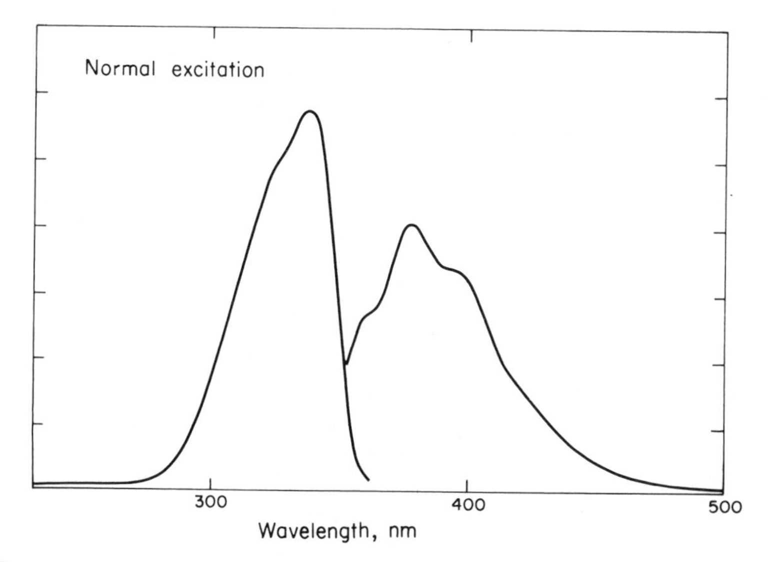

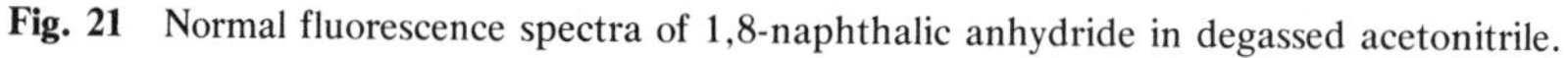
Fig. 21 Normal fluorescence spectra of 1,8-naphthalic anhydride in degassed acetonitrile.

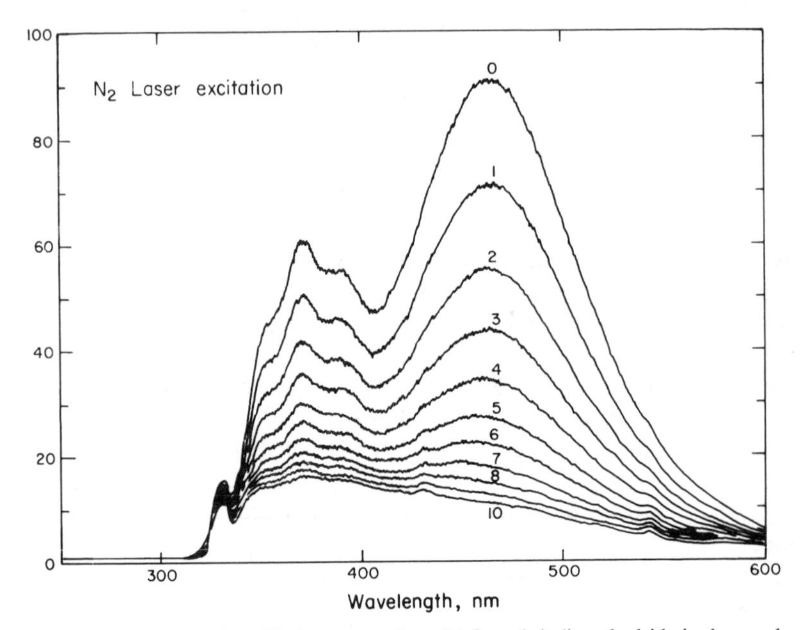

Fig. 22 Emission spectra from N_2-laser excitation of 1,8-naphthalic anhydride in degassed acetonitrile.

laser, as the intensity dependence is curvilinear (Fig. 24), which represents a saturation effect and reflects the very different powers available from the two lasers. The incident light from the nitrogen laser is about 1 mJ or less, whereas that from the ruby laser is 10 to 200 mJ/pulse. At the higher powers of the ruby laser, all the molecules in the optical path are excited and so no intensity dependence is observed.

To determine which state was leading to the delayed fluorescence, we measured the transient absorption spectrum produced by exciting **2**. We observed two species, one with a maximum at about 485 nm, which disappeared upon addition of either air or 10^{-3} M 1,3-cyclohexadiene, and another, $\lambda_{max} \leq 370$ nm, which was only slightly affected by air. We believe that the 485-nm absorption is due to the triplet of **2**, and the short-wavelength species is either a biradical or a radical ion of **2**.

The short-wavelength species decay is first order with a rate constant of $4.8 \pm 0.5 \times 10^5$ sec $^{-1}$, but the triplet decay, measured at 480 nm, obeys neither first- nor second-order kinetics. The triplet decay does, however, fit the kinetics expected for a mixture of both first- and second-order kinetics, yielding a first-order rate constant k_T of $4.4 \pm 0.2 \times 10^4$ sec $^{-1}$ and a second-order constant k_{TT}/ϵ of $5.3 \pm 0.2 \times 10^5$ M^{-1} sec^{-1}.

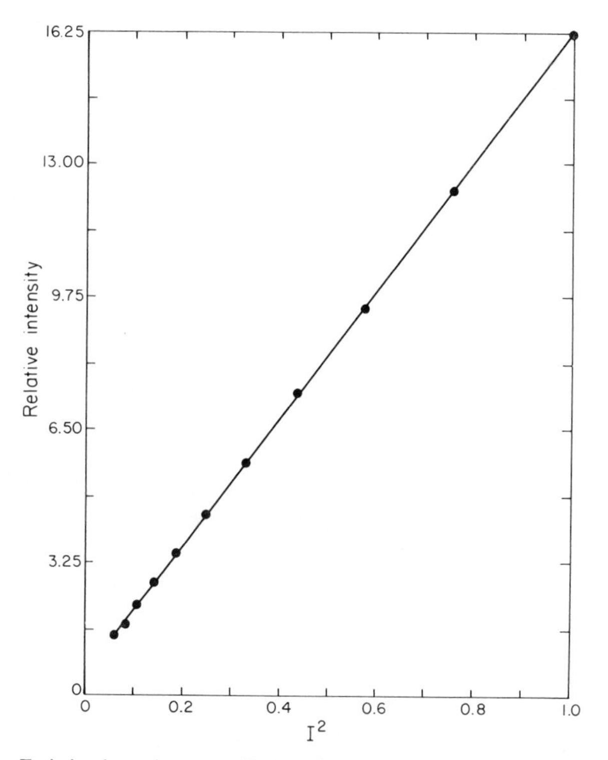

Fig. 23 Emission intensity versus N_2-laser intensity for 1,8-naphthalic anhydride.

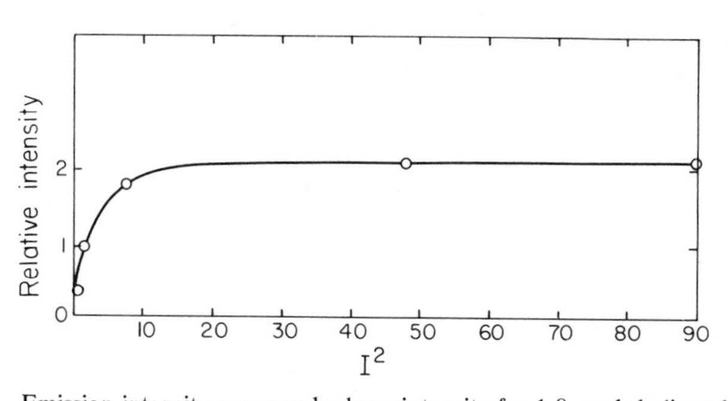

Fig. 24 Emission intensity versus ruby-laser intensity for 1,8-naphthalic anhydride.

These observations are all consistent with excimer formation by triplet-triplet annihilation:

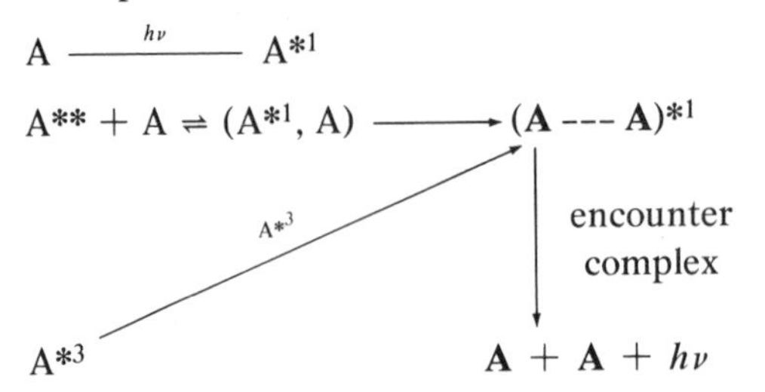

As is indicated in the above diagram, we would expect the excimer to be produced even under normal low-intensity conditions if the concentration of naphthalic anhydride were high enough to capture the photoexcited molecule. When we excite a concentrated sample of **2** with a 450-W xenon lamp, we observe a small excimer peak, ~5% of the total emission, in the same region as the long-wavelength emission from the laser-excited sample. In the solid state, at 77°K, where **2** has a nearest-neighbor relationship with **2**, we observe only long-wavelength structureless emission identical with the excimer emission observed in solution.

The kinetics for delayed emission produced by triplet-triplet annihilation are complicated but have been solved by Birks [106] and others [105]. For this system, plots of the inverse square root of the delayed emission versus time should give a straight line. The delayed excimer emission obeys this kinetic pattern, but the short-wavelength emission, delayed fluorescence, shows mixed behavior, suggesting that there is another mechanism leading to delayed fluorescence, and not excimer emission.

The usual route for excimer formation is from the singlet state, but a combination of factors prevents their observation for many compounds. If a compound has a short singlet lifetime and is only moderately soluble in organic solvents, its singlet state will not be trapped by a ground-state molecule within its lifetime. Consider an ideal case in which the excimer is formed at diffusion-controlled rates and emits with unit efficiency. The amount of fluorescence quenching, which is equal to the amount of excimer emission, is then given by the Stern–Volmer equation:

$$\mathrm{A}^{*1} + \mathrm{A} \xrightarrow{k_{\mathrm{diff}}} (\mathrm{A}\ \text{---}\ \mathrm{A})^{*1} \longrightarrow 2\mathrm{A} + h\nu$$

$$\mathrm{A}^{*1} \xrightarrow{k_f} \mathrm{A} + h\nu$$

$$\frac{\phi_0}{\phi} = 1 + k_{\mathrm{diff}}\tau[\mathrm{A}]$$

where $\tau = 1/k_f$. For many organic solvents $k_{\text{diff}} = 10^{10}\ M^{-1}\ \text{sec}^{-1}$, and for many compounds $\tau = 10^{-8}$ to 10^{-9} sec. Thus the concentration of A required for 50% excimer formation would be $\sim 10^{-1}$ to $10^{-2}\ M$. This estimate is for an ideal case, and any inefficiency will increase the necessary concentration. Many materials such as **2** are not very soluble. In acetonitrile, **2** is soluble to only $2 \times 10^{-2}\ M$.

In the triplet-triplet annihilation mechanism, the singlet state is produced in near contact with a ground-state molecule, and so concentration does not play a role once triplet-triplet annihilation has occurred. The major requirement for triplet-triplet annihilation is that $k_{\text{TT}}\ [\text{A}^{\text{T}}] > k_{\text{T}}$, where k_{TT} is the rate constant for annihilation and k_{T} is the rate constant for unimolecular decay of the triplet state. Although we do not know the value of k_{TT} (our ex-

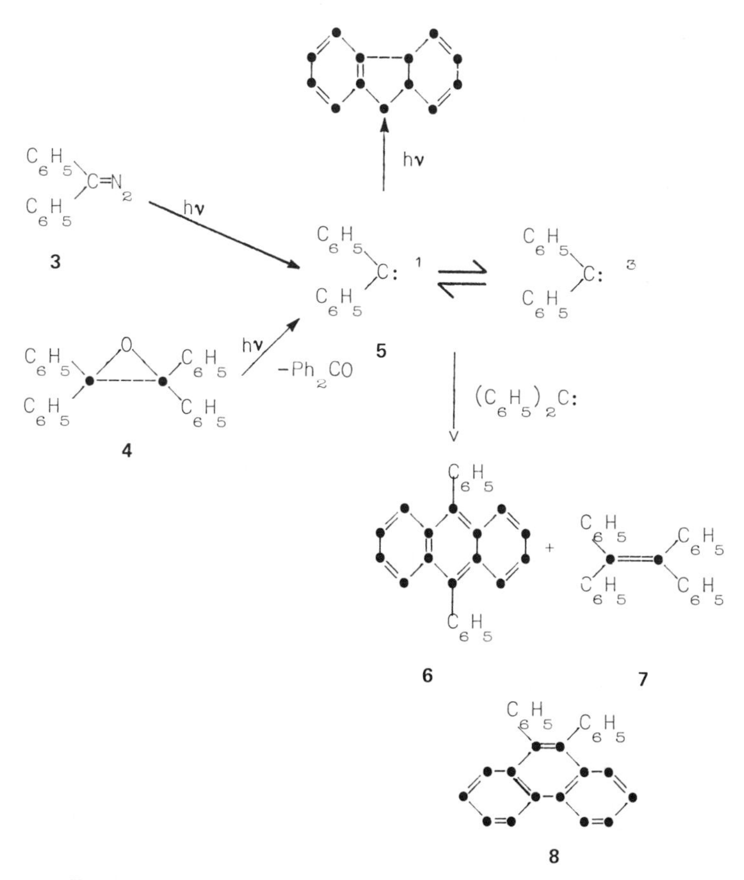

Fig. 25 Laser induced photochemistry of diphenylazomethane [161].

periments give us only $k_{TT} \ll \epsilon$), triplet-triplet annihilation is usually diffusion controlled, that is, $k_{TT} = 10^{10}\, M^{-1}\, sec^{-1}$. The measured value of k_T is $4.4 \times 10^4\, sec^{-1}$, and therefore for $k_{TT}[1^T] > k_T$ the concentration of $\mathbf{1}^T$ must be greater than $4 \times 10^{-6}\, M$. Even without focusing, the frequency-doubled ruby laser can easily produce this many excited-state molecules.

It is well established that triplet-triplet annihilation can lead to electron transfer, $A^3 + A^3 \longrightarrow A^{\dot{+}} + A^{\dot{-}}$, in addition to singlet- and excimer-state formation [143, 157–160].

A recent example of an intensity-dependent bimolecular reaction has been provided by Turro et al. [161]. The irradiation of **3** or **4** with a KrF laser (249 nm) produces the carbene **5**, which can either absorb another photon, yielding fluorene, or dimerize from one of the two spin states of the carbene to give the hydrocarbons **6**, **7**, and **8**, Fig. 25.

In the gas phase small molecules photodissociate after multiphoton absorption at 193 nm [162–170]. For example, excitation of acetylene produces excited-state CH fragments. The most likely mechanism for the various emissions observed is two-photon sequential absorption of acetylene, dissociation into CH, and then excited-CH absorption leading to the upper states (Fig. 26).

That such energetic species can be easily prepared is a great advantage in the study of their physical and chemical properties (Table 5).

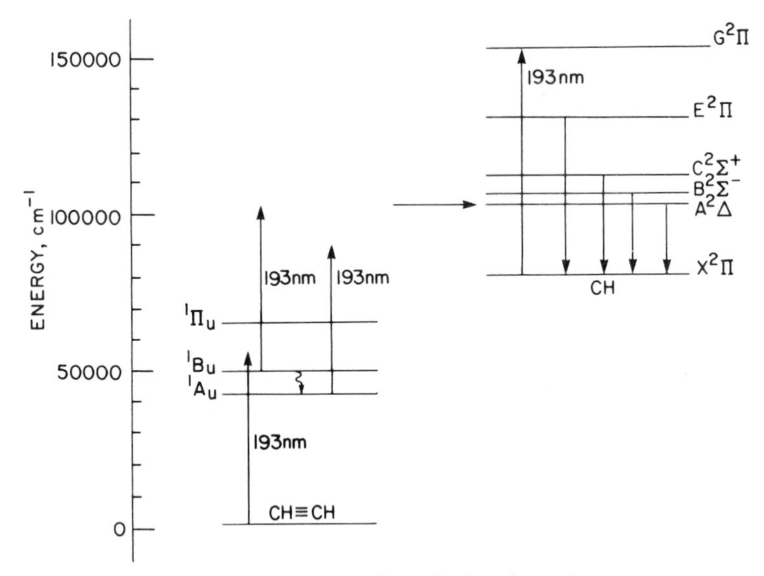

Fig. 26 State diagram for excitation of acetylene.

Table 5. Products Observed from Small Molecule Photodissociation

Molecule	*Products*	*Excitation Wavelength (nm)*	*Reference*
ClCN	CN($B^2\pi$, $D^2\pi$, $F^2\Delta$, $H^2\pi$)	193	162
$(CN)_2$	CN($B^2\pi$, $D^2\pi$, $H^2\pi$)	193	162
C_2H_2	CH($A^2\Delta$, $B^2\Sigma^-$, $C^2\Sigma^+$, $E^2\pi$)	193	162
C_2H_4	CH($A^2\Delta$, $C^2\Sigma^+$)	193	162
C_3H_6	CH($A^2\Delta$, $B^2\Sigma^-$, $C^2\Sigma^+$)	193	162
CH_3OH	CH($A^2\Delta$, $B^2\Sigma^-$), OH($A^2\Sigma$, $D^2\Sigma$)	193	162, 167
C_2H_5OH	CH($A^2\Delta$, $B^2\Sigma^-$), OH($A^2\Sigma$, $D^2\Sigma$)	193	162
H_2O	OH($A^2\Sigma$, $D^2\Sigma$)	193	162
CH_3Br	CH($A^2\Delta$)	193	167
CH_3I	CH($A^2\Delta$, $B^2\Sigma^-$)	193	167
CD_3I	CD($A^2\Delta$, $B^2\Sigma^-$)	193	167
CF_2Br_2	$CF_2(X'A_1)$, CF($X^2\pi$)	193, 248	165
CH_2I_2	CH_2	266	172
CBr_4	Br_2, C_2	248, 193	163
CF_2Br_2	Br_2, CF_2 ($X'A_1$)	248	163
$CF_3—C{\equiv}C—CF_3$	$C_2(X'\Sigma^+)$	193	173
CH_3NH_2	NH, CN, CH, NH_2	193	168
HCN	HCN(A' A'')	193	164
I_2	I	193	166

Reddy and Berry have used intracavity techniques to irradiate into highly forbidden overtone bands of gas-phase molecules [171] (Fig. 27). With this elegant technique, selected vibrational levels can be populated and the reaction dynamics analyzed. The unimolecular isomerization of CH_3NC to CH_3CN was initiated by preparing the isocyanate with either five or six quanta in the C—H stretching mode.

Molecules can interact with an intense light source to simultaneously absorb two photons. Although the basic theory for this experiment has been known for 50 years, it has become useful only since the invention of the laser. In general, the electric dipole selection rules are reversed for one- and two-photon absorption. Two-photon photochemistry has been examined for several systems [172–179]. Because of the rapid interconversion of the upper states, one-photon and two-photon photochemistry yield identical products, although differences in emission have been reported for one-photon and two-photon excitation [180–181].

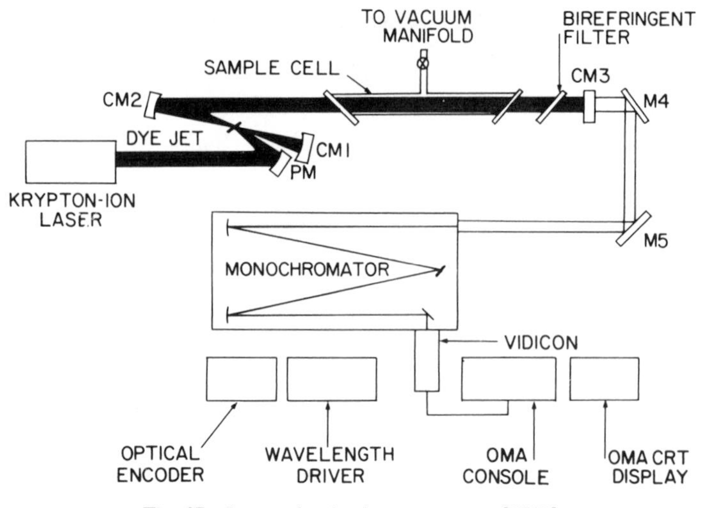

Fig. 27 Intracavity dye-laser apparatus [171b].

References and Notes

1. R. G. W. Norrish and G. Porter, *Nature*, **164**, 658 (1949).
2. G. Porter, *Proc. R. Soc. London, Ser. A*, **200**, 284 (1950).
3. G. Porter in "Rates and Mechanisms of Reactions," *Techniques of Organic Chemistry*, Vol. 7, 2nd ed. S. L. Friess, E. S. Lewis, and A. Weissberger, Eds., Wiley, New York, 1963, p. 6.
4. G. Porter and M. A. West, in "Rates and Mechanisms of Reactions," *Techniques of Chemistry*, Vol. 6, Part 2, G. G. Hammes, Ed., Wiley, New York, 1974, p. 367.
5. W. F. Kosonocky, S. E. Harison, and R. Stander, *J. Chem. Phys.*, **43**, 831 (1965).
6. G. Porter and J. I. Steinfeld, *J. Chem. Phys.*, **45**, 3456 (1966).
7. J. R. Novak and M. W. Windsor, *J. Chem. Phys.*, **47**, 3075 (1967).
8. For a review of ultraviolet gas lasers, see J. M. Green, *Opt. Laser Technol.*, 289 (1978).
9. Because of the proliferation of terminology used to describe the various possible complexes that exist only in the excited state, a decision was made in 1968 to define an exiplex as the general term for electronically excited complexes and to limit the term excimer to those formed of identical atoms or molecules [10]. It is unfortunate that the laser community has chosen to misname the rare gas halide exciplex laser.
10. E. C. Lim, Ed., *Molecular Luminescence*, Benjamin, New York, 1969, p. 907.
11. T. R. Loree, R. C. Sze, and D. L. Barker, *Appl. Phys. Lett.*, **31**, 37 (1977); T.

R. Loree, R. C. Sze, D. L. Barker, and P. B. Scott, *IEEE J. Quant. Electron.*, **QE-15**, 337 (1979).
12. L. Lindquist, *C. R. Ser. C*, **263**, 852 (1966).
13. C. R. Goldschmidt, M. Ottolenghi, and G. Stein, *Isr. J. Chem.*, **8**, 29 (1970).
14. U. Lashish, D. J. Williams, and R. W. Anderson, *Chem. Phys. Lett.*, **65**, 574 (1979).
15. D. Bebelaar, *Chem. Phys.*, **3**, 205 (1974).
16. A. van der Ende, S. Kimel, and S. Speiser, *Chem. Phys. Lett.*, **21**, 133 (1973).
17. R. Ronneau, J. Joussot-Dubien, and J. Fauve, *Chem. Phys. Lett.*, **2**, 65 (1968).
18. D. V. Bent and E. Hayon, *J. Am. Chem. Soc.*, **97**, 2599 (1975).
19. F. P. Schafer, Ed., *Dye Lasers*, 2nd ed., Springer-Verlag, New York, 1977.
20. C. Rulliere, J. P. Morand, and O. deWitte, *Opt. Commun.*, **20**, 339 (1977).
21. C. Rulliere and J. Joussot-Dubien, *Opt. Commun.*, **24**, 38 (1978).
22. V. N. Lisitsyn, A. M. Razhec, and A. A. Chernenko, *Sov. J. Quant. Electron.*, **8**, 244 (1978).
23. O. Svelto, in *Laser Handbook*, F. T. Arecchi and E. O. Schultz-Dubois, Eds., North-Holland, Amsterdam, 1972, Chap. C2.
24. J. P. Goldsborough, in *Laser Handbook*, F. T. Arecchi and E. O. Schultz-Dubois, Eds., North-Holland, Amsterdam, 1972, Chap. C4.
25. F. Collier, G. Thiell, and P. Coltin, *Appl. Phys. Lett.*, **32**, 739 (1978).
26. W. W. Wladimiroff and H. E. Anderson, *J. Phys. E*, **10**, 361 (1977).
27. H. M. von Bergmann, *J. Phys. E*, **10**, 1210 (1977).
28. For tabulation of the properties of available lasers and associated equipment see *Laser Focus Buyers Guide*, Advanced Technology Publications, Newton, MA, 1979.
29. D. Klemm, E. Klemm, A. Graness, and J. Kleinschmidt, *Chem. Phys. Lett.*, **55**, 503 (1978).
30. G. Porter and M. R. Topp, *Nature (London)*, **220**, 1228 (1968).
31. A. Muller, *Z. Naturforsch.*, **23a**, 946 (1968).
32. D. S. Klinger and A. C. Albrecht, *J. Chem. Phys.*, **50**, 4109 (1969).
33. R. H. Dekker, B. N. Srinivasan, J. R. Huber, and K. Weiss, *Photochem. Photobiol.*, **18**, 457 (1973).
34. A. Kellman, *J. Phys. Chem.*, **81**, 1195 (1977).
35. T. Hviid and S. Nielsen, *Rev. Sci. Instrum.*, **43**, 1198 (1972).
36. W. B. Taylor, J. C. LeBlanc, D. W. Whillans, M. A. Herbert, and H. E. Johns, *Rev. Sci. Instrum.*, **43**, 1797 (1972).
37. T. A. Alexander, E. R. Lawler, P. W. Wilson, and R. I. McDonald, *Rev. Sci. Instrum.*, **36**, 1707 (1965).
38. B. W. Hodgson and J. P. Keene, *Rev. Sci. Instrum.*, **43**, 493 (1972).
39. L. H. Luthjeus, *Rev. Sci. Instrum.*, **44**, 1661 (1973).
40. G. Beck, *Rev. Sci. Instrum.*, **45**, 318 (1974).
41. S. L. Olsen, L. P. Holmes, and E. M. Eyring, *Rev. Sci. Instrum.*, **45**, 859 (1974).
42. J. W. Hunt, C. L. Greenstock, and M. J. Bronskill, *Int. J. Radiat. Phys. Chem.*, **4**, 87 (1972).
43. R. D. Small and J. C. Scaiano, *Chem. Phys. Lett.*, **50**, 431 (1977).

44. J. A. van Best and P. Mathis, *Rev. Sci. Instrum.*, **49**, 1332 (1978).
45. G. Porter and M. Topp, *Nobel Symp.*, **5**, 158 (1967).
46. G. Porter and M. Topp, *Proc. R. Soc. London, Ser. A*, **315**, 163 (1970).
47. M. R. Topp, *Appl. Spectrosc. Rev.*, **14**, 1 (1978).
48. R. A. McLaren and B. P. Stoicheff, *Appl. Phys. Lett.*, **16**, 140 (1970).
49. C. Lin and R. H. Stolen, *Appl. Phys. Lett.*, **28**, 216 (1976).
50. R. F. Leheny and J. Shah, *IEEE J. Quant. Electron.*, **QE-11**, 70 (1975).
51. (a) E. Sahar and I. Wieder, *IEEE J. Quant. Electron.*, **QE-10**, 612 (1974); (b) *Chem. Phys. Lett.*, **23**, 518 (1973).
52. E. Sahar and D. Treves, *IEEE J. Quant. Electron*, **QE-13**, 962 (1977).
53. H. Schroder, H. J. Neusser, and E. Q. Schlag, *Chem. Phys. Lett.*, **48**, 12 (1977).
54. K. A. Hodgkinson and I. H. Munro, *J. Phys. B*, **6**, 1582 (1973).
55. H. Masuhara and N. Mataga, *Bull. Chem. Soc. Jap.*, **45**, 43 (1972).
56. G. Beck, *Rev. Sci. Instrum.*, **47**, 537 (1976).
57. D. Hartman, *Rev. Sci. Instrum.*, **49**, 1130 (1978).
58. A. Fenster, J. C. LeBlanc, W. B. Taylor, and H. E. Johns, *Rev. Sci. Instrum.*, **44**, 689 (1973).
59. T. M. Jedju, *Rev. Sci. Instrum.*, **50**, 1077 (1979).
60. E. T. Lynk, *Rev. Sci. Instrum.*, **50**, 1074 (1979).
61. S. Bialkowski and W. A. Guillory, *Rev. Sci. Instrum.*, **48**, 1445 (1977).
62. E. Ostertag, *Rev. Sci. Instrum.*, **48**, 18 (1977).
63. G. F. Albrecht, E. Kaline, and J. Meyer, *Rev. Sci. Instrum.*, **49**, 1637 (1978).
64. K. H. Schmidt, S. Gordon, and W. A. Mulac, *Rev. Sci. Instrum.*, **47**, 356 (1976).
65. D. Holten, D. A. Cremers, S. C. Pyke, R. Selenski, S. S. Shah, D. W. Warner, and M. W. Windsor, *Rev. Sci. Instrum.*, **50**, 1653 (1979).
66. Princeton Applied Research Corp., Princeton NJ 08540.
67. D. S. Bethune, J. R. Lankard, and P. P. Sorokin, *Opt. Lett.*, **4**, 103 (1979); see also ref. 68.
68. J. H. Newton and J. F. Young, *IEEE J. Quant. Electron.*, **QE-16**, 268 (1980).
69. R. Wilbrandt, P. Pagsberg, K. B. Hansen, and C. V. Weisberg, *Chem. Phys. Lett.*, **36**, 76 (1975); *ibid.*, **39**, 538 (1976).
70. K. B. Hansen, R. Wilbrandt, and P. Pagsberg, *Rev. Sci. Instrum.*, **50**, 1532 (1979).
71. H. Fabian, A. Lau, W. Werncke, and K. Lenz, *Chem. Phys. Lett.*, **48**, 607 (1977).
72. R. B. Srivastava, M. W. Schuyler, L. R. Dosser, F. J. Purcell, and G. H. Atkinson, *Chem. Phys. Lett.*, **56**, 595 (1978).
73. L. R. Dosser, J. B. Pallix, G. H. Atkinson, H. C. Wang, G. Levin, and M. Szwarc, *Chem. Phys. Lett.*, **62**, 555 (1979).
74. A. B. Harvey and J. W. Nibler, *Appl. Spectrosc. Rev.*, **14**, 101 (1978).
75. J. P. Devlin and M. G. Rockley, *Chem. Phys. Lett.*, **56**, 608 (1978).
76. J. Tretzel and F. W. Schneider, *Chem. Phys. Lett.*, **59**, 514 (1978).
77. W. Werncke, H. J. Weigmann, J. Patzold, A. Lau, K. Lenz, and M. Pfeiffer, *Chem. Phys. Lett.*, **6**, 105 (1979).

78. R. D. Frankel and J. M. Forsyth, *Science,* **204**, 622 (1979).
79. G. Beck and J. K. Thomas, *J. Chem. Phys.,* **57**, 3649 (1972).
80. Y. Taniguchi, Y. Nishina, and N. Mataga, *Bull. Chem. Soc., Jpn.,* **45**, 764 (1972).
81. L. C. Aamodt and J. C. Murphy, *J. Appl. Phys.,* **49**, 3036 (1978).
82. T. K. Razumova and I. O. Starobogatov, *Opt. Spectrosc.,* **42**, 274 (1977).
83. P. S. Mak, C. C. Davis, B. J. Forster, and C. H. Lee, *Rev. Sci. Instrum.,* **51**, 647 (1980).
84. J. Mizuguchi, *Chem. Biomed. Environ. Instrum.,* **9**, 219 (1979).
85. A. D. Trifunac, J. R. Norris, and R. G. Lawler, *J. Chem. Phys.,* **71**, 4380 (1979).
86. J. K. Kochi, K. S. Chen, and J. K. S. Wan, *Chem. Phys. Lett.,* **73**, 557 (1980).
87. P. J. Hore and K. A. McLauchlan, *Chem. Phys. Lett.,* **75**, 582 (1980).
88. G. L. Closs and R. J. Miller, *J. Am. Chem. Soc.,* **101**, 1639 (1979).
89. C. Davis, *J. Phys. E,* **5**, 544 (1972).
90. P. P. Dymerski and R. C. Dunbar, *Rev. Sci. Instrum.,* **45**, 1293 (1974).
91. A. L. Smirl, R. L. Shoemaker, J. B. Hambenne, and J. C. Matter, *Rev. Sci. Instrum.,* **49**, 672 (1978).
92. J. N. Demas in *Creation and Detection of the Excited State,* Vol. 4, W. R. Ware, Ed., Dekker., New York, 1976, Chap. 1.
93. H. Grueter, *J. Appl. Phys.,* **51**, 5204 (1980).
94. F. Karel and J. Novak, *Chem. Listy,* **73**, 1276 (1979).
95. H. Labhart and W. Heinzelmann, in *Organic Molecular Photophysics,* Vol. 1 J. B. Birks, Ed., Wiley, New York, 1973, Chap. 6.
96. K. R. Naqvi, *Chem. Phys. Lett.,* **63**, 123 (1979).
97. The original design was kindly given to us by K. Weiss.
98. S. L. Friess, E. S. Lewis, and A. Weissberger, Ed., "Investigation of Rates and Mechanisms of Reactions," in *Techniques of Organic Chemistry,* Vol. 8 Parts 1 and 2, Interscience, New York, 1961 and 1963.
99. S. W. Benson, *The Foundations of Chemical Kinetics,* McGraw-Hill, New York, 1960.
100. F. A. Matsen and J. L. Franklin, *J. Am. Chem. Soc.,* **72**, 3337 (1950).
101. E. A. Moelwyn-Hughes, *Chemical Statics and Kinetics of Solutions,* Academic, New York, 1971, Chap. 6.
102. G. M. Fleck, *Chemical Reaction Mechanisms,* Holt, Rinehard, and Winston, New York, 1971.
103. L. J. Andrews J. M. Levy, and H. Linschitz, *J. Photochem.,* **6**, 355 (1976/1977).
104. K. H. Grellman and H. G. Scholz, *Chem. Phys. Lett.,* **62**, 64 (1979) and references therein.
105. M. H. Hui and W. R. Ware, *J. Am. Chem. Soc.,* **98**, 4718 (1976).
106. J. B. Birks, D. J. Dyson, and I. H. Monroe, *Proc. R. Soc. London, Ser. A,* **275**, 575 (1963).
107. G. K. Heidt, *J. Photochem.,* **6**, 97 (1976/1977).
108. A. J. Street, D. M. Goodall, and R. C. Creenhow, *Chem. Phys. Lett.,* **56**, 326 (1978).

109. T. Kelen, *Z. Phys. Chem. Neue Folge*, **58**, 268 (1968); *Acta Chim. Acad. Sci. Hung.*, **60**, 87 (1969).
110. R. Sriram, J. F. Endicott, and K. M. Cunningham, *J. Phys. Chem.*, **85**, 38 (1981).
111. T. Donohue, in *Lasers in Chemistry*, M. A. West, Ed., Elsevier, New York, 1977, p. 144.
112. M. Hercher, W. Chu, and D. L. Stockman, *IEEE J. Quant. Electron.*, **QE-12**, 954 (1968).
113. C. R. Guiliano and L. D. Hess, *IEEE J. Quant. Electron.*, **QE-3**, 358 (1967).
114. G. Haag and G. Marowsky, *IEEE J. Quant. Electron.*, **QE-16**, 890 (1980).
115. K. Hamal, H. Jelinkova, A. Novotny, and M. Vrbova, *IEEE J. Quant. Electron.*, **QE-12**, 510 (1976).
116. M. Andorn and K. H. Bar-Eli, *J. Chem. Phys.*, **55**, 5008 (1971).
117. A. Zunger and K. Bar-Eli, *IEEE J. Quant. Electron.*, **QE-10**, 29 (1974).
118. (a) M. L. Spaeth and W. R. Sooy, *J. Chem. Phys.*, **48**, 2315 (1968); (b) W. Rudolph and H. Weber, *Opt. Commun.*, **34**, 491 (1980).
119. L. Huff and L. G. DeShazer, *Appl. Opt.*, **9**, 233 (1970).
120. M. M. Fisher, B. Veyret, and K. Weiss, *Chem. Phys. Lett.*, **28**, 60 (1974).
121. R. F. Leheny and J. Shah, *IEEE J. Quant. Electron.*, **QE-11**, 70 (1975).
122. R. Arsenault and M. M. Denariez-Roberge, *Chem. Phys. Lett.*, **40**, 84 (1976).
123. S. Speiser and A. Bromberg, *Chem. Phys.*, **9**, 191 (1975) and references therein.
124. E. Sahar, D. Treves, and I. Weider, *Opt. Commun.*, **16**, 124 (1976).
125. D. E. Evans, J. Puric, and M. L. Yeoman, *Appl. Phys. Lett.*, **25**, 151 (1974).
126. I. Wieder, *Appl. Phys. Lett.*, **21**, 318 (1972).
127. V. L. Bogdanov, V. P. Klochkov, and A. M. Makushenko, *Opt. Spectrosc.*, **41**, 341 (1976) and references therein.
128. T. K. Razumova and I. O. Starobogatov, *Opt. Spectrosc.*, **42**, 274 (1977).
129. K. I. Rudik, A. I. Maksimov, L. P. Senkevich, M. Y. Kostko, *Zh. Prikl. Spektrosk.*, **28**, 91 (1978).
130. T. R. Evans and L. F. Hurysz, unpublished.
131. R. Bensasson, C. R. Goldschmidt, E. J. Land, and T. G. Truscott, *Photochem. Photobiol.*, **28**, 277 (1978).
132. S. B. Babenko, V. A. Benderaskii, and V. I. Goldanskii, *JETP Lett.*, **10**, 129 (1969).
133. J. T. Richards, G. West, and J. K. Thomas, *J. Phys. Chem.*, **74**, 4137 (1970).
134. M. Ottolenghi, *Chem. Phys. Lett.*, **12**, 339 (1971).
135. A. R. Watkins, *J. Phys. Chem.*, **80**, 713 (1976).
136. G. E. Hall and G. A. Kenny-Wallace, *Chem. Phys.*, **28**, 205 (1978).
137. J. K. Thomas and P. Piciulo, *J. Am. Chem. Soc.*, **100**, 3239 (1978).
138. J. K. Thomas and P. Piciulo, *J. Chem. Phys.*, **68**, 3260 (1978).
139. U. Lashish, M. Ottolenghi, and G. Stein, *Chem. Phys. Lett.*, **48**, 402 (1977).
140. A. Matsuzake, T. Kobayashi, and S. Nagakura, *J. Phys. Chem.*, **82**, 1201 (1978).
141. G. E. Hall, *J. Am. Chem. Soc.*, **100**, 8262 (1978).

142. S. Navaratnam, B. J. Parsons, G. O. Phillips, and A. K. Davies, *J. Chem. Soc., Faraday Trans. 1*, **74**, 1811 (1978).
143. A. Bergman, C. R. Dickson, S. D. Lidofsky, and R. N. Zare, *J. Chem. Phys.*, **65**, 1186 (1976).
144. A. Kellmann and F. Tfibel, *Chem. Phys. Lett.*, **69**, 61 (1980).
145. U. Lachish, A. Shafferman, and G. Stein, *J. Chem. Phys.*, **64**, 4205 (1976).
146. G. Makkes van der Deil, J. Dousma, S. Speiser, and J. Kommandeur, *Chem. Phys. Lett.*, **20**, 17 (1973).
147. T. R. Evans and L. F. Hurysz, unpublished work.
148. J. Joussot-Dubien and R. Lesclaux, *C. R.*, **258**, 4260 (1964).
149. A. Charlesby and R. H. Partridge, *Proc. R. Soc. London, Ser. A*, **283**, 329 (1965).
150. W. A. Gibbons, G. Porter, and M. I. Savadatti, *Nature* **206**, 1355 (1965).
151. K. D. Cadogan and A. C. Albrecht, *J. Chem. Phys.*, **43**, 2550 (1965); *J. Chem. Phys.*, **51**, 2710 (1969).
152. C. Helene, R. Santus, and P. Douzou, *Photochem. Photobiol.*, **5**, 127 (1966).
153. N. Yamamoto, Y. Nakato, and H. Tsubomura, *Bull. Chem. Soc. Jap.*, **40**, 2480 (1967).
154. J. Moan and H. B. Steen, *J. Phys. Chem.*, **75**, 2893 (1971).
155. L. J. Mittal, J. P. Mittal, and E. Hayon, *Chem. Phys. Lett.*, **18**, 319 (1973).
156. H. S. Pilloff and A. C. Albrecht, *J. Chem. Phys.*, **49**, 4891 (1968).
157. L. P. Gary, K. deGroot, and R. C. Jarnagin, *J. Chem. Phys.*, **48**, 1577, 5280 (1968).
158. S. D. Babenko and V. A. Bendersky, *Opt. Spektrosk.*, **29**, 616 (1970).
159. A. Kellman, *Photochem. Photobiol.*, **20**, 103 (1974).
160. M. Yokoyama, M. Hanabata, T. Tamamura, T. Nakano, and H. Mikawa, *Chem. Lett.*, 937 (1976); *J. Chem. Phys.*, **67**, 1742 (1977).
161. N. J. Turro, M. Aikawa, J. A. Butcher, and G. W. Griffin, *J. Am. Chem. Soc.*, **102**, 5127 (1980).
162. W. M. Jackson, J. B. Halpern, and C. Lin, *Chem. Phys. Lett.*, **55**, 254 (1978).
163. C. L. Sam and J. T. Yardley, *Chem. Phys. Lett.*, **61**, 509 (1979).
164. A. P. Baronavski, *Chem. Phys. Lett.*, **61**, 532 (1979).
165. J. J. Tiee, F. B. Wampler, and W. W. Rice, *J. Chem. Phys.*, **72**, 2925 (1980).
166. H. Hemmati and G. J. Collins, *Chem. Phys. Lett.*, **67**, 5 (1979).
167. C. Fotakis, M. Martin, K. P. Lawley, and R. J. Donovan, *Chem. Phys. Lett.*, **67**, 1 (1979).
168. N. Nishi, H. Shinohara, and I. Hanazaki, *Chem. Phys. Lett.*, **73**, 473 (1980).
169. F. B. Wampler, J. J. Tiee, W. W. Rice, and R. C. Oldenborg, *J. Chem. Phys.*, **71**, 3926 (1979).
170. F. B. Wampler, W. W. Rice, R. C. Oldenborg, M. A. Akerman, D. W. Magnuson, D. F. Smith, and G. K. Werner, *Opt. Lett.*, **4**, 143 (1979).
171. (a) K. V. Reddy and M. J. Berry, *Chem. Phys. Lett.*, **66**, 223 (1979); (b) K. V. Reddy and M. J. Berry, *Chem. Phys. Lett.* **52**, 111 (1977).
172. P. M. Kroger, P. C. Demou, and S. J. Riley, *J. Chem. Phys.*, **65**, 1823 (1976).
173. L. Pasternack and J. R. McDonald, *Chem. Phys.*, **43**, 173 (1979).

174. S. Speiser and S. Kimal, *J. Chem. Phys.*, **53**, 2392 (1970).
175. Y. H. Pao and P. M. Rentzepis, *Appl. Phys. Lett.*, **6**, 93 (1965).
176. G. Porter and J. Steinfeld, *J. Chem. Phys.*, **45**, 3456 (1966).
177. H. Zipin and S. Speiser, *Chem. Phys. Lett.*, **31**, 102 (1975).
178. S. Speiser, I. Oref, T. Goldstein, and S. Kimel, *Chem. Phys. Lett.*, **11**, 117 (1971).
179. V. F. Mandzhikov, A. P. Darmanyan, V. A. Barachevskii, and Y. N. Gerulaitis, *Opt. Spektrosk.*, **32**, 214 (1972).
180. L. Parma and N. Omenetto, *Chem. Phys. Lett.*, **54**. 544 (1978).
181. Y. P. Pilette and K. Weiss, *J. Phys. Chem.*, **75**, 3805 (1971).

Chapter **IV**

INFRARED LASER PHOTOCHEMISTRY

A. M. Ronn

Department of Chemistry
Brooklyn College of the City University of New York
New York, New York

1 INTRODUCTION

Chemistry in a very real sense is the science of transformations, and an integral part of such transformations is the gain or release of specific amounts of energies. As such, the true nature of a chemical change can be understood, assessed, and employed when the specific amounts of energy are measured and controlled. Measurements of energy changes that accompany chemical reactions have formed the core of basic research in a multitude of fields for well over 40 years. The control of the specific energy changes is, however, a relatively new and as yet untapped resource that may ultimately result in major breakthroughs in experimental and theoretical chemistry. This is largely due to the laser, one of the most significant scientific developments of this century. The laser's inherent properties of monochromaticity and high intensity are tailor-made for the measurement and control of specific energy changes that induce or catalyze chemical reactions. To truly appreciate the impact that lasers have already had on chemistry and to speculate on what is to come, one needs to look closely at what a chemical reaction really is in terms of the building blocks of nature, namely, molecules, atoms, electrons, and nuclei, and their interaction with the laser radiation. Even a brief sketch, such as is given below, will immediately point out how most inherent shortcomings of conventional chemical reactions can be overcome or circumvented by the use of a proper laser system.

Chemical reactions can and do occur in a variety of media in an equal number of different conditions; reactions can occur spontaneously, that is, on mixing, the gases, solutions, or solids that are reactants require no additional energy input to transform into products. Other reactions must be driven by heat, pressure, catalysts, or combinations of these. It is the coupling efficiency of the driving force with the reacting molecules that determines the total efficiency of the reactions in terms of the desired product yields. Of paramount significance is the direct delivery of energy into the reaction pathway as opposed to the random and often wasteful deposition of energy into a large number of possible pathways. The latter situation is a common occurrence in synthetic chemistry and results in a product mixture rich in unwanted species.

It is this particular kind of chemical reaction, that requiring driving, that is perfectly suited to the application of laser excitation for consequent specific product formation. The laser with its monochromatic light and high intensity forms the perfect controlled energy depositing device, as its radiation may be made to match that desired fraction of energy required for a particular pathway to be preferentially driven. Let us take a closer look at some fundamental definitions and terminology regarding lasers, radiation, energy, and the reacting molecules to construct the foundation for the rather sophisticated technology of laser chemistry.

The study of specific control over potential chemistry and its kinetics has maintained a very large growth rate in the field of infrared laser chemistry. The basic drive for the growth of this field was the strong motivation, perhaps a dream, of the chemists to effect bond selective chemistry. The hope of achieving such bond selective or bond specific chemistry is clearly indicated by the laser's unique ability to excite a specific bond in a molecule. The translation of such specific excitation to specific reactivity has been the focus of extensive studies for the last decade.

The ultimate success of such potential control over chemistry is quite exciting, since it can offer incredible breakthroughs in such areas as synthetic chemistry, genetic engineering, and selective isotope preparation. The methodology to which chemists have been attracted to using laser techniques for experiments affecting chemical dynamics is traced historically on the next few pages.

Although great interest in the field was present from the first invention of the device (1958) it was not until the mid-1960s that the first experiments with lasers began and were centered mainly on elucidation of molecular energy transfer rates. This particular approach was tailored uniquely to the laser's abilities to excite specific motions in complex systems with very high intensity and resolution, as the application of such intense and narrow banded radiation allowed for the creation of relatively large and instantaneous nonequilibrium situations. Therefore relaxation methods could be used to study the energy flow of nonequilibrium systems back to thermal equilibrium. From such pioneering experiments in the mid-1960s a large body of data was generated by the mid-1970s, encompassing a sizable number of molecules spanning the spectrum from diatomics to polyatomics, allowing the general theoretical and experimental framework of energy transfer kinetics to be established.

Once such dynamics were amply studied the next logical progression was to apply the lasers unique properties to the possibility of catalyzing molecular rates of reaction and/or to affect unimolecular dissociations. In the period 1970 to 1973 a large number of experiments were conducted with a variety of infrared laser sources. These ranged from chemical lasers (such as HF and DF) to the carbon monoxide and carbon dioxide systems in which single photon absorptions (2.3 kcal/photon in CO_2) were utilized to enhance reactivity of desired systems. In 1973 Pandora's box was opened by the Russian announcement of successful laser induced selective unimolecular dissociations of isotope specific SF_6 species. The new era of infrared multiphoton dissociation (MPD) had begun. The literature from 1973 to 1979 is a veritable treasure of laser induced multiphoton dissociations, with reported experiments ranging in authenticity, understandability, importance, and complexity from the unique to the bizarre to the unexpected and to the predictable.

In the following pages we examine in some detail what is now routinely called infrared laser photochemistry and we approach the subject from the logical historical progression of single-photon excitation (achieved by continuous or pulse excitation) and go on to multiple-photon excitation and isotope separation, which form today's most interesting and colorful experimental systems. The specific systems discussed encompass both single and multiphoton excitation and dissociation reactions dwell on the constraints as well as the successes of such experiments in terms of model and real life systems. Present and future industrial applications of infrared laser photochemistry must indeed form a cornerstone, and thus we end with a few words regarding this subject.

2 AN OVERVIEW OF INFRARED INDUCED CHEMISTRY

The development of the laser some 18 years ago has resulted in the resuscitation of a number of experimental disciplines including measurements of energy transfer, chemical reaction rates, high-resolution spectroscopy, nonlinear effects, and laser isotope separations. The past few years have seen a tremendous interest resulting in literally thousands of published research articles dealing in one of the subjects mentioned above. The audience tuned to laser and laser related research has steadily expanded to include widely varying disciplines ranging from engineers to chemists to physicists, and laser research is beginning to have an impact even on the industrial community. The reason for such diverse interests in the new technology, and we refer to it as a new technology since 18 years defines the progression of a basic research project into an applied field, is probably traceable to the very fundamental properties of lasers themselves.

The availability of monochromatic, high-powered, short-pulse, and almost totally tunable sources is an indescribable boon to chemists and physicists alike. This new research tool offers the intriguing possibility of measuring rate constants for processes taking place between atoms and molecules that are prepared in well-defined and well-characterized quantum states. As a field within a field, the specific application of vibrational excitation (infrared energy is concomitant with vibrational excitation) of molecules and the subsequent measurements of various relaxation rates as well as chemical dissociation and reaction rates has received the most abundant attention of the community. This interest is due no doubt to the properties of the molecular infrared lasers that were available in years past, but also to the significant effect of vibrational excitation on the rate and mechanism governing chemical reactions. There is a growing feeling among scientists, engineers, and industrialists that for chemistry, the advances in laser technology and growing research development may ultimately lead to a series of major breakthroughs that are quite difficult to quantify at present, since there seems to be no end to the inventiveness of the researchers in this area. Lasers have been referred to jokingly as the answer without a question since they seem to suggest an infinite variety of experiments and technologies that are projected to function better with lasers than with conventional light or other energy delivering devices. The potential applications in the chemical fields alone have been the subject of more than a dozen review articles in the last couple of years and of an equally large number of conferences dealing with the past impact and potential impact of lasers, infrared lasers specifically, on device technology, chemical production, energy conversion, communications, and so forth. It is nearly impossible to cover all these areas in one short chapter and thus we confine our remarks to the utilization of infrared lasers in chemical transformations, whether they be synthetic, dissociative, or isotope selective reactions. However, to at least inform chemists and physicists of the potential areas that are currently being probed by dozens of laboratories around the world, it is undoubtedly significant to list a few of what are thought to be the more important areas of present and future research.

The applications of infrared lasers that have received the more significant public disclosure and that form perhaps the most bombastic of success stories are the laser isotope separations. In this context, the unique ability of lasers to separate and remove specific isotopes presents a serious contender to presently available technologies. Such isotope separations have been accomplished for a large number of nonradioactive isotopes with, of course, the potential of uranium, deuterium, and other isotope separations of paramount significance to energy problems foremost in all minds. The preparation and synthesis of unique chemicals is a subject of much current research not only to academicians, but to the industrial community at large. The perception is that lasers are uniquely suitable to the preparations of fine chemicals, biological and pharmaceutical chemicals, high-purity materials, and other specialty applications. In addition, the laser with its uniquely monochromatic radiation and high intensity in a narrow frequency range is equally suitable for separation and purification of materials. Such problems as trace material monitoring and recovery, purification in specific synthetic steps, and even removal of trace contaminants in a stream are but a few of the separation capabilities currently being discussed.

As analytical and diagnostic tools lasers have been used for many years now in absorption spectroscopy, as well as fluorescence and Raman spectroscopy. The unique combination of high power and monochromaticity allows lasers to act as superior sampling probes and excitation sources for very small concentrations (in the ppm and ppb range) in nondestructive testing. The application of lasers as photochemical or spectroscopic sources has been around for at least a dozen years and is well entrenched in many laboratories around the country. Some of the less acceptable technologies that are still receiving much attention by the well versed community are such specialized applications as rapid and localized energy deposition in a specific bond of a molecular specie. Such localized energy deposition can be a boon to high-energy reactions, can assist in the creation of high-energy compounds, and may lead to enhanced radical production and a variety of other possible chemical manipulations.

The areas of combustion chemistry, catalytic chemistry, and surface chemistry have received some attention; although not as well investigated as these areas, there are clear indications that lasers may become superior sources for effective driving of photocatalyzed reactions. Additionally, of course, since reactions are not just confined to gaseous species but can be either induced or catalyzed both in gas–solid interfaces and in gas–liquid interfaces, lasers can be utilized for either surface treatment for solid- or liquid–gas reactions, or for gas "preparations" prior to gas–solid or gas–liquid reactions. The more commonly and industrially acceptable technology of high-power laser applications for cutting, trimming, drilling, and scribing of metals and non-metal components, such as ceramics and silicon, has been around for a number of years and a large number of books and review articles have been written on the subject. There are many industries across and outside the United States that are currently utilizing lasers as sources for just such applications as welding, cutting, drilling, and even shaping of various materials. Therefore, with all these potential areas open to the scientist newly inducted to the art, the decision on which one is to be pursued can only be determined by the interest and the amounts of monies that are allocated to specific research areas. Laser photons are still costly and unless such costs are offset by some reasonably immediate gains no chemical or other industry is likely to target major developmental areas. The field is extremely young but very exciting in its potential, yet still lacks a large base of fundamental experimental data. There are perhaps two or three dozen reactions that have been well characterized, but certainly these do not form the kind of data base that has existed, for example, in conventional photochemistry and that is typical of the type of data base required for basic research before developmental aspects begin.

It is difficult to predict at present what short-term impact, if any, the laser would make on the market; there are few guiding principles and not enough data, and while many economic market studies can be made, no sufficient experimental data on large-scale processing has materialized to allow a sound economic analysis. Therefore it is perhaps instructive to go through an exercise, as is attempted in this chapter, and summarize currently acceptable constraints, pitfalls, and hopes that the more exciting research results have uncovered before any decision making is attempted. The opinions over the past decade or so are somewhat colored by the author's interests, but the writing is intended as an introduction to the field describing the experimental methodology and thus allowing the practicing chemist, physicist, or engineer to judge for his own benefit whether infrared laser induced chemical reactions can impact significantly on his own specialty.

3 SINGLE-PHOTON EXCITATION

Consider a translationally thermalized molecular gas subsequent to its being irradiated by an infrared laser (Fig. 1). If the absorption cross section in the gas is reasonably large the deviation from equilibrium of the population of the levels responsible for the absorption is considerable. The molecules return to their thermal distribution by way of collisional processes and the excess vibrational energy that was deposited in the molecules is transferred to the rotational and translational degrees of freedom (V–T/R). This results in

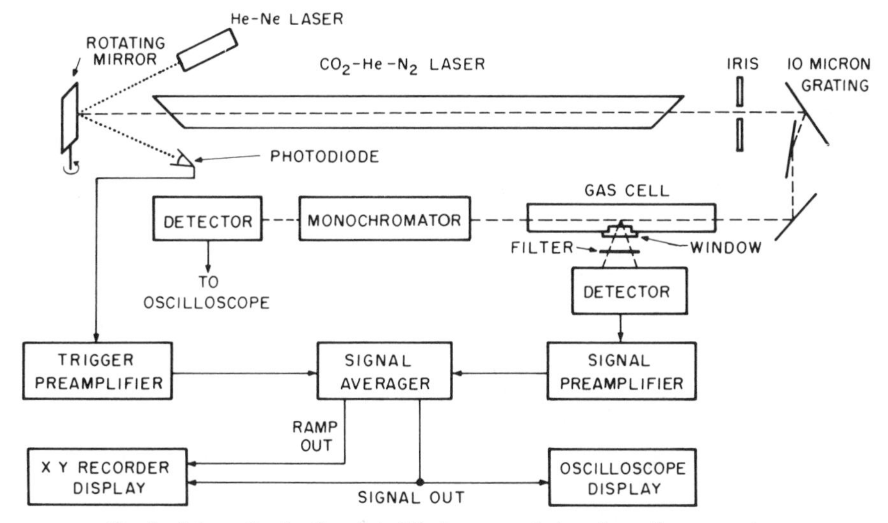

Fig. 1 Schematic of a Q-switch CO_2 laser one-photon absorption apparatus.

a new equilibrium situation at a different translational temperature. Typically, if an excitation is of short duration and modest power, say, for example, a microsecond excitation from a pulsed CO_2 laser of 10 kw power, the new temperature is some 10°K above that of the thermalized sample. However, during continuous illumination the temperature increase may reach hundreds to thousands of degrees within the illuminated volume. This new translational temperature in the gas will eventually return to the original thermal conditions by diffusion and thermal conductivity.

Most molecules studied to date, polyatomics at least, have shown large cross sections for resonant and near resonant energy transfer processes to other nearby vibrational modes, a process normally referred to as vibration to vibration relaxation (V–V). Many overtones and combination bands are thus populated with very high efficiency, thereby creating a generally elevated vibrational temperature in the sample. The time scales of such relaxation processes typically fall in a range where the time for V–V equilibration is faster than the time for V–T/R, which in turn is faster than the diffusion time (see Tables 1 and 2). Thus in a specific reactive experiment there is a set of time constraints for particular reactions under study that argues for pulse times shorter or longer than the times mentioned above. Additionally, such rapid vibration to vibration or even the slower vibration to rotation and translational relaxation processes will drain the initially excited levels in the molecule under study and thereby remove the selective excitation. A typical apparatus (circa 1970) for measurements of energy transfer rates such as

Table 1 Deactivation Rate Data

	Self	*He*	*Ne*	*Ar*	*Kr*	*Xe*
CH_3F	0.59[a]	0.78	0.065	0.039	0.022	0.019
	15,300[b]	17,300	117,000	195,000	315,000	387,000
CH_3Cl	7.5	3.1	0.51	0.54	0.54	0.53
	1,100	4,800	15,000	14,000	12,000	13,000
CH_3Br	20	4.2	1.2	1.3	1.0	1.0
	300	3,700	6,300	5,500	5,900	5,700
CH_3I	22	5.6	1.9	2.2	1.6	1.3
	500	4,000	4,600	4,400	4,800	5,600
CD_3F	0.44	1.6	0.12	0.074	0.025	0.082
	18,000	8,300	65,000	102,000	222,000	86,000
CD_3Cl	2.8	7.3	0.48	0.17	0.12	0.13
	2,900	2,000	17,000	44,000	55,000	52,000
CD_3Br	12	7.4	1.2	1.1	1.0	0.70
	600	1,800	6,400	6,500	5,700	9,300
CD_3I	27	13	1.6	1.7	1.8	1.4
	400	1,800	7,100	5,900	4,400	5,300

[a] First number is rate constant for V–T/R deactivation process in $msec^{-1}$ $torr^{-1}$.
[b] Second number is Z, number of collisions needed for deactivation, rounded off to the nearest hundred.

these depicted in Tables 1 and 2 is shown in Fig. 1. Figure 2 shows a typical infrared fluorescence signal of difluomethane obtained on such an apparatus. Note the effect of added Xe in supressing the translational temperature. Figure 3 shows both activation and deactivation of infrared fluorescence in $S^{18}O_2$. The former is shown as a risetime of a signal and the latter as a decaying exponential. In other words, the localized or the initially localized vibrational energy becomes redistributed over many other molecular coordinates and degrees of freedom, thereby contributing to the excitation of what is generally referred to as the heat bath of the medium.

Since most relaxation processes are collisional in nature, most can be surpressed or controlled by working at sufficiently low gas pressures. However, another constraint may set in at this time; this one is referred to as saturation or rotational bottlenecking. This term is descriptive of a situation encountered when an intense laser pulse is absorbed by a two-level system that becomes saturated, unable to absorb any more radiation. Since these levels contain only a fraction of the total population, a limitation is imposed on the up-pumping in molecules in which rotational line spacing is rather large. Such rotational bottlenecking or saturation phenomena can be overcome by

Table 2 Activation Rate Data

	ν_2,ν_5	ν_1,ν_4	ν_3	ν_6
CH_3F	86[a]	122		
	106[b]	77	—[f]	—[e]
	−315[c]	—[d]		
CH_3Cl	160	214	76	
	53	39	110	—[f]
	−330	—[d]	285	
CH_3Br	105	108	40	
	64	62	170	—[f]
	−360	—[d]	340	
CH_3I	101	225		
	65	32	—[e]	—[f]
	−370	—[d]		
CD_3F		680		
	—[e]	12	—[f]	—[e]
		—[d]		
CD_3Cl		490	52	52
	—[f]	17	150	150
		—[d]	260	260
CD_3Br		650	86	86
	—[f]	10	77	77
		—[d]	280	280
CD_3I	—[f]	780	92	92
		10	80	80
		—[d]	295	295

[a]The first number in each section is the rate constant for the activation process in units of $msec^{-1}$ $torr^{-1}$
[b]The second number in each section represents Z, the number of collisions needed for activation. Parameters used to calculate Z were obtained from Ref 16.
[c]The third number in each section represents the energy deficit (cm^{-1}) between the energy levels involved in the rate-determining step.
[d]No mechanism proposed.
[e]Rate constant not measurable because of experimental difficulties (laser scatter, weak signal, fast signal).
[f]Laser-pumped level.

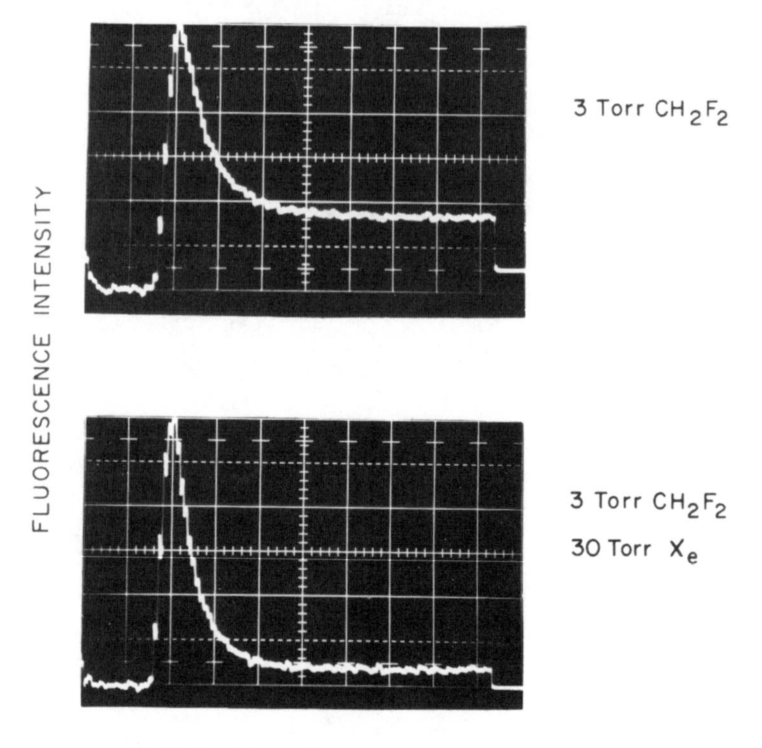

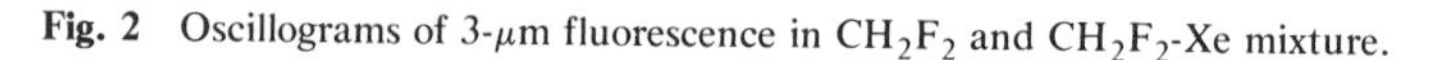
Fig. 2 Oscillograms of 3-μm fluorescence in CH_2F_2 and CH_2F_2-Xe mixture.

taking advantage of rotational relaxation processes that can be enhanced by adding an inert gas such as helium and argon, which makes the rotational relaxation faster than the laser excitation rate. A pulse duration can be selected that will avoid bottlenecking; this pulse length must be larger than the rotational relaxation time but shorter than the vibrational to translational relaxation time to avoid thermal heating. The thermal heating effects mentioned in the preceding sentence could complicate the analysis by not permitting a choice between true laser induced (catalyzed) rate enhancement and an enhanced thermal rate. Therefore such thermal heating must be distinguishable, measurable, controllable, or circumventable.

If it is assumed that selected pulse timing, absorption cross section, reacting species, and so forth have been considered carefully and all conditions have been preset, the next problem that must be addressed is the expected observation on the selected reaction rate. For example, let us consider the reaction of molecule AB with the molecule CD. Should vibrational excitation

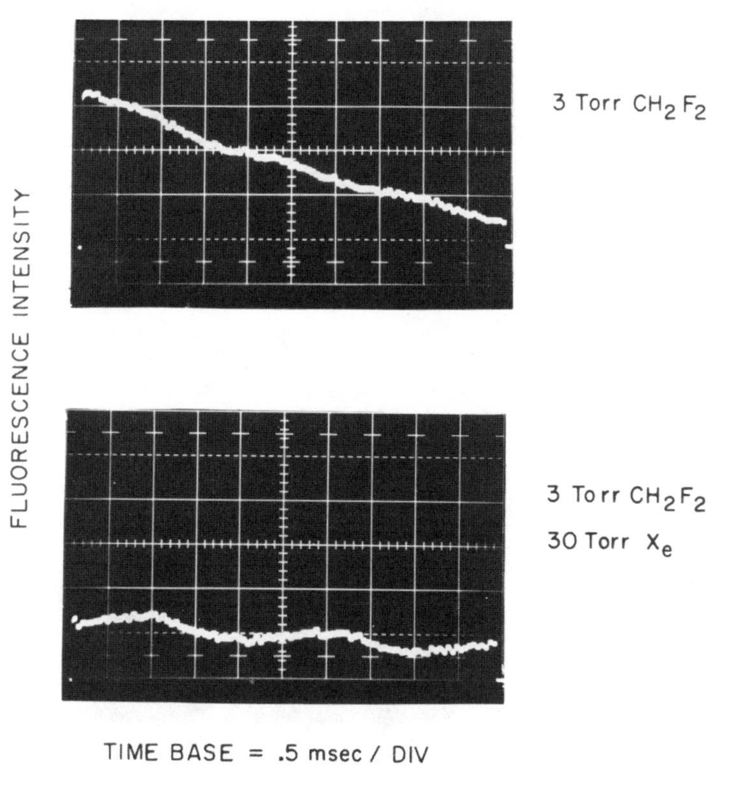

Fig. 2 (*Continued*)

be delivered to molecule AB, which is a diatomic molecule, no complications will arise if the time constraints on pulse excitation have been preset carefully, since only one vibrational state exists in this molecule and the general vibrational excitation becomes the selective excitation that was delivered to the reacting specie. Conversely, within a polyatomic molecule, say ABC, just a triatomic molecule, a vibrational excitation delivered, say, to an asymmetric stretching vibration may within a V–V transfer time to collisionally transferred to the symmetric stretching as well as to the bending vibration in the molecule. If an observation as made on the enhanced reactivity of such a system (ABC) with another system, say DE, that selective rate of enhancement would have to be extracted from a total vibrational excitation of the reaction rate. In other words, if the asymmetric stretching vibration was expected to affect the reaction coordinate in a specific sense, then the selectivity would be lost if the reaction time exceeded the rate of V–V (vibration to vibration) relaxation time, since the symmetric stretch and the bending

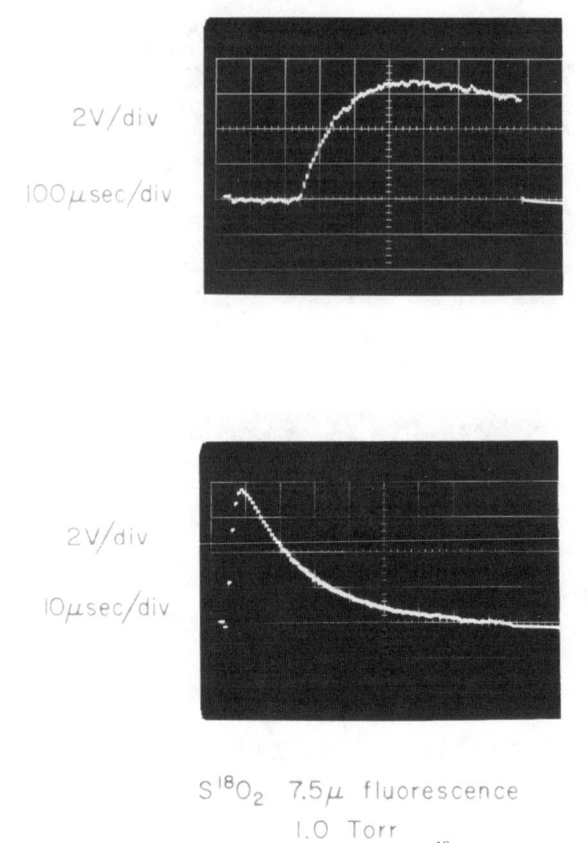

Fig. 3 Fluorescence of 7.5 μm in $S^{18}O_2$.

modes would have been equilibrated collisionally with the initially excited asymmetric stretching vibration. While such total excitation experiments are in their own rights important, they are by no means the local mode excitation that was conceptualized as a state probing experiment. Therefore careful selection of system parameters must be considered before a complete explanation for a rate enhancement is put forth.

Let us add to this set of constraints an additional one, since our choice of system to be studied may not be as straightforward as it seems at first glance. To be sure a photon carrying 2.3 kcal of energy (CO_2 laser 10.6-μ photon) cannot be expected to significantly influence a reaction having an energy of activation of, for example, 25 kcal/mole within a single absorption event. Thus it seems that one must seek out reactions with threshold energies for activation on the order of 5 or at most 8 kcal, thus ensuring the observation

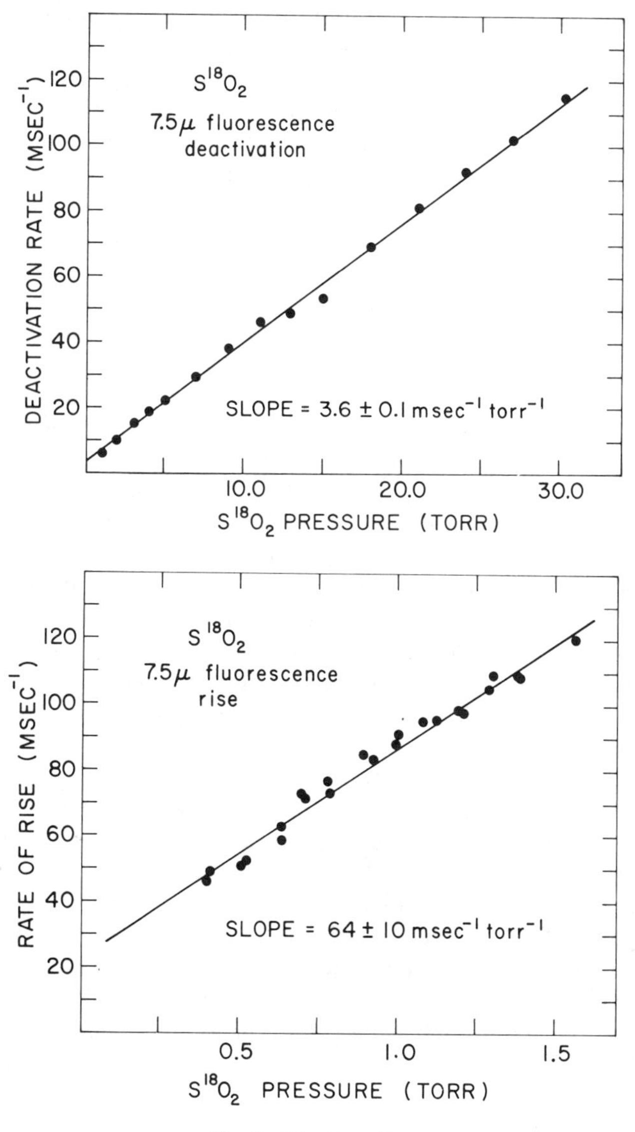

Fig. 3 (*Continued*)

of *some* rate enhancement. The problem involved in such a choice, at least for most molecular reactions, is that they would possess quite rapid rates at thermal energies and thus require experimental discrimination between a fairly fast thermal rate and a somewhat faster laser induced rate. While definitely feasible, this experiment is not easily performed by any means, since even small thermal changes may well simulate total vibrational excitation. In fact, many examples have been set forth in the literature describing the degradation of laser excitation into thermal excitation by way of V–T/R processes and a thermal rate enhancement was mistakenly taken for a vibrational enhancement of the reaction rate. To be sure, once all these pitfalls have been considered there remains the choice of a molecular system having the least and most familiar constraint, that of having a coincidence (or a near coincidence) at low pressures with a given infrared laser line. While on the surface this does not seem to be a problem since the voluminous research on energy transfer and laser spectroscopy has indeed shown hundreds of such coincidences in polyatomic species, it should nonetheless be noted that typical infrared frequencies are nominally specified with accuracies of 0.2 cm^{-1}, while laser frequencies are given with accuracies of 10 MHz (1 cm^{-1} = 30,000 MHz). Therefore to a certain extent a match requirement becomes a needle in the haystack search and while a system may look as if it possesses such an overlap at moderate laser power densities and low pressures, in reality it may not possess a sufficient absorption cross section to ensure a deviation in population sufficient to cause vibrational enhancement to give a detectable signal.

In what follows we examine two or three examples of the experiments that have been performed with single-photon absorption in the infrared and have actually demonstrated unambiguous reaction rate enhancement traceable to specific vibrational excitation. We have to discuss the bimolecular reaction first, since the first reliable experiment on laser enhancement of a reaction rate was indeed a bimolecular experiment dating back to 1971 and reported on by Odiorne, Kasper, and Brooks, who reacted potassium atoms with hydrogen chloride in the gas phase: $K + HCl \rightarrow KCl + H$. In the simplified description (Fig. 4) of such a reaction scheme, one could write an empirical Arrhenius rate equation $k_r = A \exp - (E_a/RT)$, where E_a denotes the activation energy.

For the case in point, dealing with small molecules, the dynamics of a bimolecular reaction can easily be specified in terms of such a simplified rate equation since such an abstraction reaction involves a dissociation of a single bond in the HCl molecule and the creation of a new bond in the KCl molecule. In the usual chemical fashion of defining a reaction coordinate, the vibrational coordinate of the HCl is therefore specified as the reaction coordinate. If the molecule is to be excited along this coordinate to a specific vibra-

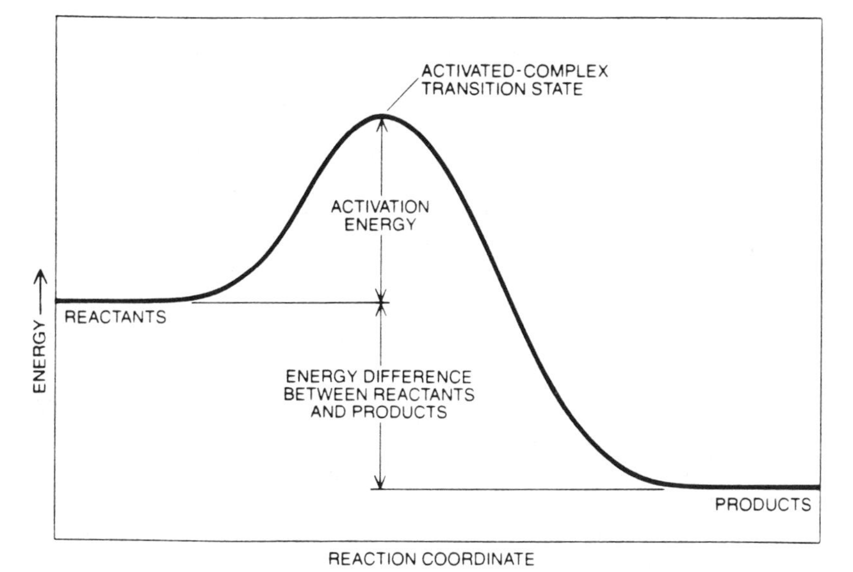

Fig. 4 Energy versus reaction coordinate for an exoergic reaction.

tional state some 2700 cm^{-1} above the ground state, then one can describe the enhanced reaction rate as a ratio between the K_r described earlier and the K_r', where the prime refers to the enhanced rate following vibrational excitation along the reaction coordinate. Such a ratio of rates translates directly to a difference in the exponential factors or the difference in the energies of activation. It is assumed, of course, in such a simplified description that the preexponential factor, A in our equation, remains a constant and consequently the ratio by which the enhanced vibrationally driven reaction will proceed relative to the ground-state reaction is expressed simply by the factor exp ($Nh\nu/RT$). That acceleration is significant since N is Avogadro's number and $h\nu$, of course, refers to the particular mode driven. The latter is equal to the laser energy and in the case of the Odiorne experiment we are speaking of an HCl molecule. Excitation was provided by an HCl laser and thus $Nh\nu$ represents an energy of about 8.3 kcal/mole. For orientation, RT at room temperature is equal to only 0.6 kcal and typical energies of activation for this reaction are on the order of perhaps 1 or 2 kcal/mole. Consequently, a rather large rate enhancement can be obtained by such an experiment.

In the Odiorne study the K atoms and HCl molecules interacted as two molecular beams crossing at 90°. An HCl laser operating at approximately 3.6 μ was made to propagate along the axis of the HCl beam. This laser operated on the $V = 1 \rightarrow V = 0$ levels in the hydrogen chloride and thus ensured a significant absorption cross section in the hydrogen chloride molecules present in the beam. Two series of experiments were performed, the normal room temperature experiment for the formation of KCl by the intersecting beams measured a rate that we call K_r, and a second experiment performed with laser excitation of the HCl species measured another rate of reaction that we call K_r^1. At the end of such a series of experiments and allowing for statistical fluctuations, the ratios of K_r^1/K_r were calculated and were shown to be slightly larger than 100, implying that the vibrationally assisted reaction was driven with a much greater efficiency than the normal room temperature reaction.

There are, however, some questions regarding such a simple interpretation of this experiment. It is reasonable to assume that the rate enhancement came strictly from vibrational excitation? Is it possible that translational heating contributed significantly to the acceleration? What is the total number of HCl molecules that are expected to play a role in this vibrationally assisted reaction? Have all constraints mentioned earlier been considered? To a certain extent this experiment is the ultimate in simplicity because of its very nature, that is, the reaction of an atom and a diatomic molecule, the latter being excited with a laser based on the same diatomic molecule. Thus exact source conditions are ensured, as collisional quenching and energy transfer are largely eliminated by working at 10^{-6} torr pressures, where the V–T/R relaxation rate of HCl is known to be quite long. Thus the very choice of the experimental partners and conditions guarantees a cleancut observation. Translational heating is the only remaining experimental problem. A simple experimental technique can be utilized to distinguish between vibrational and translational heating effects. The beam of HCl molecules can be translationally heated to simulate the temperature that a vibration to translation relaxation causes and thus a hot thermal reaction can be compared to the room temperature (thermal) reaction and both of these can be compared to the laser driven vibrationally activated reaction rate. When such a comparison is made it is found that the rate of the vibrationally assisted reaction is more efficient by more than a factor of 10 than translational energy excitation, which in turn is approximately 10 times as efficient as room temperature operation. Thus the experimental integrity prevails and a simple theoretical calculation can argue that such vibrationally assisted reactions can indeed be analyzed, even with a simple bimolecular Arrhenius theory approach. Other reactions that fall into such an unambiguous and fairly unique category are not many, but the few that do are shown in Table 3.

In these studies the HF system was excited to the $V = 1$ state and the ratio of rates of reactions was compared to the rate of ($V = 0$) HF-atom reactions. Note that all the above reactions are endothermic atom-diatomic molecule reactions. The reason for such vigorous investigations into this kind of

Table 3 Vibrationally Assisted Biomolecular Reactions

$K + HCl\,(V = 1) \rightarrow KCl + H$	$\Delta H \approx 1.7$ kcal
$Ca + HF\,(V = 1) \rightarrow CaF + F$	$\Delta H \sim 8.7$ kcal
$Sr + HF\,(V = 1) \rightarrow SrF + H$	$\Delta H \sim 6.5$ kcal
$Br + HCl\,(V = 2) \rightarrow HBr + Cl$	$\Delta H \sim -5.5$ kcal

systems dates back to the predictions of Polanyi and co-workers, who showed that vibrational excitation should preferentially drive these systems to the product side. All the work discussed above confirmed these predictions. In fact, it has also been demonstrated that the reverse case, namely, exothermic atom-diatom reactions, are not aided by depositing vibrational excitation in the reactants. We return to a discussion of the exact experimental details for one or two of these experiments, but now let us direct our attention to the polyatomic case of excitation with single photons so that we can compare the complexities that are introduced into a theoretical explanation of a diatom reacting with a triatom to that of one of an atom reacting with a diatom.

Over the last few years the laser augmented kinetics of the ozone and nitric oxide reaction has been studied intensively. This reaction, which was first investigated some years ago by Gordon and Lin (1973), continued to generate a fair amount of research activity until 1976 when another investigation was carried out by Freund and Stephenson (12). In this latter reaction scheme the ozone molecule, being a triatom with three vibrationally accessible states, is excited (13) by a CO_2 infrared laser depositing energy in the asymmetric stretching vibration and the augmented rate of the reaction is measured by observing the chemiluminescence of the ^2B states of the NO_2 product molecule (electronically excited).

Under thermal conditions the energy of activation of this reaction is quite low, 4 kcal/mole, and thus a deposition of a single CO_2 laser photon, if we are to think in terms of the model discussed earlier, would decrease such an energy of activation by 2.3 kcal, potentially causing a substantially increased rate of reaction. Indeed, the established augmented rate that was reported by Gordon and Lin (13) was higher by a factor of 20 than the thermal rate. A number of investigations of the same reaction substantiated the claim that it was indeed the vibrational excitation that was responsible for the augmented rate. What was not totally understood, or at least precipitated a reasonable amount of discussion, was the nature of the vibrational coordinate that contributed to the accelerated rate along the reaction coordinate. Although it is the asymmetric stretching vibration of the ozone that is excited initially, it is not clear whether vibrational relaxation takes place within the ozone molecule as a result of collisions with other bath molecules so that the symmetric stretch and the bending vibration both participate in the augmented laser assisted reaction rate. In 1976 Freund and Stephenson (12) decided to vary the experiment by providing the excitation not to the ozone molecule, but rather to the diatomic NO. They utilized a CO laser for the study which was Zeeman tuned into resonance with the NO single vibrational state and surprisingly enough found that the augmented rate was pretty much the same as the augmented rate that was discovered by ozone excitation.

This result, which was not at all anticipated, is suggestive of some interesting chemical arguments. The NO molecule is excited by means of a 5-μ laser and accepts 4.3 kcal of energy with such pumping whereas the ozone molecule is only excited with 2.3-kcal photons. The fact that the augmented rate remained the same while the energy content was doubled in the two extreme experiments suggests of a rather complex explanation. One such possibility is that a collision complex is established between the NO and the ozone molecule that has an inordinately long lifetime. Now this is paradoxical, to a certain extent, because there is no chemically intuitive reason to postulate an oxygen transfer reaction to be assisted by NO excitation without demanding such a long-lived collision complex. On the other hand, to say that a collision complex that contains 4.5 kcal of energy is just as probable as a collision complex formed with a 2.3 kcal in energy content necessarily means that the former must be produced at lesser efficiency since otherwise the augmented rate would be significantly higher.

The NO, ozone experiment was as tailor-made for laser studies as a reaction can be since it circumvented one of the most difficult problems of infrared investigations: namely, product state analysis. This analysis is normally carried out for dark reactions with either absorption monitoring or infrared fluorescence monitoring; both techniques are relatively insensitive to very small ratios in product state changes because of the small extinction coefficients associated with infrared transitions and the detectors that are currently available to monitor either fluorescence or absorption events. However, a chemiluminescent reaction that provides fluorescence in the visible or UV region does not suffer from either limitation, since the investigation can be carried out using photomultiplier tubes. Additionally, of course, the time resolution for such experiments is significantly enhanced, since photomultipliers have inherently better risetimes and total response curves than solid-state infrared detectors (this was particularly true during the seventies). One additional piece of information that blessed the ozone NO experiment was the *a priori* determination that the reaction that forms NO_2 in the ^2B state has an energy of activation even lower than the one reported for the normal ground-state product formation, that is, of the NO_2 in the ^2A state (the ^2B state energy of activation is 2.3 kcal while that of the doublet A state is 4.2 kcal). Consequently, one would have expected an *a priori* rate augmentation with a 2.3-kcal photon of somewhere in the neighborhood of 1 or 2 orders or magnitude. Now with all this in mind let us take a look at the possible complexities that can arise in this experiment.

In the first investigation of Gordon and Lin the reaction was indeed found to be more effectively catalyzed by a factor of 20 or so than the non-laser-induced reaction. The bending vibration was believed to be primarily responsible for the catalysis. This conclusion was based on the observation of a delay time in the reaction's inception that was attributed to the V–V energy transfer time between ozone and the rare gas diluents present in the system to suppress translational heating. This particular interpretation was the subject of a secondary investigation that verified the enhancement rate but did not in any way offer any better proof of whether it was indeed the bending vibration of the ozone that was important nor did it explain how it came to be equilibrated on the time scale of the reaction's inception. To add a bit more confusion to this seemingly simple reaction let us add that separate measurements of V–T/R energy transfer rates in ozone as a pure gas, as well as with other diluents as deactivating species, have been made as well. These established that subsequent to laser excitation of the asymmetric vibration of O_3, the symmetric stretching vibration equilibrated relatively quickly with the asymmetric stretch while the bending vibration did not. The rapid equilibration was found to be approximately 160 $msec^{-1}$ $torr^{-1}$, whereas the slower V–V equilibration with the bending vibration was found to be about 9 $msec^{-1}$ $torr^{-1}$. Based on these data one could assume that once ozone is excited to the asymmetric stretch, the symmetric stretch and the bending motion, not to mention all other combinations and overtone bands, can be equilibrated at some vibrational temperature at the pressures at which the laser augmented reaction experiments were performed. In fact, it can be shown that the induction time reported in the original experiment and verified by secondary experiments may well be explained by the equilibration of the two stretching vibrations rather than the stretching and bending vibrations. From chemical intuition it is more attractive to expect that the stretching vibration is more of a factor in determining the mechanistic route, since it is hard to conceive of an intermediate or complex that in any way depends on the bending mode, for after all this is an oxygen atom abstraction. The only piece of data that argues strongly for the long-lived complex is the 1976 investigation by Freund and Stephenson, which showed that the excitation must not be deposited at the ozone at all in order to observe the same augmentation in the reaction kinetics but that excitation into the NO stretching mode is just as effective in catalyzing the reaction. Thus at present it is accepted that vibrational excitation of either one of the reactants is as efficient as that of the other for this particular case; however a complete theoretical understanding of this mechanism has not yet been achieved.

To a certain extent this symphony remains unfinished because of the exciting discovery in 1973 (4, 5) of the multiphoton dissociation of polyatomic molecules. This discovery very quickly led to the exponential growth of two new disciplines, one that is currently known as laser isotope separation (LIS) and a second one called laser induced chemistry (LIC). Before we take a closer look at the experimental details in the single-photon absorption let me just offer that the absorption of many photons in a molecule is basically a logical extension of the single-photon absorption experiments we have been discussing. This follows since it must be postulated that more than one photon is absorbed on a per molecule basis if a dissociative unimolecular reaction such as $SF_6 \rightarrow SF_5 + F$ is driven by a moderate energy CO_2 laser pulse. This statement is based purely on energetic considerations that maintain that the sulfur fluorine bond strength is approximately 80 kcal/mole and that the CO_2 photon is equivalent to 2.3 kcal of energy and thus as many as 35 CO_2 photons must be absorbed sequentially to cause a dissociation. We return to treat the subject of multiphoton dissociations and excitations in greater detail in the following sections. Let us go back to examine Fig. 1 and discuss the experimental details that have been utilized over the years for typical one-photon laser catalyzed reaction studies.

The general methodology is reasonably simple and is depicted in the figure, which shows a source laser used for the irradiation. Typical lasers that have been used are discussed in a number of review articles as well as in this volume. The workhorses of the industry have been the CO_2 lasers, the chemical lasers HF and HCl, and on occasion HBr, and the CO laser (either chemically or electrically excited). Two types of CO_2 lasers have ben used for the study of the NO and ozone reactions. In the Gordon and Lin experiment a Q-switch CO_2 laser was utilized; typically this device is capable of delivering somewhere around 10 to 20 kw of excitation per pulse with a pulse width of roughly 1 to 2 μsec and a repetition rate that is variable within the range of 10 to 500 cps. The second laser utilized for these measurements was a continuous CO_2 laser of a fairly low power, ranging somewhere between 1 and 10 W and operated in a single-mode configuration so as to concentrate as much as possible all laser photons within the given narrow Gaussian distribution. Both of these lasers are either commercially available from a number of suppliers or are home-built devices.

The schematic depicted in Fig. 1 is an accurate representation of a complete system that can be fabricated rather easily in the laboratory with any existing vacuum and high-voltage equipment. The laser tunability, for these two particular lasers, is normally accomplished with a grating that is mounted as a termination on the cavity. Rather than using a mirror one uses a grating in a littrow configuration meaning that the first-order light reflects back into the cavity to provide the necessary optical feedback while the grating also serves as an out coupler in its zeroth order, usually with an efficiency of 10 to 20% (thus 80% of the laser output is confined within the cavity). These gratings are the normal glass substrate ruled gratings that require no special handling other than the normal care indicated for all optical components. The other mirror terminating the cavity is a 100% reflecting mirror that can be either gold coated or dielectrically coated as the case may be to provide 100% reflection and maintain total feedback. In the case of the Q-switch laser the secondary mirror, as shown in Fig. 1, is rotated about an axis perpendicular to the optic axis of the laser at variable speed, which is normally controlled by an audio oscillator-power amplifier arrangement and thus ranges anywhere from 10 to 500 cycles with reasonable ease. The rotation of the mirror is what determines pulse repetition rate but not the pulse duration; the later is determined by the rotational relaxation time of CO_2 in the laser cavity in the presence of the flowing mixture of nitrogen and helium. Typical pressures in the flowing tube are 20 to 50 torr. The laser output then is made to propagate through a cell whose construction is typically that of a cross or a tee where one axis serves as the laser entrance and exit windows and the third and fourth sides (or just the third) form a convenient observation window for either infrared or visible fluorescence. The cell is coupled to a high-vacuum system that allows for efficient and clean transfer of gases as well as mixing and pressure monitoring of the gases within the cell configuration. Should temperatures other than room temperature be desired, such a cell can easily be adapted to accommodate temperature changes from 77°K to about 250°C by locating suitable heating and cooling coils or an outer jacket around the cell. Cell materials can be chosen to be either glass or metal as dictated by the corrosivity and the reactivity of the gases under investigation. Monitoring of products that are mentioned earlier can be accomplished either by photomultiplier tubes, as in the case of the NO_2 electronically excited state, or by solid-state infrared detectors, such as indium-antimonide (InSb), gold-germanium (Au:Ge), and copper-germanium (Cu:Ge) (a detailed listing of those detecting devices commonly available today as well as a list of some suppliers from which such equipment can be purchased is found in *Laser Focus Buyer's Guide* published on a yearly basis). Additional experimental needs may require product monitoring subsequent to experimental observations. Typical techniques are gas chromatography, mass spectroscopy, infrared absorption spectroscopy, and analytical techniques; these peripheral instruments or systems are usually found in any chemical laboratory. Windows are normally made of sodium chloride since this material is transparent to practically all infrared lasers. The techniques of holding the windows on the cell have varied over the years from the simple and effective wax or epoxy seals to O ring seals. The choice is, to a certain extent, very important, since some chemicals are very reactive to the glues while others may be reactive to the O ring material. The third (and fourth) observation windows are usually composed of a material that is chosen to block any laser light scatter and transmit the wavelengths of interest. For example, if observation of a 2.5-μm fluorescence is attempted while the excitation is provided by a 10-μm laser the choice would be quartz, glass, or sapphire, since all three block the 10-μm scatter while allowing nearly 100% transmission of the 2.5-μm radiation.

The K + HCl experiment discussed earlier, and the Ca + HF, Ba + HF, and Sr + HF experiments are of a different level of difficulty, because there were laser excited molecular beam experiments, and chosen configuration for a collision free regime. In a molecular beam configuration the metal was enclosed in a specialized oven that was heated to allow the atoms to effuse through a slit to form a beam traversing a secondary chamber into which was crossed a beam of HCl molecules that were made to effuse through another slit and another entry port. The crossed geometry of the potassium and hydrogen chloride beams was then monitored by either an energy analysis device configured from a molecular beam detector or by an energy detecting device that allowed differential and total cross-section measurements on the product-state distribution of KCl formed.

While molecular beam configurations are preferable study conditions since unwanted collisions are literally eliminated or at least contained and understood with great precision, it should be emphasized that these experiments are not quite as easy as bulb or cell experiments and do require a great deal of engineering expertise. Controlled conditions must be maintained for one or both beams, and laser parameters must be carefully maintained since no pressure broadening can help the energy absorption in the system at a pressure regime of 10^{-6} torr. It goes without saying that high-flux infrared lasers are preferred to low-flux lasers, but it must be emphasized that high-flux lasers coupled with a strongly absorbing gas can also be responsible for very high thermal gradients within the sample and thus make interpretation of the vibrationally driven reaction much more complex (thermally induced rate may well overcome any vibrational enhancement). It is our opinion that pulsed lasers are much preferred to continuous illumination since the temperature profile is better understood and results have a higher degree of significance with respect to true vibrational enhancement of reaction. We return to this particular point when we discuss laser multiphoton dissociation with high-flux infrared lasers.

To conclude this section, which deals with single-photon absorption for laser assisted reactions, let us discuss briefly a two-photon absorption and consequent kinetics of a simple atom-diatom reaction.

It is rather clear from the energetics (see Fig. 4) that if a reactant can be supplied with a sufficient amount of energy, not only can the energy of activation be lowered for a bimolecular reaction, but in principle it can be totally overcome. A particular example of such an experiment is the reaction of HCl with Br atoms where HCl is excited to the $V = 2$ state.

The experiment, as described by Wolfrum and co-workers (14), was based on the earlier observation that ground-state Br atoms react very slowly with HCl ($V = 0$) at very low pressures. Additionally, the inevitable occurrence of HCl ($V = 1$) species either created by direct laser excitation or by near HCl ($V = 2$) + HCl ($V = 0$) $\leftrightarrows$ 2HCl ($V = 1$) resonant collisional relaxation did not contribute to the reaction study since HCl ($V = 1$) deactivates very efficiently to $V = 0$. Thus the reaction between Br ($^2P_{3/2}$) atoms and HCl ($V = 2$) was the only available and energetically favorable channel. The experiment then measured the reaction between these two species, Br ($^2P_{3/2}$) + HCl ($V = 2$) $\rightarrow$ H Br ($V = 0$) + Cl ($^2P_{3/2}$), and energetically was shown to overcome the barrier by some 2400 cm^{-1}. Thus a reaction occured in effect on every Br ($^2P_{3/2}$) and HCl ($V = 2$) encounter. The experimental conditions were such that HCl was present at 200 mtorr and only 10^{-1} or less of those molecules became $V = 2$ species. The rate of this reaction was found to be 9×10^{11} cm^3/(mole)(sec) with laser excitation as compared to a rate of 10 cm^3/(mole)(sec) for the vibrationless reaction. This is well in line with a bimolecular reaction occurring with every collision, since it is within the 10^{-5}-sec pulse duration of the HCl laser ($Z \simeq 2$ collisions).

This experiment was driven by a sequential two-photon absorption from a chemical HCl laser. There is clearly a great deal of research still to be performed on this class of reactions, yet the path seems clear. If sufficient energy is delivered to the system within the time constraint of the reaction rate constant the total barrier indeed can be overcome. Therefore reactions can be driven, in principle, at rates equivalent to the Arrhenius A factor, that is, the reaction scheme is collisional.

The experimental arrangement shown in Fig. 5 was taken from an identical experiment performed with a different atomic species, that is, the reaction HCl + O. The basic laboratory requirements for such an experiment are an HCl laser, a mixing cell and gas flow system, a source of atoms (Br, O, etc.), usually a microwave discharge (2450 MHz), and electronic and optical monitoring devices. High-vacuum requirements and fast flows are an integral part of such studies, since the rate of reactions must be measured prior to collisional deactivation.

It should be mentioned that while one- or two-photon assisted reactions have formed the basis of what is now commonly called laser induced chemistry there is a present lack of research and development interest in this class of studies. The thrust of research and development has shifted almost

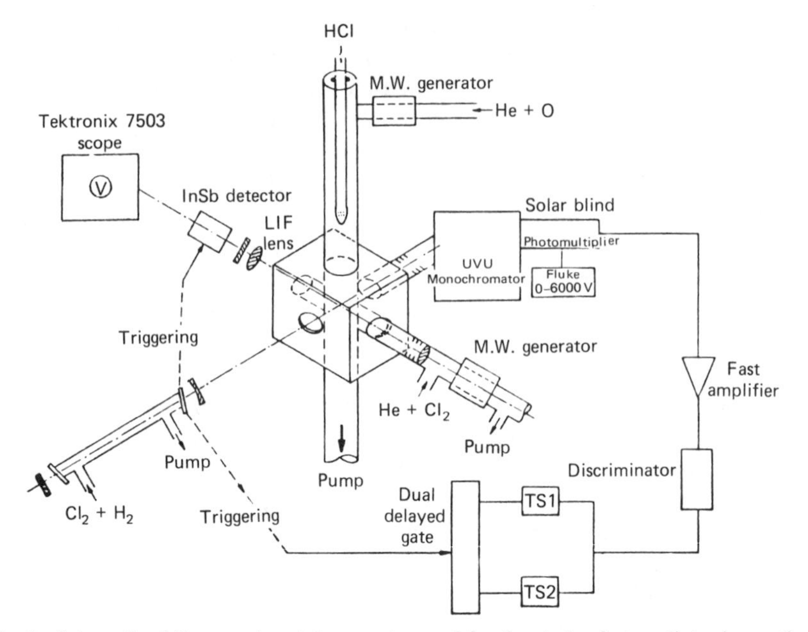

Fig. 5 Schematic of the experimental apparatus used for the study of atom-diatomic reaction: O + HCl.

totally toward multiphoton absorption (and dissociation) as a result of high flux infrared excitation. In keeping with the trend more experimental details are included in the next section dealing with high-power lasers.

4 MULTIPLE-PHOTON EXCITATION

The discovery and proliferation of research activity concerned with laser catalyzed reactions continued unabated, until 1973 dealing basically with the same single-photon absorption under different experimental conditions. By 1973 a large body of such work had accumulated establishing the technique yet uncovering no remarkable phenomena. The basic constraint, which is mentioned earlier, of having efficient lasers (CO_2) at wavelengths such that photons carry only 0.117 eV (2.3 kcal) did little to encourage photochemical and kinetic investigations of many reactions, since a majority of chemical reactions possess high energies of activation typically in the range of 3 eV.

Academic interest sharply peaked in 1973 subsequent to the Russian discovery of the collisionless isotope selective unimolecular dissociation of SF_6. Theoreticians and experimentalists alike were instantly alert to the new investigative technique of multiphoton dissociation (MPD). Clearly, a novel and unique phenomenon has been established linking unimolecular dissociation with high-energy lasers by the absorption of many photons. In the case of the SF_6 dissociation, the energetics were quite astounding, showing that the reaction $SF_6 + nh\nu \rightarrow SF_5 \cdot + F \cdot$ required an n of approximately 35. Chemists and physicists alike instantly homed in on this new discovery, since embodied within this simple experiment were hints of the new physics of isolated polyatomic molecules and their interaction with intense laser fields. Even more exciting were the chemical speculations regarding specific cleavages in complex molecules as the ground floor to new syntheses with all their implications.

It is instructive and fulfilling to realize that the last 5 years have witnessed an exponential growth in MPD research and development, that a great many questions raised within a month of the Russian disclosure are still being asked, and that feverish research activity was still as exuberant in 1981 as it was in 1973. In what follows we sketch the original experiments to present the immediate questions that they raised and attempt to answer these on the basis of more recent and consequently more carefully designed experimental and theoretical investigations.

The work of Ambartsumian and Letokhov (4, 5), which was verified by Lyman et al. (15) within a few months and by countless others within the next few years, is essentially described in Fig. 6. A high-energy CO_2 TEA laser is focused into a reaction vessel containing SF_6. This polyatomic molecule, which has been extensively studied previously, possesses a large number of coincidental absorption lines accessible to the CO_2 laser output. One such specific absorption frequency is utilized to excite the molecule under low-pressure conditions (50 to 300 mtorr). Following a predetermined number of pulses the cell contents are analyzed by infrared and mass spectroscopy and the isotopic and molecular contents are identified. The laser absorption in the work of Ambartsumian and Letokhov (4, 5) was tuned to match the $^{32}SF_6$ (ν_3) vibrational mode shown in Fig. 7 and was not at all absorbed within the $^{34}SF_6$ species since the isotopic shift was large enough. Analysis subsequent to some hundreds, thousands, and tens of thousands of laser shots indicated an almost total destruction of the irradiated species ($^{32}SF_6$), and thus progressive enrichment of the residual gas by the unirradiated and nonabsorbing species occured (see Fig. 8). Specific laser energy thresholds, pressure dependences of the enrichment factor, and its relation to the number of shots, presence of scavengers, cell, and irradiation configuration were also studied.

In the first experiments it was demonstrated quite clearly that the enrichment factor, that is, $(N_1/N_2)_{\text{product}}/(N_1/N_2)_{\text{starting material}}$, varied exponentially with the number of pulses and inversely with the initial gas pressure. Determination of the dependence of the laser parameters followed some time

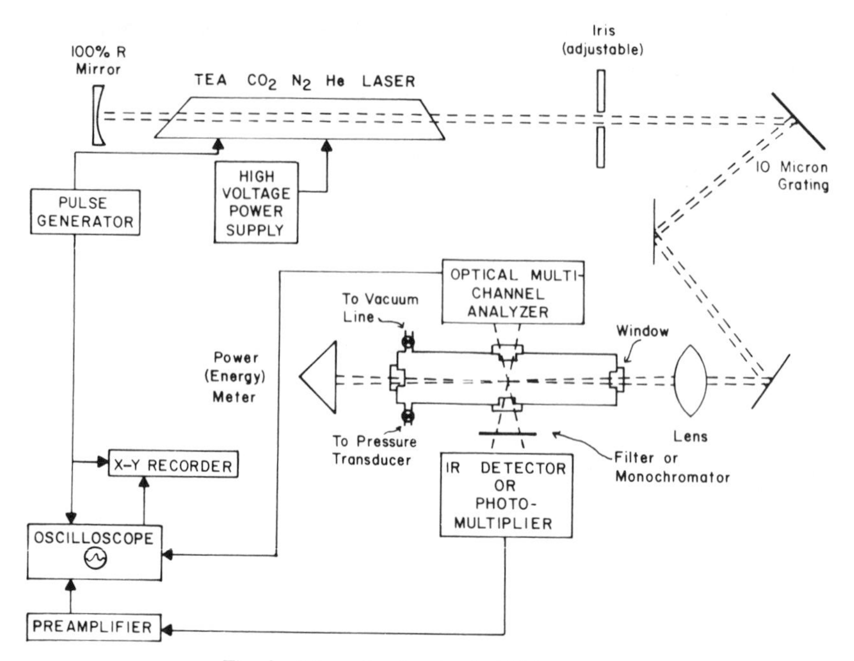

Fig. 6 Schematic of a simple MPD apparatus.

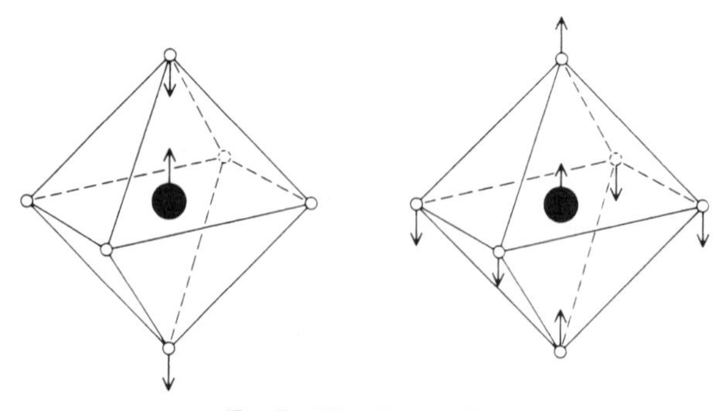

Fig. 7 Vibrations of SF_6.

later and confirmed the original observation of a sharp power threshold for dissociation.

All the experiments demonstrated an extremely efficient single-step enrichment scheme for the ^{34}S isotope that actually reached the theoretical limit. While no one could argue with such a spectacular demonstrated capability of a new isotope separation technique, many questions were instantly raised regarding the mechanism of such unimolecular dissociations.

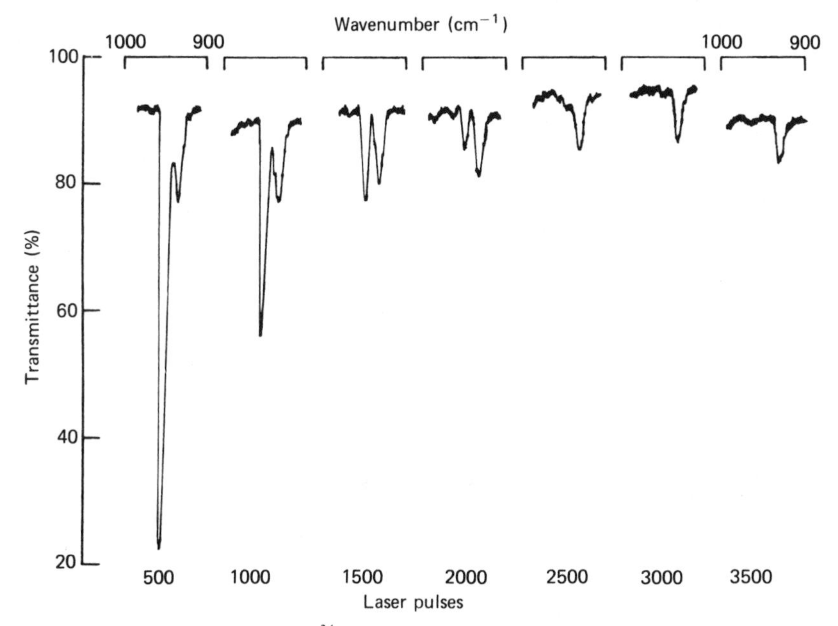

Fig. 8 Laser isotope enrichment of $^{34}SF_6$. Infrared absorption scan of a cell containing SF_6 being irradiated by a CO_2 laser tuned to $^{32}SF_6$ absorption as a function of number of pulses.

Some particularly disturbing questions for a single-photon absorption world are raised below. The S–F bond rupture requires an energy equivalent to 35 CO_2 laser photons. In view of rapid collisional and perhaps even collisionless vibrational energy exchange between isotopic species and in view of inherent quantum-mechanical anharmonicities of the vibrational energy levels we ask:

1. Is the observation unique to SF_6 (in terms of its size)?
2. How are 35 photons absorbed?
3. How long does such an absorption event take?
4. How can the excitation remain isotope specific?
5. What about intramolecular energy transfer between vibrational modes—is there a distribution?
6. Is the mode (bond) that is excited the bond that ruptures?
7. Is the process truly collisionless?
8. Is the process coherent?
9. Why does the process require a threshold?

These questions can, of course, be posed (and they are) in much more rigorous terms. Quantum-mechanical, statistical, and mathematical models can be constructed that can be used to calculate and predict the conditions that must exist within the isolated molecule to explain the observations and predict the behavior of analogous systems.

The past 4 years have witnessed an explosive exploration of the SF_6 MPD and the reaction is quickly becoming a model system for such unimolecular laser induced chemistry. Therefore in the next few pages let us attempt to address as many of the above questions as possible so as to construct a sound basis for MPD. In Fig. 9 it is evident that the rigorous resonance conditions that are quantum mechanically imposed on a molecular system interacting with a single-frequency radiation field are only maintained for the first few

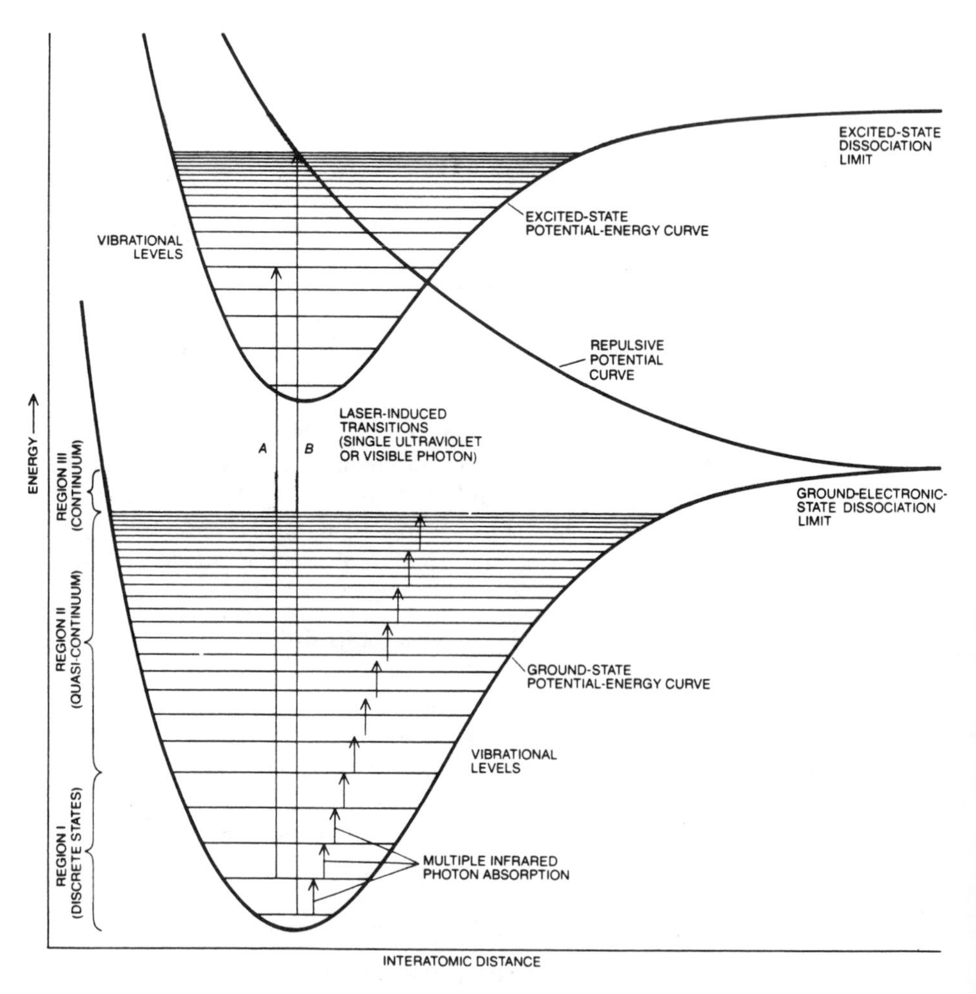

Fig. 9 Potential energy surface. Molecular fragmentation can occur as a result of one- or multiple-photon absorption. Dissociation occurs here strictly from the electronic ground state.

(four) steps. In itself, the absorption of four identical photons is rather surprising, since molecular anharmonicity normally does not allow for more than a single photon absorption. However, this anharmonicity may be compensated by either rotational transitions or by anharmonic splittings inherent in degenerate overtones. Additionally, power broadening, a mechanism unique to strong laser fields may also play a part in offsetting such deviations. While all three mechanisms are probably operative to some extent it is quite illuminating to review the rotational compensation mechanism.

If the laser photons are absorbed in the sequence P branch ($V = O, J \rightarrow V = 1, J - 1$), Q branch ($V = 1, J - 1 \rightarrow V = 2, J - 1$), R branch ($V = 2, J - 1 \rightarrow V = 3, J$), the anharmonicity is clearly offset by the rotational shift. If, in fact, the first absorption occurs from either a combination band or an overtone, a strong possibility for SF_6, the $V = 4$ level is immediately populated. The additional 30 or so photons that are needed to overcome the S-F bond energetic barrier are thus to be accounted for from energy levels above the $V = 3$ or $V = 4$ limit.

A simple statistical mechanical partition function allows a calculation of the density of energy levels above 3000 to 4000 cm^{-1}. This calculation shows that more than 10^6 levels/cm^{-1} are present above the 4000-cm^{-1} limit ($\nu_3 = 4$ range). Such an enormous level density has been termed the "quasicontinuum."

Figure 9 shows schematically that the identical wavelength photons, represented by the same length arrow, are absorbed from literally any level that is within reach (± 1 cm^{-1}). Basically this condition represents a relaxation of the rigorous quantum-mechanical resonance condition due to the incredibly high-level density. The existence of the quasicontinuum, which ends naturally in the true continuum, is well established by numerous experiments by the same Russian group discussed above, as well as by others. Thus the total phenomenon of a system absorbing as many as 35 identical photons is no longer very surprising. The basic requirements for such a molecule are reasonably small (< 10 cm^{-1}) anharmonicity and a high-energy-level density (10^4 to 10^6/cm^{-1}). Both requirements are satisfied in polyatomic molecules containing at least three atoms other than hydrogen. The questions remaining concern the times of dissociation vis-à-vis relaxation processes.

The question of a bond specific excitation and dissociation is truly one of determining the total time between excitation and fissure. In the case of SF_6, uncertainty arises since the molecule contains six chemically equivalent fluorines. Isotopic substitution is not truly feasible in this system because only ^{19}F is a stable isotope. Thus a chemical substitution of a chlorine atom for one fluorine has been carried out. An identical experiment has been performed with SF_5Cl, where a TEA CO_2 laser excited the nearly identical S-F stretching vibration and the products of this MPD were analyzed. The unexpected result was that the molecule selectively released a chlorine atom and formed the identical $SF_5\cdot$ fragment initially. The rather glaring suggestion here is that the SF_5Cl molecule was indeed raised to its quasicontinuum by means of a multiple-photon absorption but that the excitation was quickly and collisionlessly shared by the entire molecular framework. The sulfur-chlorine bond, the weakest one in the system was then cleaved preferentially to the stronger S–F bonds. The dissociation is thus clearly *not bond specific*.

Similar conclusions have been reached for a large number of polyatomic species driven to MPD by energetic TEA CO_2 lasers. The implication is that a rather severe statistical thermodynamic constraint is imposed on such experiments. As long as the excitation has sufficient time to redistribute within the system prior to bond fissure a statistical thermodynamic product is the major one. If, however, the excitation can be delivered in a very short time, say picoseconds, with sufficient intensity to cause bond breakage there is insufficient time for excess energy sharing within the molecule. Such times are indeed short, yet well within experimental ability. The potential payoff will be tremendous, since bond selective dissociation will then become a reality. An alternate approach to cutting down excitation time is to synthesize molecules that are constructed in such a way that the total excitation sharing either will not be feasible or will take a longer time than the dissociation.

What about the pressure dependence of a MPD reaction? A glance at Fig. 10 immediately shows that isotope separation efficiency is lost exponentially with increasing pressure. Vibrational energy transfer into and out of the excited vibrations is very rapid and thus every collision disrupts a selective

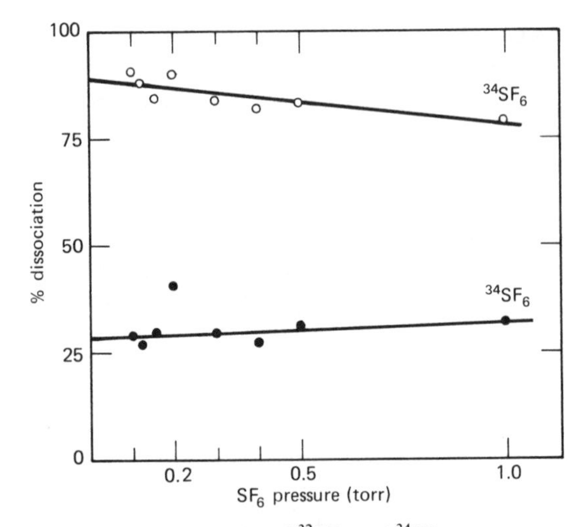

Fig. 10 Percent dissociation of $^{32}SF_6$ and $^{34}SF_6$ versus pressure.

MPD event. A more detailed discussion of the isotopic separation of $^{34}SF_6$ and $^{32}SF_6$ is found elsewhere, yet let us draw some conclusions from what we have already assessed above as a MPD reaction. It is clear that a frequency dependence is observed for all MPDs. In some cases the frequency dependence is quite sharply peaked and in others very wide. Sharply peaked dependencies on the whole indicate stringent resonance requirements that are harder to satisfy than these in the SF_6 P, Q, R, power broadening, or anharmonic, mechanism. By contrast, a wide frequency distribution indicates that one of the three mechanisms is dominating. Some laser fluence (J/cm^2) thresholds are almost always indicated for MPD reactions. This threshold is dependent on two factors, the inherent absorption cross section in the system and the efficiency of the anharmonic compensation mechanism. In all experiments reported to date, observations indicate that the reaction's yield depends on the laser fluence (energy/cm^2) rather than on the laser's peak power. Thus for a successful MPD a fairly stringent set of constraints in indicated.

The theoretical body and the experimental diagnostic tools may not as yet have reached the desired goals because indications of final reaction products are of statistical thermodynamic origin and it is by no means demonstrable that the same conditions prevail in the highly excited intermediate states. A number of calculations and experiments point out significant differences between conventional, thermal excitations and MPD via lasers. Nano- and picosecond time resolved measurements are going to be mandatory to establish the structure of the quasicontinuum. It will be of prime importance to establish experimentally whether the "transition state" achieved by laser induction is identical with a thermally generated state. In parallel, theoretical models of coupled anharmonic oscillators must be developed that will map out the merging of a localized coherent excitation into the incoherent distributions of the more highly excited states in the quasicontinuum and beyond. While many data exist for SF_6 (rapidly approaching the H atom of MPD definition) not enough is known for a general picture to emerge and the total behavior of molecular systems subjected to extremely intense laser pulse (10^6 to 10^{10} W/cm^2) is not fully understood as yet. However, the potential of utilizing these techniques to selectively deposit excitation in a given bond with its consequent fissure completely free of by-products is perhaps the most exciting chemical "discovery" of the early seventies. The research is far from complete and the number of investigating laboratories is still growing by leaps and bounds. It is presently expected that with shorter laser pulses, in the picosecond range, ultimate bond fissure control will be obtained.

Let us then discuss some recently investigated MPD reactions in terms of both the theoretical and experimental details that have shaped the interpretations of these studies. The rather voluminous body of research on MPD of polyatomic molecules naturally divides into two categories. Excitation and dissociation in a collisionally dominated pressure regime and in an isolated molecule. Since the very early days of MPD much significance has been placed on the distinction between these two classes of experiments, as it was conjectured that the presence, or absence of molecular encounters must play a significant role in determining the reactions mechanism and end products. The 1981 approach to MPD still maintains the same profile; collisionless dissociation of polyatomic species is clearly demonstrated to be the only candidate for future bond selective excitation and dissociation. Many of the hard to answer questions regarding the mechanism and selectivity of MPD became much more tractable in a collisionless regime and thus we focus on this experimental and theoretical model first by going through a specific example, that of chromyl chloride (CrO_2Cl_2).

The first reports of infrared multiphoton dissociation of molecules also stated that this process is accompanied by visible emission. Since then there has been much interest in and speculation about the origin of this molecular luminescence. The early experiments using c.w. CO_2 laser irradiation were thought to have produced luminescence through simple heating of the absorbing gas. However, this mechanism cannot explain the visible emission observed in the field of a pulsed TEA CO_2 laser at a power level insufficient to cause optical breakdown.

Much of this molecular emission appears to be caused by radical recombination or by subsequent chemical reaction although the first workers were led in many cases to believe that this emission was "instantaneous" with the laser pulse at the pressures they used.

However, in at least one instance the existence of collision free production of an electronically excited product seems well documented. Haas and Yahav studied the unimolecular decomposition of tetramethyldioxetane using a pulsed TEA CO_2 laser. On focusing the laser beam they observed a blue emission whose rise time followed very closely that of the laser pulse at pressures as low as 0.07 torr. They suggested that this luminescence was due to the formation of an acetone excimer. Only in the infrared MPD of CrO_2Cl_2 was it clearly demonstrated that collisionless production of electronically excited CrO_2Cl_2 was accomplished. This rather rare phenomenon, which was named inverse electronic relaxation, provides a rather direct probe into the dynamics of a multiphoton excitation mechanism that operates in CrO_2Cl_2. Experimentally, CrO_2Cl_2 vapor at pressures ranging from 5×10^{-4} to 3 torr was irradiated in either static or flow cells. The CO_2 lasers used for irradiations were grating tuned and allowed pulse energies of 0.2 to 2.5 J (unfocused) and pulse lengths of 0.2 to 4 μsec. Pulse shaping was accomplished with lenses having focal lengths, of 5, 10, or 12 in., and the molecular emission was observed perpendicular to the exciting beam. The time dependence, pressure dependence, and intensity dependence on laser parameters of this visible luminescence were recorded, as were its (wavelength) spectra. Conventional photomultiplier tubes and monochromators were employed (RCA C31034 and 1/4 Jarrel-Ash for example). Gated electronic detection (PAR boxcar) was also employed for the lower pressures investigated in this experiment.

The CrO_2Cl_2 system displays two strong absorption fundamentals in the 10-μ region (Fig. 11), the ν_6 asymmetric Cr-O stretch at 1000 cm^{-1} and the ν_1 symmetric Cr–O stretch at 990 cm^{-1}. Many of the higher R-branch transitions of the CO_2 10.6-μ laser band coincide with both absorption bands. Absorption experiments at a number of laser fluences yield an apparent absorption coefficient of 0.0074 cm^{-1} $torr^{-1}$ for the R(30) CO_2 laser line. Multiphoton absorption (MPA) results in the appearance of visible emission followed at higher excitation by the formation of brown particulates, identified as CrO_2. The quantity of the latter and the intensity of the former are

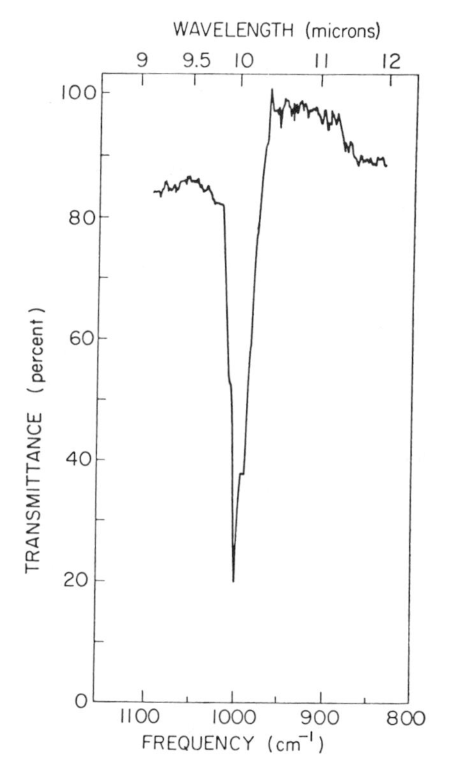

Fig. 11 The infrared spectrum of CrO_2Cl_2 in the gas phase.

closely controllable by the laser output parameters, and in fact suggest the following MPD mechanism for the CrO_2Cl_2 system:

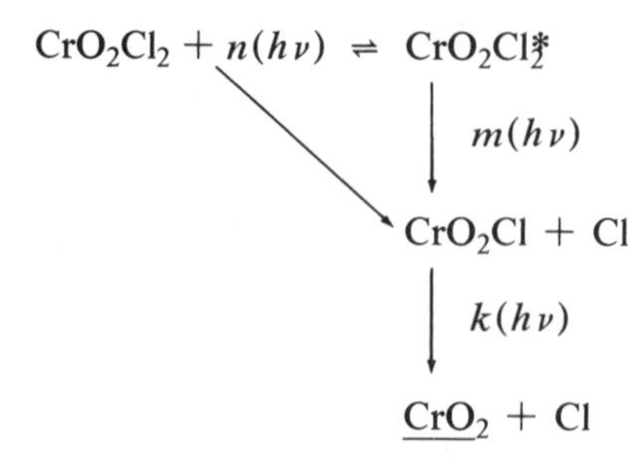

The mechanism presented above is strictly a logical interpretation of the observed facts; its verification and quantification are discussed below.

Figure 12 gives the details of the visible emission from CrO_2Cl_2. Comparison of this curve to previously published spectra of $CrO_2Cl_2^*$ in a matrix allows an unambiguous identification of the emitting species. The nature of the electronic transition is not clear, since there appear to be at least three singlet states within this wavelength region. What is interesting is the fact (from MO theory) that the transition is one of $(2a_2, 4b_1, 4b_2) \rightarrow 7a_1^*$, where the $7a_1^*$ is a chromium $d\pi^*$ orbital that is strongly antibonding for the Cr–Cl bond. The origin of this transition is calculated at 17,200 cm^{-1}, a requirement of an absorption of 18 photons (975 cm^{-1}) from the exciting laser source. By comparison, the energy required to cleave the CrO_2Cl–Cl bond is estimated as 83 kcal/mole from mass spectrometric studies and 68 kcal/mole from the appearance of photodecomposition studies. The dissociation limit thus exceeds the origin of the first excited electronic state by 19 to 34 kcal/mole.

One thus constructs a potential energy surface along the (Cr–Cl bond)

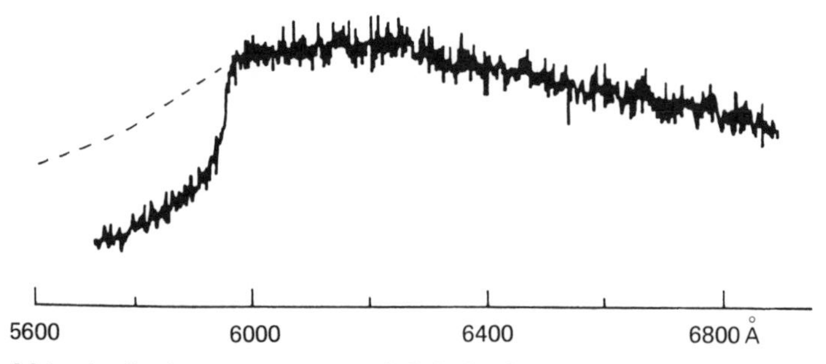

Fig. 12 Molecular luminscence spectrum of CrO_2Cl_2 following irradiation by a TEA CO_2 laser.

reaction coordinate in which the first excited electronic state origin is located substantially below the dissociation limit of the ground state. Figure 13 indicates this concept.

Figure 14 presents the variation of this luminescence intensity with laser fluence. As is clearly evident, the intensity varies linearly with laser fluence at low to medium energies. A tightly focused beam of 2 to 2.5 J/pulse promotes a significant broadening of the spectrum (dashed curve in Fig. 12) as well as an enhanced dissociation rate to CrO_2Cl and Cl. The latter can be followed exceedingly well by following the parent luminescence disappearance or the daughter's appearance. The figure presents the luminescence of CrO_2Cl and CrO_2 as well as that of CrO_2Cl_2 as a function of laser fluence. It is evident that under appropriate fluence conditions and wavelengths of observation a complete mapping of parent-daughter-grandaughter can be accomplished.

While the above observations are self-contained and form a sufficient set of conditions to firmly establish the mechanism as presented earlier, it is distinctly advantageous to probe the time behavior of the visible emission as a function of pressure, laser fluence, and wavelengths so as to obtain a true representation of the dynamics of the reaction sequence.

Three distinct wavelengths 6300, 5300, and 4000 Å, were chosen as

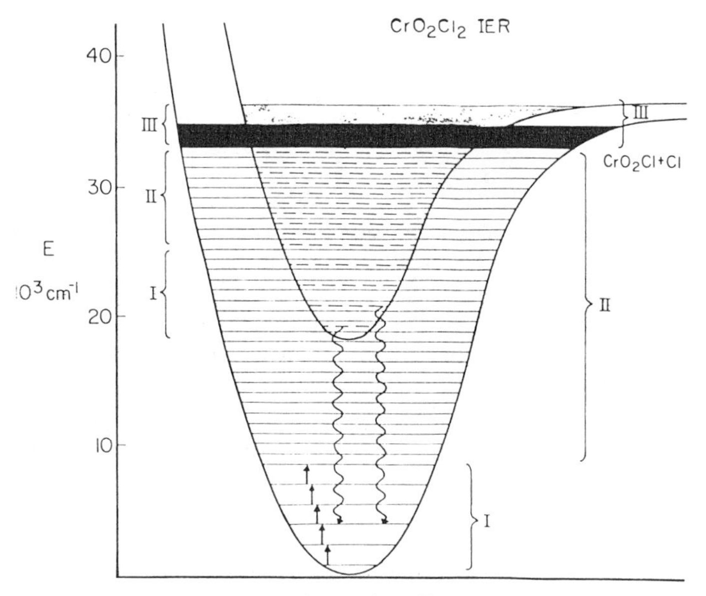

Fig. 13 Potential energy diagram of CrO_2Cl_2. Regions *I*, *II*, and *III* refer to the MPD nomenclature of the discrete, quasicontinuum, and dissociative regimes, respectively.

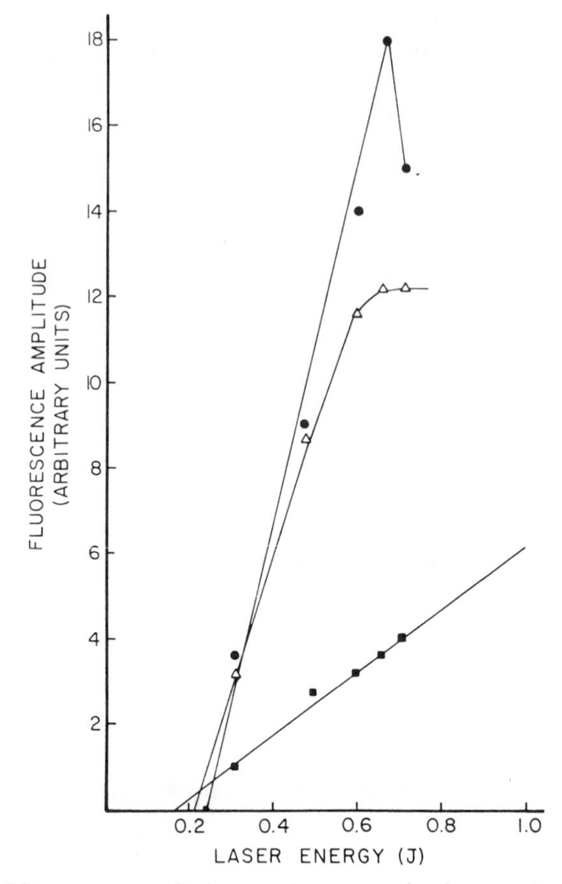

Fig. 14 Plot of fluorescence amplitude versus laser energy for three wavelength regions: (■) 6300 Å, (△) 5300 Å, and (●) 4000 Å.

nonoverlapping and are representative fluorescent wavelengths for CrO_2Cl_2, CrO_2, and CrO_2Cl. Figure 15 shows an oscillogram of the fluorescence signal at 6300 Å, that is, the parent fluorescence. It is evident that the parent's fluorescence follows the laser pulse. The risetime is independent of pressure as well as laser fluence, while fall time is both fluence and pressure dependent.

Both a collisionless and a collisional component are clearly observable at higher pressures (2 torr) and various laser pulse shapes. The bottom line indicates that the cross section for quenching the $CrO_2Cl_2^*$ luminescence is quite large and a Stern-Volmer plot as is given in Fig. 16 indicates a collision-free electronic relaxation time longer than 10 μsec.

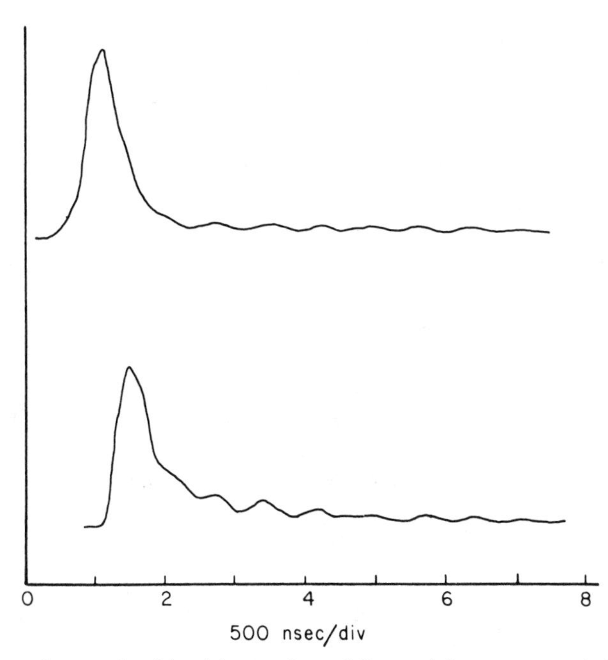

Fig. 15 Scope photographs of (*top*) laser pulse and (*bottom*) fluorescence emission at 6300 Å. CrO_2Cl_2 pressure is 0.400 torr.

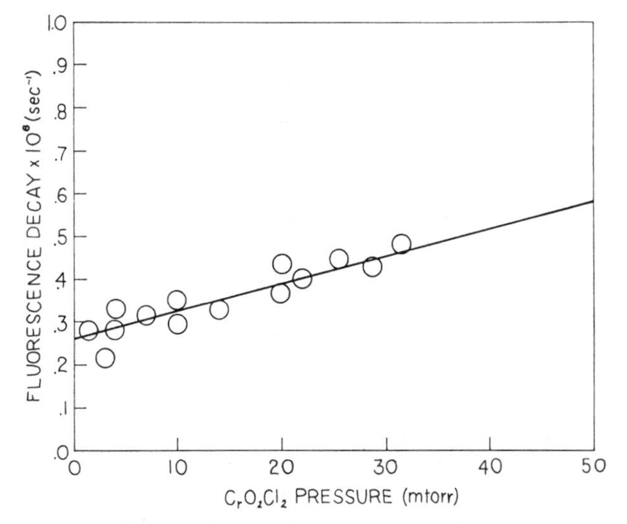

Fig. 16 Stern–Volmer plot of the fluorescence decay in CrO_2Cl_2 at 6300 Å.

The questions that are raised earlier with respect to MPD in general are thus being probed in this study with great detail. For example, while the Cr–O stretching motions are subject to laser excitation it is the Cr–Cl bond that is subsequently cleaved. This cleavage demands the absorption of some 30 photons (975 cm^{-1}) and is clearly demonstrated to occur collisionlessly. Therefore it is clear that bond specific excitation occurring on the 200 to 500 nsec time scale again does not result in bond specific cleavage.

The process of MPD in CrO_2Cl_2 requires a threshold much like that of SF_6. An unfocused beam does not result in visible emission or dissociation. The process is clearly incoherent, since the molecule attains an energy level density of $3 \times 10^6/cm^{-1}$ at 17,200 cm^{-1}. Its small size requires a fair-sized anharmonicity and it is thus clear that the first few steps only can be coherently driven and that a quasicontinuum condition is easily achieved even in this pentatomic molecule. While the experimental observations suffice to establish both the collisionless multiphoton excitation and the consequent dissociation, the channel by which the injected vibrational energy in the ground state crosses over to the excited electronic state remains to be theoretically explored.

From Fig. 13 one gets the notion of an intermediate level structure (quasicontinuum) in the ground electronic state communicating with the rather sparse level structure of the excited state. Initially, the interpretation rested on the notion of infrared active transitions between the high vibrational levels of the ground state and the low-lying ones in the excited state. Such a notion was incomplete, since it probed only interstate coupling. Intrastate radiative coupling within the high vibrational levels of the ground state should also be considered. Intrastate radiative coupling is theoretically similar to an intramolecular V-E process. Therefore in a complete treatment both intra- and interstate (radiative) coupling were considered and the total phenomenon was renamed inverse electronic relaxation (IER). In this treatment one calculates the IER rate and state coupling by considering the scheme

$$\begin{matrix} \text{groundstate} & \xrightarrow{V} & \text{single} & \xrightarrow{H_{int}} & \text{radiative} \\ \text{molecular Q.C.} & & \text{electronically} & & \text{continuum} \\ & & \text{excited bound state} & & \end{matrix}$$

where one considers nonadiabatic coupling (V) between the molecular quasicontinuum (Q.C.) of the ground state and an electronically excited state. The latter state is in turn coupled by the radiative interaction to a radiative continuum originating from spontaneous one-photon decay to low-lying vibrational levels of the ground state. It is the presence of this radiative continuum as the final decay channel that ensures the irreversibility of the IER. The practicality and generality of IER induced by MPD has been explored elsewhere and it suffices to state that this novel phenomenon fits well within the framework of electronic relaxation theory.

The collisional part of the CrO_2Cl_2 emissions and dissociation offers basically the same experimental end product (i.e., CrO_2), yet is not as rich in detail for physical analysis. Collisions erode phase coherence and provide many other deactivating channels for the excited species. As such no detailed studies in the collisionally dominated regime were made for the pressure or fluence dependence of the decomposition. With respect to fluorescence it is worth noting that the intensity of the emissions reaches a peak at 4 torr, plateaus, and finally is not discernible at all. The same observation holds for the fragment's fluorescence as well. These observations agree well with the total picture, since, as is noted earlier, the CrO_2Cl_2 cross section for quenching the electronic luminescence is quite large. The infrared V–T/R and V–V rates are not known, but they are estimated to be rather high, thus contributing strongly to the dilution of molecular excitation and the diminishing probability for either emission or decomposition with increasing pressure.

Mechanistically, this reaction is not much different than perhaps a dozen others (see Table 4). It calls for the same three distinct regions, namely, discrete, quasicontinuum, and dissociative continuum. The same sequence of events is followed, that is, the low-energy phase in which radiative coupling to the anharmonic oscillator causes resonant (compensated) transitions of a few photons in an isolated molecule and then the second phase, which is an absorption of as many as 30 photons occurring within the quasicontinuum (high density of discrete levels) and involving all states available to the fundamental pumping frequency. This second stage is followed by adiabatic crossover to the dissociative continuum resulting in the weak bond fragmentation. Alternatively, as was the case in CrO_2Cl_2, crossover to an excited electronic state occurs rather than dissociation from the ground state at lower excitation energies. This crossover requires the location of the excited state origin to be below the ground-state dissociation continuum.

Of the few dozen reactions that have been investigated to date, in which small- to medium-size molecules were employed, the basic generalities appear to support the theoretical concept that no bond or mode selectivity need be anticipated. The intramolecular (and in a collisional regime also the intermolecular) energy redistribution inevitably results in a rapid randomization. While rates of redistribution are generally inferred, experiments with rather short pulses suggest an intermode randomization time scale of picoseconds in most species. Within such a framework and given typical chemical reaction rates, no specificity is really expected nor encountered.

The lowest energetic channel, as viewed by the systems, is often predictable and equally often experimentally verified. The ongoing current preoccupation is the uncovering of a system in which a nonstatistical behavior is the

Table 4

Absorber	*Laser*	*Wavelength and Operational Mode*	*Reaction (M = absorber)*
N_2F_4	CO_2	0.3 J, 0.3μsec	$M \rightarrow 2NF_2$
			$M + H \rightarrow$
			$M + NO \rightarrow FNO + NF_3 + F_2 + N_2$
		1 J, 1 μsec	$M + (CH_3)_2C{+}CH_2 \rightarrow CF_4 + HF + N_2$
			$M + N_2O \rightarrow NO_2 + NF_3$
			$M + CH_4 \rightarrow C_2H_2 + N_2 + H_2 + HF$
NH_3	CO_2		$M \rightarrow NH_2 + H$
BCl_3	CO_2	c.w., 600W, 30 msec	$M + H_2 \rightarrow Cl_2 + B$
			$M + H_2S \rightarrow BSCl_2$
		1.5J, 200 nsec	$M + H_2 \rightarrow BHCl_2 + HCl$
			$M + O_2 \rightarrow BCl + BO$
B_2H_6	CO_2	c.w.	$M + C_4H_8 \rightarrow B_2H_4(C_4H_9)_2$
	CO_2	c.w., 1.5 W, R(16)	$M \rightarrow B_{20}H_{16}$
$B(CH_3)_3$	CO_2	c.w., 7 W, R(12)	$M + HBr \rightarrow B(CH_3)_2Br + CH_4$
H_3BPF_3	CO_2	c.w., 100 W	$2M \rightarrow B_2H_4 + 2PF_3$
D_3BPF_3	CO_2		$2M \rightarrow B_2D_6 + 2PF_3$
HCl	HCl		$M(\nu = 1) + K \rightarrow KCl + H$
	HCl		$M(\nu = 1) + O \rightarrow OH + Cl$
	HCl		$M(\nu = 2) + Br \rightarrow HBr + Cl$
O_3	CO_2	200 W, 50 Hz	$M + NO \rightarrow NO_2 + O_2$
NO	CO		$M + O_3 \rightarrow NO_2 + O_2$
CF_2Cl_2	CO_2	c.w., 400 W, P(20)	$M \rightarrow CF_3Cl + C + Cl_2$
		1.5 J, 100 nsec	$M \rightarrow CCl^*$

CH_3X (X = F, Cl, Br) F, Cl, Br)		1 J, 200 nsec	$M + Cl_2 \rightarrow CH_2XCl + HCl$
CH_3OH	CO_2	1.5 J, 100 nsec	$M \rightarrow CH^* + OH + C_2^*$ (!)
C_2H_2	CO_2	c.w., 300 W, 5 sec	$M + SF_6 \rightarrow C$
C_2H_4	CO_2	c.w., 660 W, 0.4 sec	$M \rightarrow C$
		c.w., 50 W	
C_2H_4	CO_2	1.5 J, 100 nsec	$M \rightarrow C_2^*$
			$M \rightarrow C_n$
C_2H_6	CO_2	c.w., 680 W, 1 sec	$M + SF_6 \rightarrow C$
C_2H_5Cl	CO_2	150 W, c.w., 972 cm^{-1}	$M \rightarrow CH_3 + CH_2Cl \rightarrow CH_4$
		200 W, 1 msec, 946 cm^{-1} 946 cm^{-1}	$M \rightarrow C_2H_4 + HCl$
$C_2H_2Cl_2$	CO_2	100 nsec, 60 nsec	*trans*-M → *cis*-$C_2H_2Cl_2$
C_2Cl_4	CO_2	6 W	$M + BCl_3 \rightarrow C_6Cl_6$
C_4F_6	CO_2	0.4 J, 100 nsec	$M \rightarrow F_2C = CF - CF = CF_2$
C_4H_8	CO_2	10 MW, 100 nsec	*trans*-2-Butene → *cis*-2-butene
i-C_4H_9X (X = Cl, Br, I)		0.3 J, 100 pulses	M → *i*-C_4H_8
$C_6H_{12}O_2$	CO_2	0.3 J, 1 μsec, P(20)	$M + CH_3F \rightarrow 2CH_3COCH_3$
$C_6H_{10}D_2$	CO_2	0.3 MW, 100	
SF_6	CO_2	0.8 MW, P(20)	
UF_6	CO_2	c.w.	$M + HCl$
$H_2CO(HDCO,D_2CO)$	CO_2	P(20) 2 J, 100 nsec, 0.5 Hz, 300 pulses	$M \rightarrow HD + CO$
BCl_3	CO_2	P(16), 0.1 J, 200 nsec 360,000 pulses	$M + H_2S \rightarrow B_2S_3 + HCl$
	CO_2	R(24), 2 J, 85 nsec	
	CO_2	P(20), 1.5 J, 200 nsec	$M + H_2 \rightarrow BHCl_2 + HCl$
C_2F_3Cl			$M \rightarrow C_2^*$
$C_2F_4Cl_2$	CO_2	c.w., 4 W	$M \rightarrow C_2F_4$

Table 4 *(Continued)*

Absorber	*Laser*	*Wavelength and Operational Mode*	*Reaction (M = absorber)*
CF_2Cl_2	CO_2		$M + H_2 \rightarrow ?$
SiF_4	CO_2	P(36), 0.8 J, 200 nsec	$M + H_2 \rightarrow ?$
SF_6	CO_2	P(20), 1.5 J, 200 nsec 5000 pulses	$SF_6 + H_2 \rightarrow SF_5 + HF \rightarrow SF_4$
	CO_2	P(16), 2 J, 90 nsec 2000 pulses, P(40)	
	CO_2	3 J, 90 sec, 1000 pulses	$SF_6 + H_2 \rightarrow ?$
CCl_4	CO_2	979.9 cm^{-1}, 17 J, 90 nsec	
OsO_4	CO_2	R(2), 3 J, 90 nsec	$M + C_2H_4 \rightarrow ?$

rule rather than the exception and it is believed that either shorter laser excitation times or molecular engineering of a unique system would lead to such findings.

Particularly attractive from the standpoint of utility, MPD reactions have not yet matured into industrially acceptable standards. The principal obstacle to a large-scale introduction of the laser into the chemical industry is economic feasibility. Laser photons are significantly more costly at present than "photons" that enter a reaction vessel from a jacket of steam or the direct application of a burning fossil fuel's flame. The cost of a mole of such thermal photons, if applicable to the task, is infinitesimal. The cost of laser photons varies from 2¢ per mole of CO_2 photons to several dollars for visible and ultraviolet photons. In view of the rather efficient CO_2 laser yields of about 5×10^{19} photons per pulse (1 *J/pulse*) the best that one could expect of a reaction on a mole to mole basis is an energy cost of several cents for several tens of grams of product. Thus the applicability of laser chemistry toward the production of a common chemical is not at all bright. However, specialty products such as fine chemicals, drugs, and rare isotopes typically have costs far in excess of a few dollars per pound; thus their preparation becomes reasonable for production by means of laser techniques. If the laser can be utilized as a catalyst, that is, driving a reaction not on a mole to mole basis but rather yielding many moles of product per mole of photons, the economics brighten considerably.

From both an economic model point of view and a fascinating chemical reaction, let us discuss the reaction of CrO_2Cl_2 with H_2 subsequent to MPA in CrO_2Cl_2. The experimental conditions are effectively identical to those discussed earlier for neat CrO_2Cl_2, except that the total pressure utilized is higher, 4 to 30 torr if the mixture composition is 15:1 H_2/CrO_2Cl_2. MPA at high laser fluences results in the following set of chemical reactions:

$$CrO_2Cl_2 \xrightarrow{nh\nu} CrO_2Cl + Cl$$

$$Cl + H_2 \rightarrow HCl + H$$

$$H + CrO_2Cl_2 \rightarrow HCl + CrO_2Cl$$

$$CrO_2Cl + H_2 \rightarrow CrO_2 + HCl + H$$

$$H_2 + CrO_2 \rightarrow CrO + OH$$

$$CrO + CrO_2 \rightarrow Cr_2O_3$$

$$OH + H_2 \rightarrow H_2O + H$$

The only reaction products are Cr_2O_3, HCl, and H_2O.

The reaction occurs with one laser pulse and utilizes a mere 5×10^{19} photons. At the pressure conditions discussed above, say 1.5 torr of CrO_2Cl_2

and 22.5 torr of H_2 the reaction in a 1-liter bulb appears to be complete, yielding 4×10^{-4} moles of Cr_2O_3. The photon utilization factor defined as

$$\frac{\text{moles of product out}}{\text{moles of photons in}}$$

is 4. At 4 moles of product per mole of photons the economics are quite favorable, yielding an energy cost of perhaps 1¢/mole or in the case of Cr_2O_3 in round numbers 2¢/lb. With the cost of H_2 folded in then, the price of producing a pound of Cr_2O_3 in an extremely finely divided state does not exceed \$1/lb. It is not suggested that this experiment represents a cheaper synthetic route for production of Cr_2O_3 but merely that it illustrates the notion that laser chemistry can become an economically competitive method for production of fine chemicals.

It is presently assumed that the mechanism suggested earlier is operative in this chain reaction. The reasoning behind this is based on the experimental observation of CrO_2Cl^*, CrO_2^*, and CrO^* as fluorescing (visible) species and of H_2O emission in the infrared. Equally firmly established is the absence of HCl in the $V = 1$ state with the usual behavior noted for chain branching reactions, that is, cell size dependence as well as impurity dependence point strongly toward a sequential set of events that can be described as follows:

$$CrO_2Cl_2 \xrightarrow{nh\nu} CrO_2Cl + Cl$$

$$Cl + H_2 \rightarrow HCl + H$$

These two steps as well as the next one

$$H + CrO_2Cl_2 \rightarrow HCl + CrO_2Cl$$

must lead to ground vibrational state HCl production. It is implied that the number of H_2O molecules produced is limited by the number density of CrO_2Cl_2 species. Since the pressures employed here are well within the first explosion limit known for the O_2, H_2 reaction it is reasonable to take the fundamental steps of

$$H_2 + OH \rightarrow H_2O + H$$

and

$$H + OH \rightarrow H_2O$$

as the only generating steps for the final product. Both steps may lead to the formation of vibrationally hot H_2O, since in both steps the OH radical originates from highly excited complexes of CrO_2Cl_2 or its daughters.

Thus the sequence of chain reactions is suggested in which the ultimate controlling specie is atomic chlorine, which in turn controls atomic H production, which in turn controls OH($V = n$) or OH($E = 1$) production. Of significance is the observation that parent molecule fluorescence, that is, that of CrO_2Cl_2, is believed to span the same wavelength region in the H_2 reaction as it does in the following reaction. It is difficult to completely substantiate this observation since the luminescence in the chain reaction is very strong and of brief duration. Since the experiment is basically a one-shot experiment and the Cr_2O_3 deposit must be removed subsequent to each observation, it is rather difficult to map out the spectrum as completely as was done in the case of neat CrO_2Cl_2.

5 SUMMARY

A few general comments regarding the actual experimental setup of a study such as that of CrO_2Cl_2 are now in order.

In logical progression such comments begin with a description of the laser source, which usually is a transverse electric atmospheric CO_2 laser. Such devices can be homemade or store bought quite easily and fairly inexpensively today. Desirable features are pulse duration variability (from 100 nsec to 3 μsec), wavelength tunability, and reasonable pulse repetition rates (1 pulse/sec and higher). These are necessary for most experiments since one would like to ensure excitation times that are shorter than relaxation or fragmentation times, coincidence with many molecules, and possibilities of electronic signal processing.

Energy outputs of most of these lasers range from 0.5 to 100 J/pulse. Focusing such beams results in tremendous energy densities that are responsible for multiphoton absorptions and dissociations.

It is mandatory to monitor the laser parameters before and after (even during) an experiment with energy meters, wavelength measuring devices, and pulse length detectors. All are commonplace items today, especially for CO_2 TEA lasers. The cross section of any laser beam is rarely uniform. That is, many "hot spots" may be present in a beam that represent much higher power levels than the measured average. Thus careful monitoring and tailoring of the beam is indeed a must for careful and repetitive experimentation. For experiments such as the SF_6 case described earlier, a simple vacuum flask may suffice, since the reaction can be driven to completion and the products analyzed subsequently. Midterm monitoring, as is described above for CrO_2Cl_2, requires more sophisticated cell and laser design. For example, a flow system is preferred to a stationary fill since conditions are much better controlled with respect to pressure. Also a long cell, with laser entrance and exit windows located very far from the observation port, eliminates troublesome "window fluorescence" and extraneous light. Precise fluence measurements require laser beam control and measurement.

Focusing optics for 10-μm radiation are readily available from many manufacturers and in many materials. Mirrors, spherical or parabolic (for focusing) should be metal made to withstand the high power densities of TEA CO_2 lasers. Lenses should be chosen carefully with regard to the desired versus necessary fluences (J/cm^2). Oftentimes the more costly lens system is not required and can be replaced with the more durable and inexpensive mirrors. Finally, diagnostics are truly the experimenters choice. It is, of course, always preferable to have on line diagnostics, but these are both expensive and more difficult to set up. Often postexperimental diagnostics suffice to determine the feasibility of a chosen route in a given system.

The reality of LIC is now out of the alchemy stage of the early seventies and is now a sophisticated network of government-industrial-university "hot topic." The search for economic laser isotope separation techniques and laser induced chemistry for synthetic, catalytic, and clean up preparations is vigorous, serious, and "now" oriented. The perception is that the promise has been around long enough and it is now time for the fulfillment.

Bibliography

Rather than list innumerable references in a multitude of scientific journals I list the most recent review articles and monographs. The reader is referred to the exhaustive lists of references contained in each of these for more detailed descriptions of specific experiments.

1. Boyd L. Earl, Leonard A. Gamss, and Avigdor M. Ronn, "Laser Induced Vibrational Energy Transfer Kinetics: Methyl and Methyl-d_3 Halides," *Acc. Chem. Res.*, **11**, 183 (1978).
2. C. B. Moore Ed., *Chemical and Biochemical Applications of Lasers, Vol. 1,* Academic, New York, 1974. Equally important are Vols. 2, 3, and 4, published in 1975, 1977, 1980, respectively.
3. A. M. Ronn, "Infrared Laser Catalyzed Chemical Reactions," *Spectrosc. Lett.*, **8**, 303 (1975).
4. Rafael V. Ambartzumian and Vladilen, S. Letokhov, "Selective Dissociation of Polyatomic Molecules by Intense Infrared Laser Fields," *Acc. Chem. Res.*, **10**, 61 (1977).
5. V. S. Letokhov, "Photophysics and Photochemistry," *Physics Today*, p. 23, (May 1977).
6. J. P. Aldrige III, J. H. Birely, C. D. Cantrell III, and D. S. Cartwright, "Laser Photochemistry, Tunable Lasers and Other Topics," in *Physics of Quantum*

Electronics, Vol. 4, S. F. Jacobs, M. Sargent III, M. O. Scully, C. T. Walker, Eds., Addison Wesley, Reading, MA, p. 57.

7. N. Bloembergen and E. Yablonovitch, "Infrared Induced Unimolecular Reactions," *Phys. Today*, **31**, 23 (May 1978).
8. Avigdor M. Ronn, "Laser Chemistry," *Sci. Am.*, **240**, 114 (May 1979).
9. Sol Kimel and Shamai Speiser, "Lasers and Chemistry," *Chem. Rev.*, **77**(Aug.), 437 (1977).
10. *Physics Today*, November 1980, Special Issue deveoted to "laser chemistry."
11. *Photoselective Chemistry* Advances in Chemical Physics Series Two Vols., Wiley, New York 1981.
12. S. M. Freund and J. C. Stephenson, *J. Chem. Phys.*, **65**, 4303 (1976).
13. R. J. Gordon and M. C. Lin, *Chem. Phys. Lett.*, **2**, 262 (1973).
14. D. Arnold, K. Kaufman, and J. Wolfrum, *Phys. Rev. Lett.*, **34**, 1597 (1975).
15. J. L. Lyman, R. J. Jensen, J. P. Rink, C. P. Robinson, and S. D. Rockwood, *Appl. Phys. Lett.*, **27**, 87 (1975).

Author Index

Numbers in parentheses are references. Pages in *italics* indicate where full references appear.

Subject Index